MOUNTAIN BATTERY

Mountain Battery

The Alps, Water, and Power
in the Fossil Fuel Age

MARC LANDRY

STANFORD UNIVERSITY PRESS
Stanford, California

Stanford University Press
Stanford, California

© 2025 by Marc Landry. All rights reserved.

No part of this book may be reproduced or transmitted in any form or by any means, electronic or mechanical, including photocopying and recording, or in any information storage or retrieval system, without the prior written permission of Stanford University Press.

Printed in the United States of America on acid-free, archival-quality paper.

Library of Congress Cataloging-in-Publication Data
Names: Landry, Marc, author.
Title: Mountain battery : the Alps, water, and power in the fossil fuel age / Marc Landry.
Description: Stanford, California : Stanford University Press, 2025. | Includes bibliographical references and index.
Identifiers: LCCN 2024026942 (print) | LCCN 2024026943 (ebook) | ISBN 9781503639775 (cloth) | ISBN 9781503641570 (paperback) | ISBN 9781503641587 (ebook)
Subjects: LCSH: Water-power—Alps Region—History. | Hydroelectric power plants—Alps Region—History. | Electric power production—Alps Region—History. | Renewable energy sources—Alps Region—History.
Classification: LCC TC495.A44 L36 2025 (print) | LCC TC495.A44 (ebook) | DDC 621.31/2134094947—dc23/eng/20240705
LC record available at https://lccn.loc.gov/2024026942
LC ebook record available at https://lccn.loc.gov/2024026943

Cover design: Michele Wetherbee
Cover art: Deutsches Museum, München, Archiv, BN23481

For Evelyn

CONTENTS

	List of Illustrations	ix
	Acknowledgments	xi
	Introduction	1
ONE	Mountains of White Coal	15
TWO	Carrier of Wasted Natural Forces	49
THREE	Exploiting Nature's Gifts	80
FOUR	Emergency Power	113
FIVE	Between Cooperation and Autarky	134
SIX	The Alps and the Energetic Struggle for Existence	162
SEVEN	Completing Europe's Battery	193
	Conclusion	211
	Notes	225
	Bibliography	259
	Index	283

LIST OF ILLUSTRATIONS

MAP	Ridgelines and waterways of the Alps	xvi
FIGURE I.1	Bergès's "White Coal" exhibit	2
FIGURE 1.1	Portrait of Aristide Bergès	37
FIGURE 1.2	Relief of the Grésivaudan and source of white coal	42
FIGURE 2.1	Artificial waterfall at Munich exhibition, 1882	58
FIGURE 2.2	"Gallery of Electricity"	65
FIGURE 2.3	Lauffen-Frankfurt transmission with waterfall, 1891	69
FIGURE 3.1	"Overview Map of Water Power in Bavaria"	87
FIGURE 3.2	"On the Coal Emergency"	97
FIGURE 3.3	"Load for Electric Operation"	98
FIGURE 4.1	Distribution of white coal plants	118
FIGURE 4.2	"White Coal Coming to the Aid of Black Coal"	123
FIGURE 4.3	Cross-section of the Walchenseewerk	131
FIGURE 4.4	Painting of the Walchenseewerk	132
FIGURE 5.1	High-Alpine waterpower in Central Europe	142
FIGURE 5.2	World Power Conference brochure, 1924	144
FIGURE 5.3	Badge for World Power Conference, 1926	147

FIGURE 5.4	Geographic distribution of European energy sources	150
FIGURE 5.5	Germany's power plants and high-voltage lines	153
FIGURE 5.6	Oliven's "Proposal for a European Superpower Grid"	155
FIGURE 6.1	Layout of the Tauern waterpower plant	170
FIGURE 6.2	Diagram of the functioning of the slope canals	171
FIGURE 6.3	Göring breaking ground at Tauern Works	177
FIGURE 6.4	Plans for Tauern Works Kaprun	180
FIGURE 6.5	Energy supply map of the Ostmark	185
FIGURE 7.1	Economic Commission for Europe	200
FIGURE 7.2	Why save electricity?	202
FIGURE 7.3	Electric infrastructure of Montecatini plant	205
FIGURE 7.4	South Tyrolean village of Reschen	206
FIGURE C.1	Dams and reservoirs of the Alps	216

ACKNOWLEDGMENTS

As I think of adequately conveying my gratitude to the people whose time, assistance, and support enabled the completion of this project, it seems a task almost more daunting than writing the book itself. But try I must, and the best place to start is Georgetown University, where the idea for this book took shape. While there I benefited from the mentoring of Jeffrey Anderson, Tim Beach, Richard Kuisel, Amy Leonard, and Adam Rothman. Thanks to Thomas Zeller for his valuable feedback on earlier versions of the manuscript. James Shedel and Roger Chickering showed me what it means to be a historian and improved my work with insightful advice. I owe a particularly large debt to my chief advisor John McNeill. I cannot imagine a better mentor, and I will be forever grateful for his time, advice, and support up to this day. Thanks finally to a great cohort including Tait Keller, Peter Engelke, Kevin Powers, George Vrtis, and Benjamin Francis-Fallon.

Many thanks go to the archivists and librarians whose professionalism facilitated this research. In particular I would like to acknowledge the staffs at the Bayerisches Hauptstaatsarchiv, the Bayerische Staatsbibliothek and the Deutsches Museum Archive and Library (and particularly Wilhelm Füßl, Daniel Gebauer, and Alexander Rieppenhausen) in Munich; the Firmenarchiv der Wacker-Chemie in Burghausen; the Salzburger Landesarchiv; the Schweizerisches Bundesarchiv in Bern; the Südtiroler Landesarchiv-Archivio provincial in Bozen/Bolzano; the Archives du Musée de la Houille Blanche in Lancey (where Frédérique Virieux graciously hosted me); the Bundesarchiv Berlin-Lichterfelde; and the Archiv der Republik in the Österreichisches Staatsarchiv. Thanks also to the dedicated professionals at the libraries at Georgetown University, Utah State University, and the University of New Orleans.

A project like this requires resources, and I am thankful for the support of a number of organizations and institutions. The History Department at the College of William & Mary got me started on historical research with scholarships for overseas archival visits. The Georgetown History Department and Graduate School provided assistantships, travel funds, and summer research support. The German Historical Institute in Washington afforded me the opportunity to serve as a research assistant and allowed me to participate in their Archival Summer Seminar. Grants from the German Academic Exchange Service (DAAD) and the Deutsches Museum allowed me to conduct research in Europe, and the Rachel Carson Center provided backing and intellectual community there as well. The IEEE Life Member Fellowship in the History of Electrical and Computing Technology enabled me to finish writing my manuscript, and the Fulbright-Botstiber Award in Austrian-American Studies helped me to refine one of the later chapters by financing a stay at the University of Innsbruck. The University of New Orleans Office of Research greatly facilitated the completion of this manuscript with support from a CEO Award, and by helping me receive an ATLAS grant. The final writeup occurred with the support of the Louisiana Board of Regents through the Board of Regents Support Fund (LEQSF(2021–22)-RD-ATL-10). Many thanks to Matt Tarr and Rebekah Cossaboom, and the ATLAS anonymous reviewers.

Since leaving Georgetown, I have had the good fortune to be part of three fine institutions. The Department of History at Utah State University was a wonderful community to land in after completing my PhD. I was lucky to learn from and exchange with Lawrence Culver, Victoria Grieve, David Rich Lewis, Dan McInerney, Colleen O'Neill, Ross Peterson, Len Rosenband, and James Sanders. Particular thanks to Tammy Proctor for the institutional support and Chris Conte for the fellowship in the forests. The Department of History and European Ethnology at the University of Innsbruck was an idyllic place for a visiting professorship. Thanks to Patrick Kupper for making it possible, and to Elisabeth Dietrich-Daum, Stefan

Ehrenpreis, Ute Hasenöhrl, Maria Heidegger, Odinn Melsted, and Irene Pallua for the collegiality. Since then I have had the pleasure of working with Dirk Rupnow and Eva Pfanzelter in the Department of Contemporary History.

At the University of New Orleans, I have benefited greatly from the support of the History Department. Thanks especially to Kathryn Dungy, John Fitzmorris, Max Krochmal, Allan Millett, Jim Mokhiber, and Andrea Mosterman. MA student Devin Sweet has been extremely helpful in preparing the manuscript. I have had the great fortune of being part of UNO's Austrian Marshall Plan Center for European Studies. Gertraud Griessner has facilitated all of my work and made my family feel at home in the Big Easy. Günter Bischof welcomed us too, and generously read parts of the manuscript.

Along the way I have been fortunate to cross paths with numerous other individuals who have supported and sustained me. Laurie Koloski provided crucial feedback and assistance that put me on the path to becoming a historian. Margaret Davidson, Rolf Steininger, and Lonnie Johnson introduced me to the German language and Austrian history. In Washington, I had the pleasure of getting to know Christof Mauch, who has enriched my scholarship on two continents. Todd, Dave, Seamus, Maggie, and Angus made living in DC as a doctoral student possible and can never know what their kindness has meant to me and my family. In Munich thanks to Drew Eisenberg, Martin Geyer, Ulf Hashagen, Martin Knoll, Uwe Luebken, Reinhold Reith, Helmut Trischler, and Frank Uekoetter. In Vienna, Verena Winiwarter and the group at the Institute of Social Ecology (Simone Gingrich, Gertrud Haidvogl, Fridolin Krausman, and Martin Schmid) welcomed me during several visits.

I am obliged to numerous colleagues who—at conferences, workshops, symposia, in editorial offices, and elsewhere—allowed me to refine my work: Brian Black, Christophe Bouneau, Lisa Brady, Jeroen Dewulf, Matthew Evenden, Alison Frank, Samuel Gladden, Chris Jones, Christian Karner, Philipp Lehmann, Kim Martin Long, Howard Louthan, Geneviève Massard-Guilbaud, Charles-François Mathis, Martin Melosi, Sara Pritchard, Georg Rigele, Caroline Schaumann, Victor Seow, and Igor Tchoukarine. Julie Cohn provided feedback on several chapters. Friends like Luke Johnson, Karl Pfefferle, Sabine Barcatta, Andrew Denning, Bradley Coates, Stephen Scala, Casey Cater, Charles Hadley, and Jana Lipman have provided fellowship and advice along the way. It has been a pleasure working with Margo Irvin and the team at SUP. I am thankful she saw potential in the manuscript, that she engaged two extremely thoughtful anonymous readers, and for her support during the entire process.

Finally, I would like to thank my family. Every opportunity I've been afforded ultimately comes down to my parents, Marc and Pamela. My saintly sisters, Erin and Meghann, have graciously put up with me all these years. As this book has taken shape, I have acquired not only knowledge about the Alps but also an extended Alpine family who have kindly embraced the American in their midst. Between the idea for this book and its publication, the gang has been joined by Dennis and Eleanor, who will never know how much joy they have brought into my life. Lastly, all thanks to Evelyn, to whom this book is dedicated. It simply would not exist without you.

MOUNTAIN BATTERY

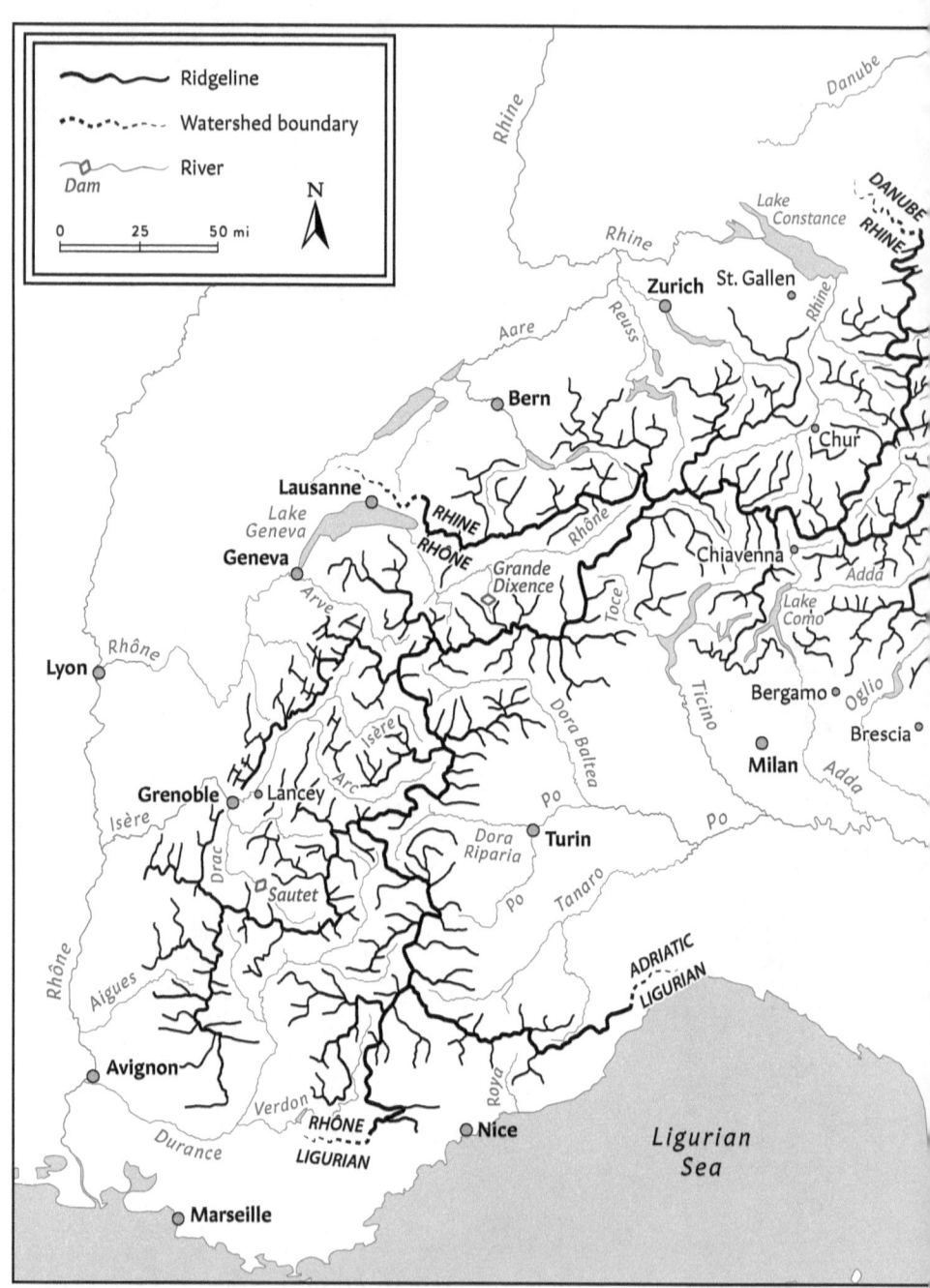

MAP Ridgelines and waterways of the Alps.

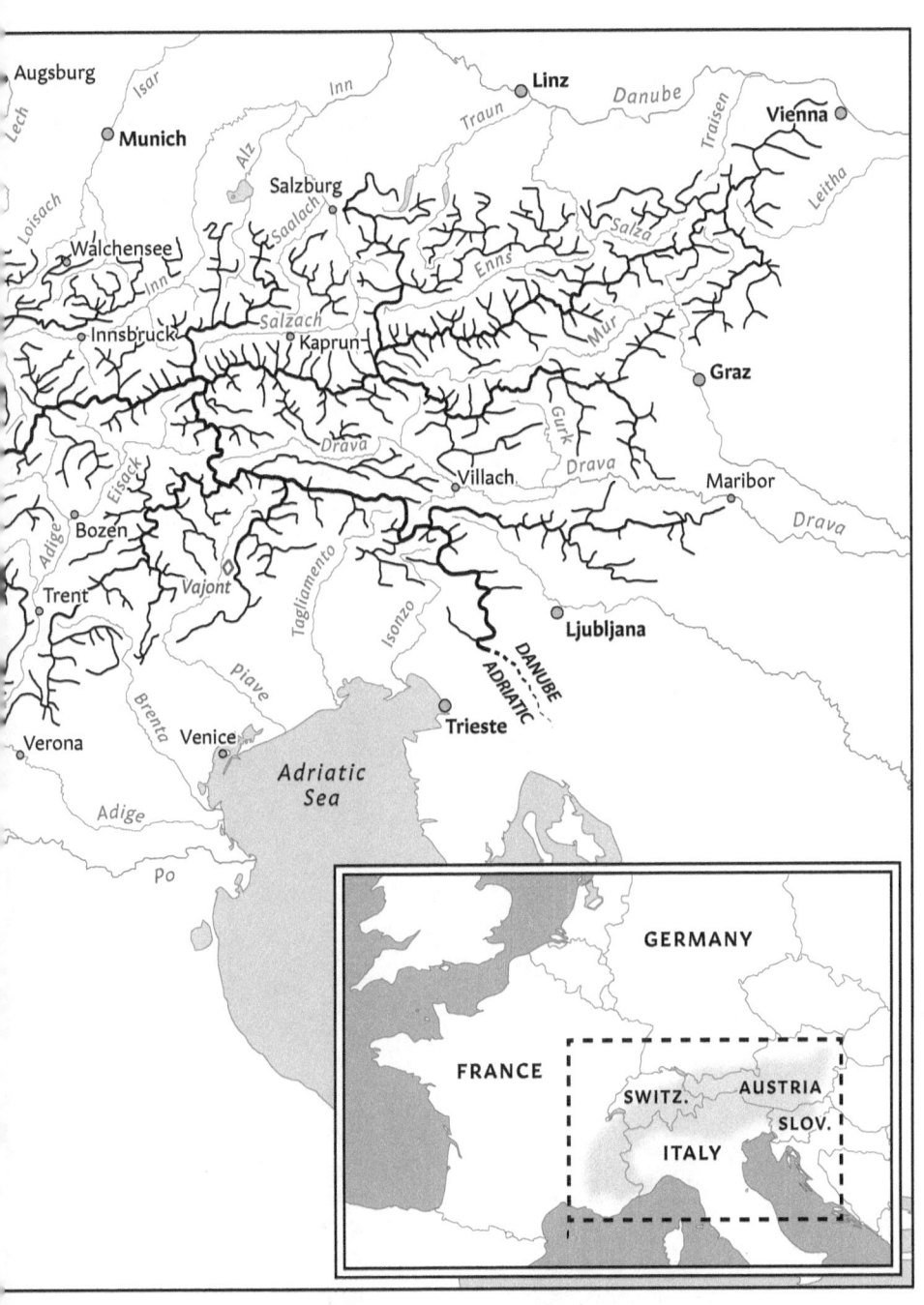

Introduction

"But suppose I treated these very fissures and crevasses as letters, attempted to decipher them, shaped them into words, and learned to read them, would you have any objection to that?"
—*Johann Wolfgang von Goethe,* Wilhelm Meister *(1795–1796)*

IN 1889, A FRENCH ENGINEER named Aristide Bergès attended the Exposition Universelle in Paris to inform the world of a new form of energy he had discovered. The site for this revelation was the gigantic glass and iron Galerie des machines, located across the Champ de Mars from the main attraction of the fair, Gustave Eiffel's tower. Here, the graduate of the prestigious École Centrale des Arts et Manufactures and proprietor of a successful paper mill, set up one of the numerous exhibitions on French industry that filled the sprawling display area. Bergès's exhibit displayed a device he had been instrumental in developing, as well as some fairly bold claims about its potential.

The display consisted of a metallic turbine measuring two meters in diameter. This innovation of the early nineteenth century amounted to a reinvention of the wheel—the waterwheel in this case. Unlike the waterwheel, turbines could cope with the destructive force of water falling from great heights. They had been developed as part of an effort to help waterpower stimulate the same sort of commercial explosion occasioned by rapidly spreading steam engines. While Bergès had not invented the turbine himself, he was among the first to adapt it to handle high-pressure waterpower. Emphasizing this fact, the turbine was placed above a plaster relief map depicting a small section of the French Alps (Figure I.1). On a plaque attached to the turbine stood a long inscription that began with the words "exploitation of the WHITE COAL (*HOUILLE BLANCHE*) of the glaciers by the creation of chutes between 500 and 2000 meters in height."[1]

By "white coal," Bergès meant the waterpower of France's glaciated Alpine peaks. With the help of the turbine, he had managed to draw power from several torrents emerging from these white glacial snowfields high above the Alpine city of Grenoble. The energy derived from the creation of artificial "chutes"—in this case, iron pipes—through which the water

FIGURE I.1   Bergès's "White Coal" exhibit.
Courtesy of the Maison Bergès - Département de l'Isère (France).

from high-altitude streams plunged several hundred meters before being conducted to the turbines that powered Bergès's paper mill and a growing number of industrial establishments.

In a pamphlet accompanying the exhibit, Bergès explained that white coal was of course a metaphor. But he used the term to emphasize that glacial runoff could become "riches just as precious as coal for their region and for the state." Bergès predicted even more from this new energy source. While coal produced the iron that had "made our civilization and our century, of which the Eiffel Tower is in a sense the poem," white coal also possessed great potential. For white coal, "the economic force of electricity," could produce aluminum, a metal that was "light as glass, firm as iron, rustproof like silver, and certainly the secret of the future." Due to these advantages, aluminum would become prevalent in vehicle construction, in "tools of war," and would "open aerial navigation to us." Bergès concluded that aluminum derived from white coal "will give its name to the new century if it comes at a sufficiently good price."[2]

Bergès's words were prophetic, even if aluminum did not quite put its stamp on the twentieth century in the same way iron did for the previous one.[3] From its debut at the World's Fair, white coal would go on to have an impressive if not widely appreciated career. Not only would the label become a common way to describe hydropower the world over, but the actual waterpower of the Alps was put to mighty use by Europeans. The new energy source not only provided power for a new sort of industrial economy, as the Frenchman had predicted, but actively helped shape it. For reasons this book will explore, hydroelectricity proved to be the major alternative to coal in generating electricity on the European continent in the early twentieth century, and the Alps became one of the global centers of its production. Understanding why and how the Alps became one of the foremost landscapes of electricity in Europe has much to tell us about modern energy use and the history of the lands that drew this power from the mountains.

In the wake of the Exposition, interest in white coal exploded. So lucrative was the potential energy of the Alps that efforts to tap it became one of the main drivers of innovation in the new field of electrification. Up until the First World War, the belief in the superiority of Alpine waterpower led Europeans to make it the foremost substitute to coal in electricity production. Many of the world's largest hydroplants sprang up within the mountain arc, making the Alps a center of the modern industries associated with the "second industrial revolution."[4]

When in the fall of 1914 it became clear that Europe's soldiers would not be home before the leaves fell, white coal emerged as the most important alternative to coal in powering war production. In the interwar years, white coal became entangled in competing visions of energy. One party saw large-scale exploitation of Alpine hydropower as a means to produce international cooperation and economic growth. The other spied in the mountain heights the key to autarkic self-sufficiency and even racial survival. As the 1930s wore on, the latter view increasingly won ground. Though it has not been fully recognized, Nazi planners in particular saw Alpine waterpower as essential to their expansionist program. In part because such energy took time to develop, the fascist war effort came up short. But after the war, Europeans returned to the earlier vision of developing Alpine hydropower as a means to rebuild devastated societies and foster the international cooperation and prosperity that would prevent a repeat of the supremacist violence. By the early postwar period, over half of western Europe's electricity—and a quarter of the Continent's total—was generated by Alpine water. Of course, all of this new energy development necessitated dramatic interventions into politics, society, and culture. It also remade the Alpine environment. In exploiting white coal, Europeans completely replumbed the Alps, the "water tower" of Europe, with consequences—both anticipated and unintended—to this day.

Energy is one of the most pressing current issues for global society. In light of global warming and a cluster of other environmental problems, transitioning away from fossil fuels to other "renewable" or "green" sources of energy is viewed by many policymakers and experts as essential. Given the centrality of renewables in our current energy discussions, understanding of their development is key. Since the 1980s there has been a growing interest in the history of energy use, particularly for the modern period characterized by unprecedented energy consumption.[5] However, there are still important gaps in our knowledge. To date the history of fossil fuels has received the lion's share of attention.[6] There have been far fewer attempts to write comprehensive histories of renewable energy sources as such.[7] This imbalance means that our understanding of a fundamental aspect of modern energy history—and the context for current energy policy—is distorted.

This book tells the history of Alpine hydroelectricity. It is a story about mountains, water, and power in Europe from the middle of the nineteenth century until the present. In seeking to explain how the Alps became one of the most important energy landscapes worldwide, I argue that we should take seriously the insight of Aristide Bergès in styling the energy "white coal." The history of Alpine waterpower—and, I argue, of so-called renewable energy sources in general—cannot be understood without grasping the assumptions that lie behind this label. First, coal was the backdrop to all of these changes. The book's subtitle refers to the "age of fossil fuels" to signal my belief that the fundamental framework for understanding what happened in the Alps—and modern energy development as well—is that it has occurred under the sign of coal, oil, and gas.

The movement to exploit Alpine energy on a grand scale took place in the context of the epochal transition to intensive coal use in the early nineteenth century. This transition, which first took place in England and from there slowly spread across the globe, was so momentous that it deserves to stand alongside the discovery of fire and the shift to agriculture as the most consequential in human history. For reasons that I detail in the body of this book, coal removed previous limits on both energy use and, consequently, economic growth. For now it is enough to say, as many have, that coal's property as a *fossil* fuel meant that it had potent short-term advantages compared to alternatives. The ancient sunshine stored within could be accessed immediately. It did not require, as other fuels did, large amounts of land to be cultivated over a period of years, sometimes decades.

The prospect of making coal the cornerstone of a new energy era, however, first arrived with a machine that could take this thermal energy and make it lift, rotate, work. This was the steam engine. When fed with coal, it launched an energy revolution still underway today. In many ways it powered the industrial revolutions that remade economy and society worldwide.[8] Coal-fired industrialization in some parts of the globe also touched off a scramble for new sources of energy in other places. The goal was to replicate the economic dynamism of steam power. In this context, Europeans could begin to spy potential energy in mountain torrents. A change in energy systems spurred new environmental perceptions.

Second, ideas about the virtues of coal as an energy source decisively influenced the development of Alpine waterpower. The Alps became one of the global centers of hydroelectric production because, in the eyes of contemporaries, mountain waterpower was the form of natural energy that could be forced to behave most like coal. Coal, the argument went, was not subject to the geographic limitations or seasonal fluctuations that plagued water, wind, and solar power. It was energy that could be used anywhere in the world a steam engine could be brought—and at any time. This viewpoint ignored the herculean efforts it took to wrest the mineral from the ground, ship it around the world, and feed it to hungry boilers. Perhaps this omission emanated from an unspoken assumption that the necessary human labor could be wrangled more reliably than the forces of nature. The mountains, however, had an advantage in this regard. Above all, the lakes and glaciated valleys of the Alps represented opportunities to store fickle waterpower. In this manner, more so than other flows of natural energy like wind or even the waterpower of the plains, white coal could be made to act like a fuel. Over the course of the twentieth century, the argument that the Alps should be exploited for this feature won the day. The result was the creation of what I call *Europe's battery*, a gigantic system for storing hydraulic energy and counteracting renewable energy's Achilles heel.[9] The Alps have become a hybrid landscape, where the line between the natural and the technological has become blurred.[10]

To understand the historical development of a renewable energy source, I chose not to frame this study as a history of hydroelectricity, but of Alpine hydroelectricity. This, then, is a history not of just an energy source but of an energy landscape and its modification by humans. The concept of energy landscapes has received some attention in the past, both as a way to grasp the environmental implications of energy use and to understand the broader dynamics that have led to energy transitions.[11] Implicit in the concept is that we need to consider more than just an energy source itself to understand its development and the wider impacts of its use. Using it here made sense for a couple of reasons. First, as Bergès has shown us, contemporaries actually distinguished between Alpine waterpower and other forms. That is to say that at the time, Europeans themselves perceived the Alps as a unique energy landscape. More importantly, waterpower is not simply inherent to water but depends on its environmental context. It was not only the water that mattered, but the reality that it flowed down steep slopes that imparted energy and through narrow canyons that seemed to make for ideal places to impound and store water. Where I am writing near the mouth of the Mississippi River, the sheer volume of water flowing represents considerable potential energy. The problem is trying to create a fall large enough to harness the flow. Such a dam would flood an enormous area and be difficult to anchor it in the landscape. In the Alps, Europeans supposed that nature had already done much of the necessary work by creating lakes and valleys that could be conveniently closed off. Landscape decisively influences the physical characteristics of flowing water—and humans' ability to manipulate it. In dealing with water in particular, its location—the mountains, the plains, the delta—matters.[12]

The Alps have become a landscape of electrification, and for that reason the mountains are an important part of that broader story. It is a story whose significance for modern energy history is vastly underappreciated. Modern energy histories tend to privilege the intensive use of fossil fuels as the most momentous occurrence. As noted, coal, oil, and gas represent a true energy windfall for societies and offer unparalleled versatility. It is also true that in absolute terms, the majority of energy deployed by humans since 1800 has come from fossil fuels. Several realities often go unrecognized in this regard, however. First, statistics that show the dominance of fossil fuels rarely note that most of the energy unleashed was not harnessed by humans but lost in the form of wasted heat. The eye-popping numbers can be misleading in this regard. Second, a large and steadily increasing share of fuel has been burned to produce another form of energy: electricity. Most of the world's electric current comes from fossil fuels.[13] Only electricity, that most versatile of energy sources, has made possible our modern, high-tech society. As one historian has noted, we can construct electric cars, trains, and to some extent aircraft. But can you picture a world run by steam-powered computers?[14] And I would add that it is electrified factories that produce the cars, planes, and ships that burn hydrocarbons.

Not surprisingly, then, historians have produced excellent research on the emergence and spread of electricity. With some important exceptions, however, most studies have focused on the history of electrification in urban or national settings.[15] As hydroelectricity played an important role in the electrification of many countries, some scholars have also focused on the significance of water in this development. The overwhelming majority of these works, however, conceive of electricity in terms of either a commodity or a technology—not as an energy source. They also treat water more as a static natural resource than a dynamic product of earth's life support systems.[16] The rise of environmental history since the 1970s has increasingly produced analyses that also consider the environmental sources of hydroelectric energy and its impacts.[17]

Something new is added by tracing the history of the Alpine energy landscape: the central role that the mountains played in the development of the electrical technology that continues to undergird our modern grids. As I will show, many of the decisive innovations required to transmit electricity over long distances arose out of efforts to bring the power of the mountains to the plains. Seen from the Alps, then, early electrification was as much about remote rural waterfalls as it was about Edison light bulbs in posh city quarters. From this perspective, the problem of renewable energy has not been peripheral but at the heart of modern energy concerns. The effort to utilize renewable energy led to the promotion of a technology—electricity—that enabled the dramatic expansion of fossil fuel use. The Alps and the globe's coal seams and oil patches are all part of the same story.

The Alps were a landscape of energy transitions as well. For historians looking at energy use in the past, one of the primary questions is how changes in energy use occur.[18] It is not hard to see why. Explaining why they happen is not only of historical interest. At a moment when shifting away from fossil fuels seems necessary to avert catastrophic global warming, it would be helpful to know how transitions have functioned in the past. The concept of energy transitions is not without detractors.[19] But a generally shared conclusion is that energy transitions do not happen overnight. Most large-scale transitions have taken decades, and in this regard white coal is no exception.[20]

Still another question is *why* societies adopt new energy sources or—as is more frequently the case—ways of converting or storing them. One group represents what might be called the necessity school. This point of view squares with the adage that necessity is the mother of invention. So, for example, the shift to intensive coal use in England and elsewhere was encouraged by increasing wood shortages.²¹ These are rationalist economic arguments that as one resource becomes scarce, movements to substitute another resource gain momentum. Another school has emerged suggesting that recent energy transitions in particular have been driven by entrepreneurs who have identified vast supplies of potential energy resources and then attempted to convince others of the wisdom of using them.²² Still another group argues for the unintended nature of energy developments and the importance of political expediency.²³

If one understands an energy transition not as a change in primary energy supply, but simply as a change in the ways that societies generate, store, and transport energy as I do, then the following pages document multiple substantial transitions. First, the introduction of the turbine finally enabled the full exploitation of the great falls of mountain waterpower. Next, coupling turbines to generators gave rise to a new energy source—hydroelectricity—with all of its peculiarities. The use of high-voltage alternating current to transport this energy overland also represented a crucial change in energy use. From today's perspective, these developments are interesting because they also signify a transition to an intensive use of renewable energy. In general, the experience in the Alps suggests that transitions defy simple categories. In the early years, market considerations played a crucial role in the spread of mountain waterpower. Alpine hydropower's initial success had much to do with its "discovery" and promotion by businessmen like Bergès. White coal, in other words, was initially slick marketing. However, the ultimate boom in Alpine hydropower owed its rise to contingency—the World Wars—and the decision on the part of state entities across the Alps to take an active role in the energy sector and force its expansion.

A major recent debate in energy history has centered around whether particular energy regimes have led to specific political outcomes. The theory that dependence on oil revenue makes for undemocratic states—the "oil curse"—is one aspect of this discussion. An influential recent thesis argues that democratization in nineteenth-century Europe was fundamentally connected to its emerging coal regime. According to this view, the physical nature of this fuel—that it is a bulky solid requiring considerable effort to remove from the ground and transport—made the region's economy vulnerable to labor disruption. Since coal will not move on its own but requires people to move it, workers had leverage to demand political reforms. The twentieth-century corollary is that liquid oil defies worker control, and states built atop the production of hydrocarbons have escaped comparable pressures to democratize.[24] In the Alps, the connection between politics and energy turns these equations on their head. Here political circumstances shaped energy development. The legal reality that water was in many cases a common good ensured that the state would play a major role in adjudicating its use. So too did the fact that modifying flows of water had wide-reaching societal effects. After the emergency of the First World War, governments throughout the Alps capitalized on their sovereignty over water resources and drove a surge in white coal development to secure domestic energy supply. The public hand would remain the dominant force in Alpine hydropower development until the liberalization of European energy markets at the end of the millennium. One major lesson from the uplands, then, may be that large-scale transitions to renewable energy require an active state role.

Writing a history of the transformation of the Alpine environment is also a different way of looking at modern European history. The history of this region and most others continues to be conceived primarily on a national basis.[25] There are good and understandable reasons for this reality. For one thing, the history profession as we now know it first developed in order to tell the story of the rise of nations. The academic discipline of history as it is currently practiced emerged in nineteenth-century Germany, and its acolytes consciously took the history of the nascent German "nation" to be its object of focus.[26] Historians ever since have almost reflexively defaulted to concentrating on the history of nations. When one trains to be a historian these days, it is still most common to become an expert in French, or US, or Chinese history. In the meantime—in part because of the efforts of national historians—people throughout the world often think of themselves in national terms. National identity is one of the most prominent forms of collective identity. And most people view world affairs through the lens of nationality. It follows then, that for many types of history, the national scale makes most sense.

But for many environmental topics, stopping at national borders promises to miss critical aspects of the story.[27] Environmental phenomena often operate without respect for lines on maps. The geologic processes that created the Alps did so eons before anything like national boundaries existed. The waterways that carve the peaks blithely transgress borders. Taking a geographical landscape like the Alps as the framework for a history shifts our perspective on these lands during this time period and challenges the assumptions that underlie some national histories. What does it mean, for instance, that monarchies in Italy or Bavaria built similar dams, as a republic, in France? Or that the same dam in the Austrian Alps was welcomed by liberal internationalists and National Socialists?[28] By following the changes in the mountain landscape, we can see that various nationalities and political groupings were united more by their commonalities than separated by their differences. In place of fragmented histories of national development, a story emerges about the interplay between environmental conditions and societies in a peculiar part of the world.[29] The timing of the modifications of the Alpine landscape, moreover, suggest the centrality of international factors—the World Wars and their aftermaths—for the fate of Europe's environments.

Turning our attention to this region, often considered peripheral, also reveals an alternative history of modern European economic development. The traditional story of industrial Europe focuses on steam-driven, heavy industrial development.[30] The changes to Europe's most iconic mountain range are a testament to the existence of another Europe, one that tried to compensate for its lack of fossil fuels through intense manipulation of water.[31] This history of a European energy landscape also suggests a possible "special path" for the region's energy past.[32] Almost forty years ago, one historian defined the history of American energy use as "coping with abundance." What he meant above all was that the United States had the uncommon luxury of having many energy options to satisfy demand. This has been a considerable boon to American businesses and entrepreneurs, but it has also often led to profligate energy use.[33] In many ways, continental Europe's energy past is and has been the opposite phenomenon. Western and Central Europe are composed of medium and small states with far narrower resource bases than countries such as the United States, China, or Russia. There is much greater pressure in these areas to economize with energy. The transformation of the Alps is the physical manifestation of West-Central Europe's coping with scarcity.

Since Fernand Braudel began his epic postwar history of the Mediterranean world with a section on mountains—"Tout d'abord les montagnes"—there has been increasing interest in the uplands of the past. Braudel's conception of the mountains has also proved durable. For him, upland regions have by definition stood apart from the more crucial developments on the plains. Due to their unforgiving terrain, Braudel supposed, mountains could not support the development of urban civilization that has been the primary driver of history. The French historian saw the Alps as somewhat of an outlier in this regard, a *montagne exceptionnelle* due to undeniable traces of high culture.[34] In the intervening time, numerous historians have put mountains first by writing histories of the Alps.[35] Studies have shown us how the mountains have figured centrally in the formation of national identities and even what it means to be modern.[36]

A recent history takes up Braudel's question of the specialness of the Alps as its main theme.[37] It finds that, in many ways, the Alps are exceptional. From a comparative perspective, these mountains have been more heavily populated since the late medieval period than other European or global highlands. But this characteristic has had much to do with the peculiarity of its surrounding areas. These regions have long been the seats of epic empires, temporal and spiritual, and some of the most economically dynamic spots in recent history. It was their dramas that led personages such as Hannibal and Napoleon to attempt legendary crossings of the mountains, bringing notoriety to the Alps by association. Since the Enlightenment and industrialization visited Europe in the last three centuries, tourists have begun flocking to the mountains en masse. They came seeking, in the words of an early British Alpinist, "the playground of Europe."[38] And their tourism has transformed society and economy in the Alps as for no other range in the world. In terms of energy, this book will show, the Alps are exceptional as well. No other upland region on the globe has been modified to store power as have the Alps.

Historians, it has been said, can be "splitters" or "lumpers." That is, when trying to reconstruct the past, they can tend to emphasize the peculiarity of phenomena or the commonalities and connections between them.[39] *Mountain Battery* falls decidedly into the lumping category. This is in part because the lower resolution of histories of a larger scale causes some objects to blend together. It would not be possible to write histories beyond the individual subject without generalization. But it also stems from my assessment that, across space and time, significant consensus existed about the making of the Alpine energy landscape. At times in the book, I will refer to things that "engineers" or "Swiss" or "Europeans" believed or did. This is not meant to suggest the absence of dissenting opinions or actions in a given group. But in general, I do recognize more similarities across groups. Europeans in this region ultimately transformed a shared mountain landscape in comparable ways. This was in part because the various states were constantly taking note of the others' progress. But above all, modern energy-intensive society presented states with similar pressures and choices. The record suggests a convergence of outcomes.

In seeking to understand how and why Europeans modified the landscape in a certain way, this book focuses primarily on the actions and thoughts of engineers, politicians, and economists. There would be other ways to tell this story, ones that focus more on the role that consumers of energy played, for instance. Those energy historians who point out the necessity of paying more attention to the demand side of the power equation are surely correct.[40] But it is a contention of this book that elite decision-makers in positions of power possessed outsized influence in determining the shape the Alps ultimately took. That these were mostly well-to-do white men does not mean that categories like class, race, and gender played no role in this story. On the contrary, questions about what it meant to be bourgeois, masculine Europeans in the early twentieth century were inextricably linked to the choices they made and how they understood them. The historical sources that they left behind—in government archives, in the pages of journals, and the protocols of international conferences—make up the bulk of the evidence for this study. If elites did carry more weight in reality, it has important consequences for how we understand phenomena such as energy transitions and environmental change.

The history of white coal also sheds light on the malleable nature of energy. Though few things could seem more material, what exactly composes the stuff is hard to define. Energy as a concept, moreover, is a rather recent one. Tracing how Europeans defined what constituted acceptable uses of hydropower, moreover, must leave one with the impression that a fair amount of imagination went into determining where, when, and how it should be used. The key is present in the very distinction that Bergès made between his energy source and others. White coal was superior because it derived from glaciers, meaning the flows upon which it depended were more regular than for other waterways. The towering falls that could be exploited in the Alps supposedly made Alpine waterpower cheaper too. Since the great relief multiplied the effectiveness of falling water, one could harvest more energy per drop of water than in the plains. This meant that all the necessary equipment—canals, pipes, turbines, and powerhouses—could be of smaller dimension.[41]

All of the distinctions imply that there would be many different ways to create the hydraulic pressure needed for useable energy. Rarely did a white coal proponent note that some of these advantages might be nullified by the reality that construction projects are more difficult in the mountains, and maintenance more frequently required. Remoteness from centers of consumption also increased the cost of transport for materials and of electricity too. The history of white coal shows just how much human culture goes into determining where useful energy can be found, how it should be harnessed, and for whose benefit. To speak with the literature on the history of nationalism, energy is very much "imagined" or "invented."[42] The outsize role that imagination can play in siting hydropower plants may be a peculiarity of the energy source.

Finally, a word about the narrative arc of this book. One way of reading this history of a renewable energy source is as a triumphal success story. In these pages you will read how scientists and inventors made new discoveries and created new, seemingly miraculous technologies. At the behest of leaders and politicians, engineers designed new systems that laborers implemented with their sweat and toil. Oftentimes these efforts were portrayed in heroic terms, with humans mastering the powers of nature for the common good. In reality, the process was much messier. Plans to modify Alpine hydrology met with opposition always and everywhere. Bold promises of superlative energy bounties went unfulfilled, dams sometimes failed. Where possible, I try to show how the confident pronouncements associated with white coal were always based on particular points of view. I also try to do justice to the messiness by zooming in on cases at the local level. Hydroelectric development in the Alps—like all human actions—failed to realize its lofty goals and resulted in unintended consequences. I hope the reader will keep these caveats in mind.

ONE

Mountains of White Coal

> The immeasurable height
> Of woods decaying, never to be decayed,
> The stationary blasts of waterfalls,
> And in the narrow rent, at every turn,
> Winds thwarting winds bewildered and forlorn,
> The torrents shooting from the clear blue sky,
> The rocks that muttered close upon our ears,
> Black drizzling crags that spake by the wayside
> As if a voice were in them, the sick sight
> And giddy prospect of the raving stream,
> The unfettered clouds and region of the heavens,
> Tumult and peace, the darkness and the light—
> Were all like workings of one mind, the features
> Of the same face, blossoms upon one tree,
> Characters of the great Apocalypse,
> The types and symbols of Eternity,
> Of first and last, and midst, and without end.
> —William Wordsworth, "The Simplon Pass," The Prelude (1850)

IN 1837, THE TOWN OF St. Blasien in southwest Germany witnessed a crucial—but often neglected—energy transition. Located in the hilly Black Forest region, St. Blasien was home to several waterfalls including one that measured over one hundred meters in height. The town also boasted its share of industrious inhabitants, and several had attempted to wring useful energy out of this power source. Taking advantage of higher falls required millwrights to build larger waterwheels. The dimensions of the waterwheel that could harness a hundred-meter drop, however, were beyond all reasonable economic considerations. All attempts to tap into this fall, therefore, necessitated the construction of a series of waterwheels situated one above another. Converting the entirety of the falls remained out of reach until the late 1830s, when the Frenchman Benoît Fourneyron installed a new instrument he had created in a cotton spinning mill at the fall's base. Measuring just fifty centimeters in diameter and weighing eighteen kilograms, he called the device a turbine. It was fed water from the top of the falls through a 500-meter-long iron pipe. The new prime mover managed to convert approximately 80 percent of the water's kinetic energy into 60 horsepower of usable energy.[1]

The ability to draw useful energy out of a force of nature such as a gigantic waterfall astounded observers at the time. The engineer Moritz Rühlmann, who visited the facility, recorded his awe in the presence of the turbine:

> On entering the wheel-room, one learns that what had been heard at a distance about this place was not merely mystification, but reality. One then feels seized with astonishment, and wonders, more than in any other place, at the greatness of human ingenuity, which knows how to render subject to it the most fearful power of nature. At every moment the powerful pressure appears likely to burst in pieces the little wheel, and the spiral masses of water issuing from it threaten to destroy the surrounding walls and buildings. Often when I went out of the wheel room, and looked at the enormous height from which the conducting tubes brought down the water to the wheel, the idea forced itself upon me, "that it was impossible," but the idea passed away when I went back into the little room.[2]

Beyond provoking amazement, the development of the turbine had very real material consequences, particularly for the Alps, Europe's foremost mountain chain. From this point onward, it became possible to create waterpower from falls over a hundred meters—what would later be called high-pressure waterpower. Over the course of the mid-nineteenth century, the implications of this fact became clearer. Since waterpower is a function of the volume of water multiplied by the distance of its fall, height itself became a sort of energy resource. And no region possessed more of this newly valuable commodity than the Alps. Eventually, in the 1860s a confluence of circumstances led a group of entrepreneurs in the French Alps to begin a sustained experiment in using high-pressure water to power their industrial establishments. These men, led by the entrepreneur Aristide Bergès, sought to outdo one another by creating higher and higher falls, what they called *hautes chutes*. Together, they established a template that would later be followed in other corners of the mountains. Bergès, part of a new class of engineers dedicated to modernizing the French nation, succeeded more than anyone else in marketing this new energy as white coal.[3] In stressing what was unique and advantageous about using Alpine waterpower, he laid the foundation for seeing the Alps as a landscape of new energy resources.

This chapter tells the story of how Alpine water came to be perceived as white coal after the middle of the nineteenth century. In a certain sense, then, it is a history of the discovery of a renewable energy source. The case of Alpine waterpower shows a number of features of interest to those concerned with energy transitions. In general, it must be said that like other energy transitions explored by historians, the shift to white coal was complex. And, like many energy transitions, white coal was not actually a new energy source. Rather, the change lay in a new way of harnessing waterpower, at least on the surface. While contemporaries chalked up the new energy source simply to the innovation of a new technology to convert waterpower—the turbine—in reality the picture was more complicated. Alpine water had been in use as an industrial energy source for centuries.[4] Traditional technology, moreover, was capable of putting Alpine water to work in a variety of ways and potentially on a large scale.[5]

What changed was a revolutionary shift in the energy regime of western Europe and other parts of the industrializing world. For the first time in human history, economies had begun to exploit fossil fuels on a grand scale. First in England, around 1800, and then in other localities, the adoption of the steam engine allowed societies to utilize coal deposits to burst the limits on economic growth. Fossil fuels represent a one-time energy windfall. States using them gained decisive economic and military advantages—as evidenced by the United Kingdom's rise to a globe-spanning empire. Many, in regions like the Alps that did not possess abundant fossil fuels, understandably feared being left behind. In this way, the fossil fuel revolution prompted an intense search for new energy sources that might compete with coal. This is the context in which entrepreneurs sought new ways to access traditional energy sources like flowing water. The discovery of white coal did require technological innovation. But it also demanded a society that permitted its implementation. Finally, the discovery of white coal demonstrates another peculiarity of what we have come to call renewable energy resources: namely, that there is a fair amount of imagination that goes into their construction.

The Nature of White Coal

In order to understand how and why contemporaries spied unique energy in its waters, it is first necessary to understand the nature of the mountains from which they sprang. As simple as it may seem, however, defining what and where the mountains are is not so straightforward. This is because the definition depends on who is looking. From a geographer's perspective, only those areas defined by high-mountain geomorphological processes are considered Alpine. This view of the Alps limits the mountains to an archipelago of zones above 2,000 meters or so. This territory is not inhabited by humans year-round. Politicians of various stripes, on the other hand, have their own definitions. In the context of European regional politics, leaders seek to ensure that Alpine issues are accorded serious consideration by expanding the mountain region to include powerful urban centers on the outskirts—places like Vienna, Munich, Zurich, Lyon, and Milan. Including them in the Alps creates a much larger region with nearly 100 million inhabitants and biomes little affected by upland dynamics.

This book uses a definition of the Alps that lies in between these two extremes. It is a combination of the general view on where the Alps begin—where the land begins to rise—with a hydrological definition of the mountains as the area where river dynamics are affected by the mountains. This results in an area about 200,000 square kilometers in size with a population of 15 million or so. These Alps stretch in the familiar arc from the Riviera in the west to the doorstep of Vienna and the head of the Adriatic in the east. They pass through the present-day countries of France, Switzerland, Italy, Liechtenstein, Germany, Austria, and Slovenia.[6]

The first mention of the Alps in European letters stemmed from Herodotus, writing in the fifth century BCE. Herodotus wrote of the Alpis, which did not refer to a mountain chain, but a tributary of the Danube. About a hundred years later Aristotle called the massif by its earliest known name, the Arcynian Mountains, perhaps alluding to the Hercynian Forest that once covered much of southern Germany. It is in the work of an obscure Alexandrian poet named Lycophron that we first find the name Salpeis in the third century BCE. The precise meaning of the word remains unknown. The Romans associated it with *albus*, the Latin word for white. In all likelihood the term is Celtic for lofty, as cognates still appear in Gaelic tongues.[7]

The loftiness that so impressed the Celts was the product of plate tectonics. In fact, since the 1960s the Alps have been used by geologists as a textbook example illustrating the validity of the theory.[8] Scientists believe the folding that created the Alps began some 100 million years ago. At this time, the northward drifting African continental plate collided with the European plate. The impact dramatically shrank the size of the Tethys Sea—the colossal sea that once separated the European and African continents and whose remnant can be seen in the Mediterranean. It also vaulted the African plate on top of the European, folding together continental crust in the process. This folding took place mostly in the horizontal dimension, creating a low mountain range. It has only been in the last 20 million years that this mountain range was uplifted to resemble the high peaks we recognize today. The ongoing process of mountain building and the erosive forces of water and ice have for the most part held each other in check, meaning that the Alps were likely never much higher than they are today.[9]

At the height of the last glacial period, 22,000 to 20,000 years before present (BP), Scandinavia and northern Europe were completely covered by enormous ice sheets; the Alps, Pyrenees, and the Caucasus sat under smaller ones. This most recent ice age ended between 17,000 and 10,000 years BP. The work of the glaciers expanded and widened valleys and created some of the most important mountain passes in the Alps. The retreat of the glaciers left behind moraines and transported sediments that aid in the formation of soils and therefore agriculture. The withdrawing ice also created high-altitude cirque lakes and carved out characteristic U-shaped valleys with very steep sides. All of these effects would have important consequences for hydropower development. The valleys would later appear as ideal places to store water. Many of their steep slopes were and remain sites of extreme events such as landslides.[10] But they have also left behind countless so-called hanging valleys. These hanging valleys are often the site of the spectacular waterfalls that have drawn visitors to the region. They were also seen as ideal spots to harness waterpower.

Hydrologically, the Alps serve as "Europe's water tower."[11] The nickname alludes to the fact that several of the continent's most important rivers have their sources in the massif. Alpine water drains via the Rhine into the North Sea, via the Rhône and Po into the Mediterranean, and lastly via the Inn and Danube into the Black Sea. While the Alps compose a relatively small area of these streams' watersheds, they nonetheless contribute a disproportional amount of water to their channels. In the case of the Rhine, the Alps comprise some 23 percent of its catchment area but deliver about half of its total annual discharge. The figures for the Rhône and Po are similar. In this way, the Alps have an importance that reaches far beyond their borders.

Thanks to a host of environmental factors, the Alpine rivers possess an abundance of water that is almost unmatched on the European continent. The mountains function as a "rain catcher" (*Regenfänger* in German). The lofty Alps drive moist air masses from both the Mediterranean and Atlantic upwards and force them to precipitate their water freight in the form of rain or snow. The mountains as a whole then receive much more precipitation than the rest of Europe. An average of 1,450 millimeters falls on the Alps annually, while western Europe receives about 800. The difference is greater still if one considers all of Europe, which averages only 600 millimeters of precipitation.[12] The high altitude of the Alps also decreases the amount of water lost due to evaporation or transpiration, meaning a higher percentage of its precipitation drains away than in surrounding areas. A large portion of Alpine river drainage basins lie above the tree line. Over half of the watersheds of the Rhine, Rhône, and Inn rest above 2,000 meters. This means that factors of altitude—especially the development of glaciers and frost—play a significant role in influencing the water volume and sediment loads of Alpine watercourses.

Of crucial importance for the ecology—and waterpower development—of Alpine streams is the seasonal distribution of water flows and the storage of water in lakes and glaciers. This is called a "flow regime." In the Alps, geographers have identified glacial, nival, and nivo-pluvial/pluvionival types. In watersheds with glacial flood regimes, a large proportion of the catchment area (greater than 30 percent) is glaciated. Glacial watersheds thus have a high median altitude. The daily and yearly fluctuations in glacial melt give glacial regimes a unique discharge rhythm. Typically, maximum flows occur in the summer months of July and August, when melting glaciers account for approximately 60 percent of all annual discharge; 90 percent of the total discharge takes place from May through September. Accordingly, streamflow in the winter months is low. The catchment areas for strictly nival regimes are devoid of glaciers. Snowmelt begins in April for lower locations, with the main phase in May and June.

The prevalence of substantial summer floods and pronounced winter dryness would prove to be the feature of Alpine hydrology that most concerned energy experts. Between glacial and nival regimes, there exist a number of transitional zones that describe a fair portion of Alpine waterways. In contrast to glacial and nival regimes, pluvial discharge patterns post several maxima over the course of the year. The first comes in spring with the snowmelt, and the second in fall due to precipitation. Purely pluvial regimes are rare in the Alps. Only very low-lying tributaries possess strongly pluvial characteristics. Even in the large river valleys, the Alpine snowmelt dominates flow regimes far into the plains. Glaciers currently store about a year's worth of precipitation in their ice fields and even out the differences between dry and wet years considerably. Some 216,200 million cubic meters of water regularly drain out of the Alps, year in and year out.[13]

While the mountains may appear unmoving and timeless, the defining feature of the Alpine environment is its labile, even volatile natural dynamics. For the Alps are a young chain whose mountain-building process is ongoing.[14] Dynamism characterized Alpine hydrology, especially in the time before hydropower development. From an ecological standpoint, scientists would have characterized undammed Alpine streams as nearly "wild river ecosystems." Alpine waterways continually changed their gradient and velocity and experienced strong seasonal variations in streamflow and sediment load. During flood periods, the carrying capacity of high-energy Alpine waterways increased. This process constantly created new microenvironments, forcing both flora and fauna to evolve special survival strategies. Certain plants, for instance, acquired flexible stalks that yield to flowing water and whose growth was stimulated by damage to its shoots. Successful riparian grasses also evolved to be able to quickly populate freshly created sandbanks. Alpine riverbeds, floodplains, and valleys also served as migration corridors for plants and animals traveling both upstream and down. Floodplain environments were (and continue to be) home to some of the greatest biodiversity in Europe.[15]

The waterways of the Alps, however, were by no means untouched entities by the nineteenth century when the story of white coal begins. Humans had significantly modified Alpine hydrology for centuries. In late-nineteenth century Switzerland, for example, a community completely drained the Märjelensee, an upland lake that had first emerged earlier in the century thanks to glacial advancement. On several occasions, the Märjelensee broke through its glacial dam, flooding the entire Rhône valley from the mouth of the Massa to Lake Geneva. To remove this threat for posterity, locals drilled a drainage tunnel in the lake's empty basin.[16] Another example of the types of modifications made to Alpine rivers comes from the French Alpine piedmont, where inhabitants thoroughly rearranged the Isère by the early nineteenth century. The Isère went from being a volatile braided channel to a heavily diked single-channel river, engineered for the purposes of agriculture and flood control in the early nineteenth century. Many might assume that the braided, unpredictable Isère of the early modern period was a waterway untouched by humans. But even the braided Isère system was at least partially the result of human activity. Anthropogenic deforestation teamed with increased Little Ice Age precipitation to give the Isère its multichannel look.[17]

Besides interventions like these, the Alps have also long been a center of waterpower usage in Europe. Traditional waterpower utilization was based on the employment of waterwheels in various types of mills. The Alps, like a number of other regions throughout Europe, possessed environmental characteristics that were well-suited to turning waterwheels. Traditional watermill technology, however, tended to foster a more decentralized landscape of smaller-scale energy production.

The use of waterpower first began in antiquity, although just when remains unclear. Different scholars suggest an origin in fourth-century BCE India or the Near East around 200 BCE. From this point, waterpower usage spread throughout the ancient world. Far more certain is the technology that harnessed the power of water for nearly two millennia, from its beginnings until the nineteenth century. The earliest converters—or prime movers—were predominantly waterwheels. These came in two broad varieties, horizontal and vertical waterwheels, each of which operated best under different hydrological conditions. The earliest wheels were utilized in grain mills.[18]

Like nearly all energy on earth, waterpower is in fact solar power. It is the useful energy that humans are able to harness by intervening in the hydrological cycle. Waterpower is the product of two primary factors: volume of flow and height of fall (sometimes called "head" in hydraulic parlance). In the waterpower equation, a small amount of water with a very high fall can create the same amount of energy as a large amount of water falling a shorter distance. The availability of waterpower is partly determined by geographic factors, and its utilization is thus regionally limited. Streamflow is influenced by geographic variables such as seasonal rainfall distribution, temperature, humidity, evaporation, air currents, vegetation regime, and geological circumstances.

The Alps possessed many characteristics that made it an ideal setting to harness waterpower. Chief among these were an abundance of water and a surfeit of slope. The mountains are rich with water because low evapotranspiration rates—a function of the colder temperatures at higher altitudes—mean that a large portion of the region's abundant precipitation remains surface runoff. Just as important as the quantity, the quality of Alpine waterpower also facilitated utilization. While Alpine watercourses demonstrated marked differences in streamflow according to season, they generally retained a minimum flow that was available year-round. The regularity of Alpine streamflow was in many cases aided by the presence of lakes and glaciers. The great relief of the mountains imparted added energy to the water coursing down its slopes, but it was not relief alone that made the Alps ideal for waterpower. The broken topography of the mountains also ensured more opportunities to concentrate falls than in the lowlands.[19]

As with the emergence of waterpower usage in general, it is difficult to say when Europeans first began harnessing the power of Alpine water. Pockets of southern France and northern Italy took advantage of waterpower as far back as Roman times. After the collapse of the empire, the watermill seems to have spread from these bases into the Alpine region. The earliest confirmed mill in Switzerland dates from the sixth century. Southern German legal codes attest to the existence of waterpower usage in that region during the eighth century; the earliest Austrian watermill seems to have appeared a hundred years later. By the early medieval period at the latest, mills were an increasingly familiar sight on Alpine waterways. Like other types of early mills, the primary purpose of these facilities was flour production. The period from the tenth until roughly the sixteenth century CE, however, witnessed a rapid expansion in the applications for waterpower in the Alpine area. A slew of new uses for waterpower originated in or near the Alpine regions of southeastern France, northern Italy, Switzerland, and southern Germany.[20] In addition to powering the first hemp mills, fulling mills, oil mills, and paper mills, Alpine waterways provided the energy for metallurgical industries such as iron and wire mills, hydraulically operated ore stamps, and metallurgical bellows. In the estimation of one historian, this industrial blossoming made the Alps—alongside a similarly innovative region in northern France—one of the centers of a "Medieval power revolution" based on waterpower.[21]

To tap into the energy of flowing Alpine water, traditional waterpower usage required modification of the local environment. Generally speaking, producing waterpower entailed the impoundment of water and its channeling onto a waterwheel and back into another watercourse. The primary method of impoundment was the construction of a weir across the river—a small dam whose function was not to completely impede flow, but to redirect a portion of water into the millrace, an artificial channel leading to the wheel. Where waterwheels were not installed directly in streams or canals, the millrace often took the form of a wooden flume or chute. After driving the wheel, the water was then led out of the mill into another artificial channel—the tailrace. In most cases, these impoundments and diversions occurred on a relatively small scale. But more serious dams and diversions took place during the medieval period on some larger Alpine streams such as the Lech in Germany, the Durance and Drac in France, and the Ticino in northern Italy. Where conditions permitted, artificial reservoirs known as millponds were also likely employed to ensure a more consistent water supply.[22]

The power of Alpine water was used outside of mills as well. While navigation never really played much of a role in the Alps, the timber floating and rafting industry have a long and important history. Wood has long been both an important source of fuel and a construction material, and Alpine watercourses offered a convenient means of accessing mountain forests. The tradition of floating individual logs and rafts from the Alps to urban markets is—like watermilling—documented all the way back to antiquity. By the thirteenth century, the Alpine wood export took on such dimensions in some areas that cities tried to stanch the flow. Zurich, for example, issued two decrees limiting timber rafting in its dominions in the late thirteenth century. Later the nineteenth-century iron industry's insatiable demand for wood caused many an Alpine forest to disappear and prompted renewed concerns about timber rafting from Alpine governments. The Swiss city of Bern, for example, lodged an outright ban on the water transport of timber in 1870. By that time, however, economic changes were already beginning to make the industry obsolete. The growth of rail networks enabled access to forests and thus sounded the death knell for this exhausting and at times dangerous type of work. Nevertheless, timber floating and rafting continued on some Alpine waterways until after the First World War.[23]

The Fossil Fuel Transition and the Search for New Energy

Up until the nineteenth century, then, Europeans had already utilized the power of Alpine water to a considerable extent. At that point, however, important developments in the broader energy regime of Europe and other industrializing regions began to change ideas about energy usage. The intensive use of fossil fuels allowed some regions to break the usual boundaries on economic growth. Compared to the revolutionary effects of fossil fuel use, traditional waterpower technology began to be seen as lacking in many respects. A push began to improve upon the waterwheel, to make a device capable of wringing more energy from water. The result was the development of a new prime mover, the turbine, and the device would have particular importance for the Alps.

At roughly the turn of the nineteenth century, one of the two or three most significant energy transitions in human history—on par with the adoption of agriculture—took place. In England around 1800, a human society began to acquire the bulk of its energy from a fossil fuel—in this case, coal. While cultures had long employed this mineral as a fuel, the introduction of the steam engine in the eighteenth century enabled unprecedented uses. From the perspective of energy historians, the transition to fossil fuels established the physical basis for a new type of economy and society. To understand the revolutionary implications of the fossil fuel transition, it is necessary to consider the material itself.

Coal is the residue of forests that existed millions and millions of years ago. In a sense, it is fossilized solar energy that has survived through the ages. Coal formed where the remnants of extraordinarily biomass-rich forests accumulated as vegetable matter in low-oxygen aquatic environments such as lakes or coastal swamps. Such conditions limited bacterial activity and thus slowed decomposition. Considering that the intensive burning of fossil fuels is a relatively recent phenomenon, coal use has a much longer history than one might expect. When humans first encountered coal, it was valued foremost for its color and was sometimes even used as jewelry. The origins of coal consumption date back to antiquity, where it was used as fuel for iron production under the Han dynasty in China. Around a millennium later, ironmongers and miners in Song dynasty China used coal to launch an abbreviated industrial revolution almost 800 years before the western European version. The Song coal economy placed the Chinese well ahead of any society in the world in this regard. At that time European coal mining was just getting its start in Belgium. Britain, however, soon emerged as the center of European, indeed worldwide, coal consumption. Shortages of wood encouraged Britons to switch to coal, both as an industrial and domestic fuel source. By the end of the seventeenth century, coal consumption in London had progressed to the point that diarist John Evelyn could compare the city to the "Suburbs of Hell" among other dismal localities, primarily because of the excess of smoke.[24]

Up until the eighteenth century, coal use in England (and elsewhere) remained a component of traditional energy systems.[25] The principal prime movers remained the muscles of humans and animals. Animate power was later augmented by the addition of devices that harnessed the flowing power of wind and water: mills and sailing ships especially. Biomass fuels—wood, charcoal, peat, and animal dung—provided heat for all household and industrial uses. Although these components remained remarkably stable over time, technological innovation greatly improved their efficiency during this time period.

Traditional energy systems were subject to stringent limitations. One of the primary problems of energy use in this period was that it was inextricably bound to land.[26] Societies had to make land use choices that had important effects for the available types of energy. Generally speaking, communities could choose to employ their land as arable, pasture, or woods. Arable provided crops, the food energy needed to feed a population. Pasture fueled livestock, a source both of food and animate power. Finally, the woods were a primary source of thermal energy for cooking and domestic heat. Production in one area could only be achieved by an increase in the overall territory exploited or at the expense of another type of energy. Transport problems, however, limited the effect of gains in territory. At a certain point one expended more energy to move resources than one gained.

In addition to these basic components, traditional energy systems had recourse to other energy sources. They too possessed limitations. As we have seen in the Alps, inanimate flows of natural power provided supplementary industrial power. Seasonal variations in water (and wind) flow, however, confined surpluses to certain parts of the year. Natural flows of energy congregated only in certain geographic locations and defied easy transport and storage. Moreover, water and wind alone could only provide kinetic energy, not heat.

Without intending it, British inventors of the eighteenth century developed a device that would help explode—or, better said, combust—these limits. If ever a technological innovation deserves credit for changing the course of history, then it is the steam engine. The new machine made its debut at the edge of a coal mine in Staffordshire, England, in 1712. It was there that Thomas Newcomen (1663-1729), an ironmonger from Dartmouth, began operation of the world's first steam engine. The engine was constructed to pump water from one of England's many deep coal shafts. Using the heat gained from burning massive amounts of coal, the device boiled water to create steam and drive a piston attached to a bulky wooden beam that rocked back and forth, raising and lowering a series of buckets. Newcomen's machine operated with a strength of nearly 6 horsepower and raised 120 gallons of water per minute. It was also wildly inefficient. To perform work, it heated water into steam, then condensed that steam by spraying water into the cylinder. Each stroke required immense amounts of fuel just to reheat the cold cylinder. More than 99 percent of the energy of these engines' fuel was lost as heat.[27]

Despite its profligacy, the steam engine revolutionized economic possibilities. In the short term, it saved the British coal industry. Its main claim to fame was that it was the first prime mover that could convert heat into motion—first reciprocal and later rotary. It was, as many in England called it, a "heat engine." Using steam-powered pumps, English miners could get around what had been an intractable problem. On the well-watered island, a pit of any size would frequently fill up with rain. Digging eventually hit the water table. But Newcomen engines at the edges of mines allowed access to virtually unlimited stores of energy.

In the long term, the steam engine inaugurated a new epoch of human economic activity. Steam engines allowed the heat of burning coal to power machines. By tapping the power of the "subterranean forest," the accumulation of hundreds of millions of years of biomass, they liberated vast amounts of energy and freed up land that would otherwise have been devoted to producing fuel in the form of food or wood.[28] With the advent of the steam engine, coal could also be used to supply mechanical energy in addition to heat, opening up a plethora of new applications. Over the years, individuals such as James Watt refined the original blueprint, resulting in more efficient and versatile models. The new machines could now be used anywhere and even installed on wheels—locomotives—and ships. The result was a positive feedback loop: steam engines permitted unprecedented speed, range, and freedom of transport, enabling the distribution of more coal for ever more steam engines. By the year 1800, single engines could perform as much work as 200 men—though they still only managed to convert 5 percent of coal's energy into useful work. A century later, steam engines had become thirty times as powerful.[29]

The most dramatic early effects of the transition to coal came in Britain. There the new and improved Watt engine revolutionized the textile industry, which exploded with the availability of new power. While Britain had imported 5 million pounds of raw cotton between 1771 and 1775, the number rose to 58 million pounds in 1834 alone. Coal also transformed transportation. In 1830, the pioneering Rocket locomotive made its inaugural route traveling from Liverpool to Manchester. By the 1840s, thousands of miles of networked rail systems emerged not only on the British Isles, but across Europe and North America. Steam power aided a similar expansion of navigation. Fossil fuels made land and sea travel not only quicker and cheaper, but they went a long way towards overcoming the seasonality of transport. Thanks to steam-powered transport, muddy roads and lack of wind were no longer impediments to travel.[30]

These new economic possibilities also reshaped political relationships the world over. On a global scale, the rise of the industrial British economy helped that nation eclipse competitors in India and China. Shifts in the balance of power occurred within Europe as well. On the continent, no state profited more from the new circumstances than Prussia. Much has been made of the significance of Prussian military prowess in its wars of unification against Denmark, Austria, and France. But Prussia also enjoyed the economic benefits of controlling some of Germany's most important coal reserves in the Ruhr basin. The Prussian statesman Otto von Bismarck famously asserted that "blood and iron" would decide the great political questions of his day, and his state had a leg up on at least the latter part of this equation thanks to the coal-rich Ruhr. It is no coincidence that the high-energy societies based upon coal that were confined to Europe, North America, and to some degree Japan until about 1950, dominated the international arena in this time period.[31]

Contemporaries recognized a connection between coal supply and economic might. Perhaps the best known of these observers is the British economist W. Stanley Jevons. Jevons, one of the founders of marginal utility theory, called early attention to the critical economic importance of English coal supplies in his 1865 treatise *The Coal Question*. In it, he demonstrated the importance of coal consumption for the British economy and attempted to ascertain how long supplies could last. Jevons devoted part of his study to an investigation of alternatives to coal. But after considering a list of potential energy sources, he concluded that the British "must not dwell in such a fool's paradise as to imagine we can do without our coal what we do with it."[32]

Coal transformed economic and industrial prospects where it could be had cheaply. However, like most energy sources, coal was unevenly distributed across the Continent. The hard coal that fueled nineteenth-century industrialization was the product of a process that had begun 300 million years ago, at a time when Europe was located in equatorial climes. Certain lakes and coastal areas that accumulated vegetal material developed the primary seams that would change geopolitical fortunes in the nineteenth century. The offshore deposits proved to be most plentiful, and they extended in a belt from what became Ireland, across Britain and northern Europe, to the Silesian fields in Poland. Another major deposit was located in the Asturias region of northern Spain.[33]

As this list indicates, the Alps possess no major pockets of coal. The mountains were not entirely devoid of fossil fuels. The so-called Briançonnais basin, which runs in an arc extending from northwest Italy through Briançon to the Valais in Switzerland, is a product of the very same geologic forces that endowed Great Britain and northern Europe with its coal stocks. Nevertheless, the orogeny that created the Alps also considerably deformed these outcroppings, making them very difficult to mine.[34] In addition to scanty local coal supplies, transport conditions made fossil fuels relatively more expensive to procure in Alpine regions than in other areas.

The fossil fuel transition, however, occasioned a reassessment of the waterpower potential of the Alps. As a waterpower landscape, the Alps had something going for it that few other places did. In no other region in Europe was it possible to harness falls as large as those in the Alps. In later years, hydraulic experts would make the case that high-pressure waterpower represented a superior form of power and supplied mathematical proof to back up this claim. In the mid-nineteenth century, however, these theories had not yet been developed. All that mattered at this point was that waterpower was a product of a given volume of water falling from a given height. By increasing the fall, one needed a correspondingly smaller amount of water to generate the same energy. A little bit of Alpine water could go a long way. But until the mid-nineteenth century or so, much of this potential energy from Alpine peaks lay beyond human grasp. Looking past the issue of accessibility, harnessing high-head waterpower exceeded the capabilities of traditional waterpower technology. The waterwheels that generated the power to drive mills could not handle the force of water falling much farther than fifteen meters.[35] In the Alps, this limitation left a lot of unused waterpower on the table.

No natural phenomenon embodied the unused potential energy so much as the plentiful waterfalls that exist throughout the Alps. The Alps, like many mountain regions, are home to an abundance of these landscape features. Waterfalls in the Alps are the result of two broad sets of conditions native to the mountain region. One group of these chutes derived from a grab bag of geologic phenomena that created steep stretches in the profile of many watercourses. Geologists call these sites "knickpoints," and they can occur thanks to changes in a channel's bedrock, local tectonic action, changes in stream discharge, or even global changes in sea levels.

Many waterfalls were also the product of glacial erosion. During ice ages, massive tongues of ice reached down the mountains, often into the larger main valleys. When the glaciers receded, they left behind mounds of debris known as moraines. Sometimes, these heaps constituted barriers to flowing water that created abrupt slopes. But these moraines sometimes turned watercourses out of their old beds and onto new terrain where the new river's course was almost always punctuated by intermittent falls. It is this phenomenon that is responsible for much of the profile of the Drac River in the French Alps, for example. It also created the Niagara Falls. The most extreme falls, however, were result of the lower erosive power of the ice in the secondary valleys, which left them stranded higher above the newly dug out main valleys. Geomorphologists refer to these higher secondary valleys as hanging valleys, and the heavily glaciated Alps are full of them. Hanging valleys often tower hundreds of meters above the main valley, in some cases reaching over 800 meters in relief. They posed serious obstacles to transportation in valleys, and they were also the sites of some of the most impressive waterfalls in the region.[36]

In the eyes of a traditional millwright, a waterfall would have in many ways appeared to be the ideal site to harness waterpower. To take greatest advantage of the power of falling water, it is necessary to concentrate the fall at a specific point. This concentration was achieved in two general ways. The first method involved creating a fall solely by obstructing the regular flow of a river. By throwing a barrage or a dam across a river, water could be impounded to create a head between the surface of the captured water and the surface of the water just below the mill. The other method utilized the natural gradient of a longer stretch of a river's course. Using another obstruction, water could be diverted into a headrace canal and led to a suitable point before falling onto the waterwheel. Waterpower, it was argued, was all the more cost-effective if the natural environment could be relied upon to help create hydraulic head, and waterfalls were the most extreme example of this phenomenon.

Unlocking the latent power of Alpine water required the development of a new prime mover. Movement on this front began with the development of just such an implement in the early nineteenth century. The new device was the turbine, and it represented, in the words of one historian, "the first radical improvement of water-driven prime movers since the introduction of the vertical wheel centuries before."[37] Turbines differed from traditional waterwheels in a number of respects. Unlike the waterwheel, the turbine could operate while submerged, meaning it could utilize more of the available fall. It also converted the power of falling water with greater efficiency than most waterwheels. Most importantly for mountainous areas like the Alps, the turbine could withstand the destructive force of high-pressure waterpower.

As with many technological innovations, it is no easy task to assign a single inventor responsible for the development of the turbine, which was the outcome of incremental improvements. But historians generally give credit to the Frenchman Benoît Fourneyron (1802–1867). Fourneyron created his new prime mover in response to a 6,000-franc cash prize offered by the French Société d'encouragement pour l'industrie nationale in the early 1820s to the person who could produce a more efficient industrial waterwheel, and especially one that would not cease to function when floodwaters downstream impeded the wheel's forward motion. This contest must be viewed too as at least a partial attempt to help waterpower make up some ground to the newly dominant fossil fuels.

In 1827, Fourneyron came up with a device that fit the society's bill. Fourneyron's invention was composed of a centrally fixed disk replete with iron compartments that captured the incoming water and guided it to buckets located on an outer wheel—or runner—mounted on a vertical shaft. The name for the device came from the Latin stem, meaning to spin or whirl. Throughout the late 1820s and early 1830s, Fourneyron demonstrated the improved efficiency of the turbine at both experimental and industrial sites. Satisfied that he had fulfilled their contest's requirements, the society awarded the engineer its grand prize in 1833. Fourneyron himself recognized that his contraption addressed the tricky issue of handling high pressure:

> The problem to be solved in order to obtain the greatest possible effect from falling water, is how to receive the water without shock in the apparatus which is supposed to transmit its force and expel it without speed.[38]

By the end of the decade, Fourneyron's turbine installation in the Black Forest, described at the beginning of this chapter, pointed the way to the future exploitation of the mountain heights.[39]

By midcentury, the ability to harness high falls led to a dawning recognition of the awesome potential power slumbering in the Alps. In northeastern Switzerland, a textile factory utilizing a 170-meter fall stood out as the highest chute in the Alps. During these same years, an elementary school student in the mountains of Croatia encountered a model of a turbine in his classroom. Fascinated, he began to design and test his own creation in local streams. He also began to dream of using such a device to harness the power of the Niagara Falls in the United States. Some thirty years later, Nikola Tesla would assist the Westinghouse Company in doing just that.[40]

Another region particularly interested in the prospect of Alpine waterpower was the Italian state of Piedmont. Piedmont—as its name implies—lay at the foot of the Alps in the northwest Italy. At the center of the Italian *Risorgimento*, Piedmont's leaders viewed industrial development as paramount, but the kingdom was decidedly lacking in industrial resources (especially coal). Thanks to the hydropower revolution inaugurated by the turbine, some observers believed Piedmont's industrial luck had changed. For the kingdom dominated the lion's share of the western Alps—the highest, wettest, and most glaciated region in the entire massif.

The awesome potential power of Piedmont's mountains was the subject of an alleged 1847 conversation between the Italian statesman Marquis Massimo d'Azeglio (1798–1866) and British politician Richard Cobden (1804–1865). During a visit by Cobden to Piedmont, d'Azeglio supposedly confided his lack of faith in the industrial potential of his country to the Briton. Cobden, whose nation's industrial pedigree presumably lent weight to his opinion, pointed to the snow-covered peaks of the Alps and declared "therewith you can secure the economic future of your beautiful country."[41]

Whatever the veracity of this story, by the mid-1850s the subject of Alpine energy potential preoccupied the highest echelons of Piedmontese government. At this time Count Camillo Benso di Cavour (1810–1861), the future architect of Italian unification, became convinced of a bright future for Alpine waterpower. Prime minister Cavour voiced his outlook in an 1854 session of Piedmont's "Sub-Alpine" parliament in Turin. In June of that year, Cavour found himself compelled to defend his engagement of an engineer to explore the prospect of transmitting waterpower overland by means of compressed air. The idea met with astonishment and not a little mockery in the engineering community, but Cavour insisted on the importance of the venture. Finding a way to transport waterpower beyond the site of its generation would remove one of the primary limitations of hydropower development and "could do for our country what steam engines have done for England." For, Cavour noted, in "chutes of water" Piedmont possessed "more motive force than England has with all of her steam engines combined." The prime minister's appeals helped win parliamentary support for the compressed-air initiative, even though the technology never became a comprehensive solution to the transport problem of waterpower (that distinction remained for hydroelectricity, the subject of the following chapter). Nevertheless, hydro-pneumatic power did prove its worth in at least one monumental undertaking. Water-powered pneumatic drills replaced steam-powered ones in the drilling of the Mount Cenis tunnel—the first large-scale "piercing of the Alps"—in the 1860s. Cavour's words also helped the count go down in history as the first major political visionary to have grasped the new energy potential of the Alps.[42]

Discovering White Coal

As French nationalists would later be at pains to emphasize, the true "discovery" of white coal actually took place at the other end of Cavour's Mount Cenis tunnel, in the French province of Dauphiné. It was there in a valley directly to the northeast of Grenoble that Aristide Bergès became one of several entrepreneurs who exploited the relief of the French Alps to produce high-pressure waterpower. Beginning in the 1860s, this group succeeded in harnessing the power of higher chutes than ever before. In the mid-1880s, when the emergence of electricity presented another possible use of Alpine waterpower, Bergès proved especially adept at promoting and publicizing the development of Alpine waterpower resources—his biographer aptly dubbed him an "apostle of white coal."[43] After coining the phrase and introducing the technology behind it at the Exposition Universelle in 1889, both caught on in the Dauphiné. From its beginnings there, white coal would go on to conquer the rest of the Alps and indeed the world as a popular way to describe all waterpower.

The Dauphiné is a historical region sandwiched between Savoy, the Rhône, and Provence, which occupies a large portion of the westernmost Alps. In the fourteenth century, the Dauphiné passed from the house of Savoy to France, where it has remained ever since. After the French Revolution, the province was divided into the three *départements* that comprise the region to this day: Isère, Drôme, and Hautes-Alpes. Grenoble, location of one of the oldest universities in Europe, was the historic province's capital and the region's foremost city.[44] The province had a history of waterpower innovation. The first water-powered hemp mills and wood lathes are attributed to the region during the medieval period.[45]

In the mid-nineteenth century, the Grésivaudan valley of the Isère River to the northeast of Grenoble became a center of waterpower innovation. Near Grenoble, the eastern flank of the valley is dominated by the Belledonne, one of the so-called granite massifs that make up some of the highest Alpine peaks including Mont Blanc. Unlike most of the Alps, these mountains originated not from the sea, but deep below the earth. Formed some 280 million years ago, most of the granite massifs—like icebergs—remain hidden below the surface. Ironically, the granite massifs have reached such towering heights because of their relative lightness. These blobs of igneous material float like buoys upon denser molten rock. When the sedimentary strata that once covered them eroded away, they rose even higher. The peaks of the Belledonne all range between 2,500 and 3,000 meters above sea level, and some of the area above 2,650 meters was glaciated as well. The valley below, on the other hand, rests at roughly 300 meters above sea level.[46] The great relief of the Grésivaudan lent considerable power to the torrents streaming down the sides of the Belledonne. The glaciers, by melting in the summer heat, also caused them to carry considerable water even in the dry part of the year—a boon to waterpower users. It was these watercourses that French engineers tapped into using the newly developed turbine, modifying torrents to create higher and higher chutes beginning in the 1860s.

The idea to use turbines to exploit gigantic falls in the Grésivaudan emerged in the context of the industrial expansion occasioned by the fossil fuel transition. In this case, it was the explosive growth of the paper industry in the Dauphiné during the period of the Second Empire. By 1860, the region found itself at the head of French paper production thanks to a major shift underway in the industry. For centuries, the trade had relied upon the use of rags for the fibers that composed paper. As papermaking became mechanized at the turn of the nineteenth century, production increased dramatically lifting the price of rags as well. It became clear that fortunes awaited the finding of a cheaper substitute. Not for nothing did Balzac make one of his main characters in *Lost Illusions* obsessed with this very question. Wood pulp emerged as the answer. Since the mountains of the Dauphiné possessed both an abundance of wood and the water necessary to power pulping machines, the region witnessed a flood of new paper mills. Given its proximity to the Grenoble market, Grésivaudan was particularly well placed to benefit. A first step occurred in 1860, when a local entrepreneur planned to build a paper mill in the valley powered by a 150-meter drop. His notion had been inspired after observing high-fall waterpower during a trip to the German Alps. The idea was not realized, however, for financial reasons.[47]

At this moment, Aristide Bergès entered the picture. While he would not prove to be the first person to harness the great relief of the Alps, he was the innovator most responsible for spreading the idea about the unique energy source. The son of a bourgeois paper manufacturer in the French Pyrenees, Bergès had demonstrated a knack for science at a young age. Based on the recommendation of his professors, he sat and passed the entrance exam for the prestigious École centrale des arts et manufactures in Paris. This "industrial Sorbonne" had been founded in 1829 by Saint Simonians looking to produce civil engineers on the British model in France. *Centraliens*, its founders hoped, would disperse throughout French society and push the country's industrialization. In the person of Bergès and others (Gustav Eiffel was a graduate), this calculation paid dividends. Most students entered the school at age nineteen or twenty, so it was something of an anomaly when Bergès gained entrance having barely celebrated his sixteenth birthday in 1849. While at school, Bergès devoted a great deal of thought to the economics of the technology at his father's paper mill. Drawings in a notebook made during summer vacation show he was preoccupied not so much with the paper machines, but in particular with the turbine—a modified version of Fourneyron's original model.

Like most *centraliens*, Bergès hoped to take over the family business after graduation. His difficult father, however, resisted the son's efforts to modernize the outfit. The young engineer was especially excited about the possibility of installing a mechanical pulp machine in the family factory. Bergès's drive to innovate pushed him to strike out on his own. After brief stints working for French railroads, Bergès hung a shingle as a civil engineer in the mid-1860s (Figure 1.1). During this time, he fused his two primary interests and patented a new, hydraulically powered wood grinder to produce pulp.[48]

FIGURE 1.1 Portrait of Aristide Bergès with an allegorical figure of white coal. To the right one can see the turbine that made harnessing large falls possible, Bergès's pulping machine, and the factory at Lancey situated below the towering mountains.

Source: Alfons Mucha, ca. 1905. Courtesy of the Maison Bergès-Département de l'Isère (France).

As his device was designed to turn mountain forests into profitable paper, Bergès traveled to the Grésivaudan in the autumn of 1866 in search of clients for his machine. A later biographical sketch would suggest that Bergès was particularly taken with the beauty of the area. Beyond its aesthetics, he was also struck by a feature held by many of France's Alpine valleys: their U-shape. The result of glacial carving, the trough-shaped profile ensured the presence of waterfalls. This geomorphology combined with the steady glacial runoff regime set the engineer to dreaming. Bergès had visions of tapping into this reservoir of energy for so little cost that he would be able to dramatically lower the price of paper (and he mused, put more books in the hands of the French people).[49] The young engineer quickly came into contact with valley entrepreneurs and received a contract to design a factory to power one of his pulping machines. He used this first engagement as an opportunity to create a high-fall plant. Unlike most industrial facilities in the area, Bergès proposed to use the force of an upland stream as motive power. After capturing it at a higher altitude than was common, he would create a significant fall down to the factory located at the bottom of a sizable gorge. The Toulousian sought not just to build a pulp factory but to build one with an innovative power source.

It is significant that the driving force behind his effort to tap into a new source of energy was not scarcity of other energy sources. The power demands of Bergès's machines did not exceed the capacities of traditional mills. It seems that the engineer was attracted instead by the efficiency of using the relief of the Alps so that small amounts of water could create large amounts of energy. The possibility of opening up unprecedented industrial capacity for the mountain region—and the fame that would come along with it—surely also played a role in the entrepreneur's calculations.[50] In this way, the transition to using Alpine waterpower fits an important pattern noticeable in many modern energy transitions. It was less a case of necessity being the mother of invention than of an attempt to create a new market for an untapped resource.[51]

That opting to utilize glacial runoff streams as power sources was a choice and not a requirement is evidenced by his business partner's response to the idea. In rejecting the engineer's proposal, the associate made clear that the enterprise's finances did not permit experimentation.

It would be necessary to negotiate with the peasants for the purchase of the source, the right of way, a difficult thing when one is in a hurry. The combination of expenses would rise to a very high figure. What's more, these waters are generally turbid (*tufeuses*), they deposit a lot of sediment in the conduits. The upkeep would be costly. Obliged as I am to limit my expenses, I must request that you only do what's strictly necessary.[52]

Clearly, hydropower innovation seemed excessive. Moreover, the list of the pitfalls of generating high-altitude waterpower would prove to be very prescient: similar problems have accompanied the development of Alpine waterpower to this day. The rosy projections of engineers often left out the difficulty of construction projects in mountainous areas. Furthermore, precisely because Alpine streams flowed so swiftly, they often eroded their beds and carried with them large amounts of sediment. These deposits then wreaked havoc on pipes and turbines alike. Thanks to these reservations, this first venture did not result in any energy innovation.

Undeterred, Bergès held fast to his energy visions. The booming paper business soon offered him another chance. In 1869, local notable Dr. Joseph Melchior Marmonnier became interested in a contract to produce wood pulp for the French market and beyond. Marmonnier, like Bergès, was a liberal republican who saw no conflict of interest between his involvement in local politics and business. The doctor reached out to Bergès for ideas on how to launch this new venture. This time Bergès would succeed in persuading others of his ideas. The factory that would produce the pulp was located in the Combe de Lancey. This glacially formed gorge near the valley floor in a hamlet of the same name stood less than twenty kilometers to the northeast of Grenoble. Through it flowed what the locals called the "torrent of Lancey." Fed by the runoff from the Freydane glacier, the brook derived from a handful of mountain lakes high up in the Belledonne range. At an altitude of 700 meters, the torrent plunged into the gorge before joining the Isère. Later in his life, Bergès would describe the torrent as an "insignificant stream, carrying at most less than a hundred liters per second and managing only with great pain to drive some grist and hemp mills of three or four horsepower each."[53] Bergès convinced his associate that creating a 200-meter chute out of the Lancey torrent represented the best means of powering the two wood grinders of Bergès's designs at the new factory.

Given the height of the fall, the small amount of water available would be sufficient to generate an impressive 500 horsepower and drive the two machines. Marmonnier agreed and the two formed a company in which the doctor supplied the capital and local connections, and the engineer the technological equipment and know-how.[54]

Setting aside the purchasing of land and rights of way, the most difficult part of creating high-chute waterpower was technical. Finding a firm that could construct equipment capable of withstanding the immense pressure being planned was no easy task. The issue caused some mild panic for Dr. Marmonnier. In a February letter to Bergès, the doctor confided that he and others in the valley feared that pipes would not be able to handle such extreme forces. He suggested lessening the pressure by diverting the water into two separate pipes.[55] The engineer was not deterred, and after considerable negotiations the partners chose a Lyon company that also specialized in making gas conduits to provide the iron pipes sealed with lead joints. The search for the turbine took a little longer. In terms of type, Bergès sought one very similar to the original turbine that had been invented by Fourneyron and installed in the Black Forest in 1837. He turned down an offer for one, however, from the Fourneyron workshop in Lyon as being too pricey.[56]

Construction began in late January 1869. Just upstream of the gorge, the project called for the diversion of the torrent into a canal nearly half a kilometer in length. The last section, made out of masonry, traversed the property of the mayor of the village of Combe de Lancey and required a 3,000-franc payment for the concession. From there the water entered a surge tank and then the iron pipe—a penstock in hydraulic parlance—450 meters in length and half a meter in diameter. This penstock then snaked its way to the edge of the ravine, where it dove down into the canyon and the factory. Within the complex, the water pressure created by the 200-meter difference in height spun a single turbine attached to two of Bergès's pulping machines. At the end of September, a milestone was reached when Bergès received word that the pipes withstood pressurization "without a lot of leakage, save for some joints that were lost, everything went well."[57] After seven months in October of 1869, the factory came online.

Though Bergès has been lionized with hindsight as the far-seeing inventor of a new form of energy, his project was far from a sure thing or an unmitigated success. The first months of the venture were extremely precarious both financially and technologically. Immediate problems with one of the pulping machines prevented full-capacity production. Occasional lack of water also slowed progress. In October, the doctor reported that none of the factory owners in Lancey could remember a time when the torrent ran so dry. It was not such an issue, ironically, since only one of the grinding machines was functional.[58] In January 1870 Marmonnier complained that frost in the valley had arrested waterflow and almost forced a complete factory shutdown.[59]

Over the course of the factory's existence, it would be plagued by irregularities in stream flow. This, as we shall see, came to be perceived as the Achilles heel of waterpower (and all renewable energy resources). Bergès's response—to make even more modifications of the local hydrology—would also become the consensus reaction. Above all, this involved converting some of the higher-lying lakes of the Belledonne massif into reservoirs in an attempt to control flows. Coal, in his own words, allowed itself to be stockpiled "but water stockpiles itself in lakes."[60] The hydrological interventions required to store water would form the basis of numerous lawsuits from locals aggrieved by the impacts. Nevertheless, the machines began to operate twenty-four hours a day, seven days a week. The rerouted energy of the stream of the Combe de Lancey now directed its efforts toward the consumption of the surrounding forests. Newly available power created new job opportunities. In the first year of operation, sixteen workers on two twelve-hour shifts disposed of the rerouted energy. Two hands ran the machines, and the rest moved, cut, and stripped the wood. From the regular reports of the first foreman, we know the factory shredded around fifty cubic meters of fir and aspen per month. Just short of two decades later, a larger, more seasoned outfit minced a record 500 tons of wood. The rhythm of production reliably matched the seasonal availability of water, with logs and workers receiving a temporary reprieve in the winter months.[61]

The eventual success of the pulp factory inspired growth. Over the years, Bergès intensified his use of the local hydrology, and his factory expanded with the control of ever greater flows of energy. In 1882 he equipped the mill with another chute, this one with a record-setting head of 500 meters. Dwarfing existing falls, the move was again hailed as a pioneering feat. In the time between the creation of the Lancey facility and this expansion, no fewer than twelve other chutes of greater than sixty meters were installed by French industrialists (Figure 1.2).⁶²

Beginning in 1881, another potential use for *hautes chutes* occupied Bergès's mind. In that year, Thomas Edison made a splash with his new system of electric light on the Continent at the first electrical exhibition in Paris (see Chapter 2). This was a year after the American inventor debuted his first central stations in London and New York. All of these events received heavy coverage in the French press, and Bergès followed the de-

FIGURE 1.2   Relief of the Grésivaudan and the source of white coal.
Courtesy of the Maison Bergès-Département de l'Isère (France).

velopments closely. He began formulating his own enlightening plans, first for Lancey and later for the entire Grésivaudan. It would take over a decade, however, before he succeeded in introducing the new electric light to Lancey. In January 1893, his waterpower began to generate current for lamps that lit streets and public buildings. It was great advertising for his product and paying "subscribers" followed. The next goal became the illumination of the broader valley, and the engineer entered into negotiations with neighboring communities. As with most new technologies, however, numerous obstacles stood in the way. These ranged from securing rights of way for wires and poles, to struggling against existing economic interest groups, to overcoming general resistance. It took until Christmas 1897 for a portion of the valley to be ready. At 4 pm on January 3, the engineer gathered a small group, including local notables, to witness the inaugural lighting. As if to prove to the public the essential wholesomeness of both the undertaking and the mastermind, Bergès brought his wife and children along to the machine hall for the ceremonial flipping of the switch. Madame Bergès herself opened up the valve, allowing water to spin the turbine and generate the electricity. A local newspaper noted how impressed tourists, traveling on the rail line from Grenoble to Chambéry, were when they saw thousands of lights twinkling on the mountainsides of the newly "luminous valley."[63]

The potential of using Alpine water to produce aluminum, however, is what truly pushed Bergès to become an apostle of white coal. Aluminum smelting represented the most important innovation in nonferrous metallurgy since the transition into the fossil fuel era. This lightweight, flexible, corrosion-resistant metal exists in practically unlimited supply—it is the most abundant naturally occurring ore in the earth's crust. Prior to the end of the nineteenth century, however, there was no practical way to mass-produce the substance. The electrolytic process—developed independently by the American Charles Martin Hall and the Frenchman Paul Héroult in the 1880s—changed that, but it came with a catch. It required enormous amounts of direct current electricity (six times more energy than required for smelting iron) to separate the aluminum element.[64] From the beginning then, the industry was dependent on extremely cheap electricity. Hall found the necessary energy in Appalachian coal. His counterpart Héroult would do the same in the Alps. The first attempt came in

Switzerland, where the Société métallurgique Suisse utilized the power of the famous Rhine Falls at Schaffhausen to operate Europe's first aluminum factory in 1887.

From there the destinies of aluminum and the water of the French Alps became entwined. In the spring of 1888, Héroult and his business partners set out to find a suitable waterpower site in France for the creation of a French aluminum industry. While they had initially set their sights on the Moselle River in northeastern France, the group soon focused on the Grésivaudan due to the inexpensiveness of *hautes chutes* waterpower. For a time, Bergès negotiated with the Societé élctrométallurgique française to lease one of his falls for their factory. This would necessitate raising a dam he had built on a high-altitude lake. In his letter seeking approval from the prefecture, the engineer took the first step in his project to spread the word on his new form of energy. The aluminum industry, he explained

> is perhaps a veritable industrial revolution in the world of metals and there is a veritable regional public interest as the Isère department could be the center of this development in France. There are in fact thousands of horsepower to be utilized therefore and an unexpected prospect of exploiting this facet of our local riches, which I have dubbed the white coal of our glaciers.[65]

Ultimately Héroult and his colleagues opted to use the power of the neighboring torrent in the town of Froges for their venture. Bergès's assertion about the importance of aluminum for the regional economy, however, proved correct. Thereafter, a number of aluminum factories popped up in the French Alps, making the region one of the early global centers of aluminum production. The other Alpine hotspot, on the Rhine above Basel, made Switzerland the first European country to mass-produce aluminum, which it sold primarily to Germany. There, West German firms turned the aluminum into flatware that it sold mostly in the United States (Swiss consumers apparently wanted nothing to do with eating utensils "made out of mud").[66]

In marketing white coal, Bergès would find his calling. As depicted in the Introduction to this book, the engineer unveiled this concept to the wider French public at the Paris Exposition Universelle of 1889. There he boasted that white coal and aluminum might possibly define the twentieth century in the same way that coal and iron had for the nineteenth. Throughout the 1890s, Bergès continued to preach this gospel. In addition to presenting the new technology at other industrial fairs, the Frenchman launched a publicity offensive with local politicians and newspaper editors, attempting to convince them of the importance of white coal for the future of the region. In retrospect, we can read these efforts as an example of how a renewable energy transition was marketed to the public.

On the occasion of his negotiations with the city of Grenoble to transport hydroelectricity from a nearby *haute chute*, Bergès spoke with a reporter for a local paper on the merits of white coal. His comments succinctly summarize the supposed virtues of mountain hydropower. Dramatically underplaying his own tribulations, Bergès made harnessing Alpine waterpower seem simple, cheap, and completely reliable:

> The conquest of the chutes of the high peaks is generally "easy," the costs "reduced"; maintenance "almost null"; their mode of construction "elementary," and their continuity and regularity can be "absolute" over the 365 days of the year.[67]

Bergès's argument was not yet formalized—that development would come later. Nevertheless, we can see the template emerging for why Alpine waterpower should be seen as different from—and superior to—other forms. First, the height of the mountains made white coal an extremely potent form of energy. The existence of great falls in the mountains—provided by nature without human intervention—meant that a little bit of water could go a long way. Second, the source of this waterpower in the glaciers and snowfields supposedly guaranteed a more reliable flow of water over the course of a year. For these reasons, and because of the lower value of property in the mountains, white coal did often cost less than waterpower on larger streams. In contrast to coal, which necessitated constant labor and running costs for maintenance and transport, the engineer countered that white coal only required low initial outlays and—once the hydraulic infrastructure was built—was practically free. And what stood in the way of realizing these riches made possible by the wonders of technology? Only the masculine, domineering will to take it: "Open your eyes and want it!"[68]

Where entrepreneurs were enchanted by the economic aspects of the new energy source, the broader public was equally interested in its social impacts.[69] Particularly as hydroelectricity became a commodity available for domestic consumption, these other virtues came to the fore. For the educated and literate classes, electricity from white coal clearly represented social progress. We can see this in the pages of the handsomely illustrated weekly *Revue des Alpes* that appeared across the Francophone Alps. Amid advertisements for bars, restaurants, hotels, watches, and hiking shoes, treatises promoting white coal could also be found. One such article celebrated the illumination of Lancey by waxing poetic about benefits of hydropowered light as compared to existing gas and petroleum lights. It brought progress for the farmers of the valley, who would find their work facilitated and threatened by fewer accidents and fires. Such electric light would even permit them to work longer hours without worry of the impending end of day—a potentially dubious benefit from the point of view of the farmers. How it would impact the domestic sphere was illustrated with a scene from later in the day when the men could finally retire:

> Later, it's the family reunited around a bright light: the child does its homework, the women work on the clothes, without tiring their eyes and without the wobbling light of bad petrol lamps. The housework is done peaceably in the gentle and healthy light.

While in the countryside women's work was never done, electric light would at least spare their eyes the ill-health effects. The author even noted that electric light improved the lot of livestock! They too would presumably benefit from a form of light that burned cleaner than oil. The vision laid out here is one of social progress derived from the cheaper energy that came along with technological innovation.[70]

But it is important to note that it came in the form of an argument. The author, who explicitly adopted the perspective of a "civilizer" in laying out these advantages, was clearly trying to persuade his audience. For, although it usually went unsaid, technological innovation also brought with it disruptions to existing social and environmental circumstances. As proof we need to look no further than Bergès's later efforts to illuminate the entire valley. Though he managed to gain the support of many municipalities, the commune of Domène refused to allow the right-of-way for wires to cross its boundaries. The mayor did this, it turns out, to spite the neighboring town for an earlier conflict over the location of a tram line. A local newspaper referred to it as "the electric war of Grésivaudan."[71] What for some was progress, was for others a question of mundane local politics. Where larger interests were at stake, more significant resistance was possible.

Conclusion

Setbacks such as these notwithstanding, Bergès's proselytization efforts paid dividends. No other person was so associated with the new energy source. In 1902, when French industrialists and entrepreneurs organized an international meeting to oppose legislation to nationalize waterpower (see Chapter 2), they christened it the "Congrès de la Houille Blanche." In addition to holding the proceedings in Grenoble, the organization honored Bergès's role in the development of white coal by scheduling an excursion—one might say a pilgrimage—to his pathbreaking factory in Lancey. The congress provided ample opportunities for the French to celebrate the energy developments in the Alps as a triumph over nature and a point of national pride. In the opening speech, the president of the congress argued that harnessing the new power of the Alps had made Grenoble the "uncontested capital of white coal and of the latest rebirth of our national industry." For this reason, the president declared, Grenoble was proud of the role that its "children" had played "in the army marching toward the conquest of the energy of the mountains."[72] When the participants made their way to Lancey, however, Bergès could only greet them weakly from the balcony of his home. Less than two years later, he would pass away in his Grésivaudan residence. The contents of his briefcase on the day of his death have been preserved through the decades. In addition to a copy of the original white coal leaflet from the Exposition Universelle can be a found a playbill for the opera *Faust*—a Goethe drama in part about the unintended consequences of its protagonist's efforts to tame the unruly power of flowing water.[73]

Thanks to an epochal shift in the energy regime of West-Central Europe, the French engineer had played a crucial role in the subsequent scramble for alternative sources of power. Turbines made it possible to perceive the Alps as a new energy landscape. The advent of electricity just a few years later enabled water to power entirely new fields of activity like metallurgy and chemical production that had previously required fuels of some kind for heat. Early electric technology managed to support new industries in the mountain valleys close to the energy source. It would take another technological innovation to bring this new power not just to the uplands, but to the cities and industrial centers of Europe. Though it is not widely recognized, the history of white coal is deeply connected to that transition as well.

TWO

Carrier of Wasted Natural Forces

> Great things are done when men and mountains meet;
> This is not done by jostling in the street.
> —William Blake, "Gnomic Verses" (ca. 1793)

IN THE TURN-OF-THE-CENTURY BOOK *Anticipations,* author H. G. Wells turned his considerable intellectual powers to reflection on the fate of humankind. In Europe, in particular, the Englishman foresaw inevitable conflict. To his mind, the forces of scientific and technological progress were radically restructuring traditional communities, creating new economic connections and friction as they transgressed old boundaries of nation and language. On the Continent, Wells envisaged a coming clash between German and French for "the linguistic conquest of Europe, and perhaps of the world." The setting for this conflict would be a vast emerging urban region that Wells predicted would become one of the greatest in the world. This territory centered on the Rhine, stretching from Paris in the west to Prague in the east, and would become the "industrial capital of the Old World." Wells believed this region would remain intact despite a new, crucial development he perceived:

> Even when the coal-field industries of the plain give place to the industrial application of mountain-born electricity, this great city region will remain in its present position at the seaport end of the Old World. Considerations of transit will keep it where it has grown, and electricity will be brought

to it in mighty cables from the torrents of the central European mountain mass.[1]

In many respects, Wells's vision of the future was quite prescient, even if he got some of the details wrong. The French and Germans did indeed engage in an epic clash, but their conflict proved to be more than just one of language. Still, Wells's prophecy illuminates an energy connection that, while strange to modern eyes, was clear to many at the time: the close connection between mountain waterpower and electricity. In his vision, electricity went hand-in-hand with mountain water. Coal-fired industry was something else entirely. What he called "mountain-born electricity" emanating from the "torrents of the central European mountain mass" was nothing other than the newly valuable waterpower of the Alps. And as he correctly foretold, electricity would eventually allow this energy to be brought down to the plains, complementing the fossil-fueled heavy industry that existed there.

This chapter tells the story of that forgotten bond between Alpine hydropower and electrification. It focuses on some of the most important figures and debates surrounding the use of electricity to tap into the energy riches of the mountains. It shows that it was not merely electricity that changed the value and fortune of Alpine water, but a type of electricity. White coal could not be a truly revolutionary energy source until it was mobile like coal.[2] This required electrical technology that, unlike Thomas Edison's direct current, could be transported longer distances overland. The promise of being able to harness white coal was one of the main drivers of the development of this new technology, which we now know as alternating current. It has been electricity, in its alternating current form, that has conquered the globe, transforming economies and societies in the process.

The intertwined history of white coal and electricity, then, suggests an alternative view to the dominant narratives of both energy and electrification in the modern period. Starting with the most influential early histories of electrification, the story of Thomas Edison and his urban illumination systems have stood front and center. Hydroelectric development has been interpreted as an interesting side note occurring in peripheral regions.[3] But the history of electricity shares a similarity with the history of oil, in that its earliest use as a source of illumination was eclipsed by a later role. Whereas oil found its ultimate calling in the internal combustion engine, electricity has ascended as a carrier of energy. The story of how electricity assumed this function played out to a significant degree in the Alps.

The Advantages of Electrical Current

The forgotten connections between the history of Alpine water and electrification are best illuminated in the lives of a network of engineers whose careers centered on developing electrical technology to access this power. No single person did so much in this respect as Bavarian engineer Oskar von Miller (1855–1934). Born the tenth son of a bronze founder, Miller's family made the jump in the nineteenth century from small-town artisans to well-to-do burghers of Bavaria's capital city. His father established this position by becoming the court sculptor for the Wittelsbach dynasty and an artist of international renown. Among his most well-known works are the colossal figure *Bavaria* on the festival grounds for the Oktoberfest and the portals for the American Capitol in Washington, DC.[4]

Miller the son took the family a step further when his exploits in the energy field gained him admission into the Bavarian House of Lords. He did so by taking advantage of good educational opportunities to join the engineering profession. After attending secondary modern school (*Realgymnasium*) and the technical university in Munich, Miller began an apprenticeship as a civil engineer in the Bavarian state service. Like Aristide Bergès, his initial activities centered on projecting new railway routes. Also like his French counterpart, his restless personality chafed at this stopover with the railroads.

Miller's entrance into the field of electrical engineering, and his role in opening up vast untapped stores of energy worldwide, owed a great deal to chance. On a Sunday morning in 1881, as Miller was following good Bavarian bourgeois tradition and enjoying some sausages and a pint at a local haunt, he learned of an upcoming electricity exposition in Paris. The news came from Miller's companion, no less a personage than the director of the Bavarian state railways, who joked that the event would be something for the young civil engineer. Miller took the remark seriously and determined to be present in Paris. By the afternoon he had already received the approval of his father. His conviction to attend was only strengthened by a conversation with a friend who urged him to go in order to learn more about electricity, which in his estimation would play a "leading role in the world" in the near future. Getting permission to take leave of his duties from the Bavarian state, however, proved more difficult. During a series of hard negotiations, Miller eventually suggested that a trip to Paris would provide an opportunity to evaluate the potential of using Bavarian waterpower to generate electricity. This approach finally reaped dividends, for in hydroelectricity, the state saw a potential means to improve its economic situation. With the task of investigating the potential for combining the new technology with the harnessing of waterpower, the state of Bavaria granted Miller a short leave from his apprenticeship. In September 1881 Miller departed for Paris in the role of Bavarian commissar.[5]

In the topic of waterpower, Miller had hit on a subject that resonated with his superiors. Over the course of the nineteenth century, alarm had been growing within some circles in Bavaria that the state—like most regions in and around the Alps—was falling behind other parts of Europe in the industrial sphere. Most chalked this up to the fact that Bavaria did not possess the necessary coal reserves. On the other hand, Bavaria possessed the greatest waterpower supplies in all of Germany, with the bulk of this energy being located in the southern part of the state where water flowed down off the Alps towards the upper Danube River. As in other Alpine regions, Bavarians had been exploiting this waterpower since the medieval period for mining, metallurgy, and mills of various kinds.[6] But traditional waterpower had not enabled the state to industrialize to the same extent as other German regions. Many began to see waterpower as inferior to coal. Unlike fossil fuels, the argument went, hydropower had to be used more or less where it was generated. It suffered from the eternal energy problem of being difficult to get in the right place, and therefore was only of limited value.

This viewpoint conveniently left out that some regions, Switzerland for example, had industrialized intensively based on traditional waterpower usage. It also passed over the many tried-and-true methods for transporting waterpower—albeit over relatively small distances. The most common one—one that had existed since the beginning of waterpower usage—was to transport waterpower via a canal. How effective this method could be is proved by the existence of the industrial center in Lowell, Massachusetts. There a dense network of hydraulic canals provided the energy for the mills powering early industrialization in America.[7] As late as 1889, engineers floated suggestions to harness the Niagara Falls via miles of similar canals—but, in this case, underground.[8] Another strategy was to use waterpower to generate compressed air, which could then be used to run machines and tools. Water-powered pneumatic drills were used in this manner in place of steam-driven ones in the drilling of the Mont Cenis tunnel—the first large-scale "piercing" of the Alps in the 1860s. Cable transmission also found application in certain situations—for example, at the famous Rhine Falls in Schaffhausen, Switzerland.

But all of these approaches had their drawbacks. Generally speaking, they all led to considerable loss of power during transmission and could only be carried out in certain conditions. Cables in particular wore themselves out very quickly. Also, traditional transmission methods could not be used to make waterpower provide heat for industrial processes. Most importantly, they were all relatively limited in reach. Exploiting Alpine waterpower on a large scale would have required the migration of industry and population into the mountains. As an Austrian industrialist at the time explained the problem, "Waterpower has always been something valuable, and whoever could use it for their factory did so. Except that you had to locate your factory where the waterpower existed, up in the uneconomical valleys, far away from transport and people."[9] Shifting industry to the source of energy would not have been an impossibility, but it would have upset existing economic arrangements. And it was a particularly hard case to make in an era where coal could be brought more or less anywhere on the planet. By the time Miller traveled to the Paris fair, it was clear that waterpower might be used to generate electricity and transported in this form. But electrical technology, too, suffered from the difficulties of delivering it long distances. Whether electricity could be used to overcome what was seen as white coal's greatest drawback was what motivated Miller to attend.

The Paris International Electrical Exhibition in 1881 is rightly remembered as one of the key events in the history of electrical engineering. It marked the introduction of Thomas Edison's legendary "central station" system for powering incandescent light bulbs to European audiences.

Edison did not attend the exhibit himself, but his "jumbo" steam generator, which powered some 1,200 lamps hanging from the ceiling and in the stairwell, dominated the fairgrounds.[10] Years later, Miller described the impression left by the exhibition as "overwhelming." The greatest commotion, he recalled, was caused by the one Edison bulb that could be lit and extinguished with the flip of a switch. Hundreds of people stood in line for hours to do just that.[11] Many of the fair's nearly 1 million visitors during its three-month span left convinced of the superiority of the Edison system. Among them was the German mechanical engineer Emil Rathenau, whose experience convinced him to acquire the German rights to the Edison franchise. The company he formed, the German Edison Company, would go on to become to the Allgemeine Elektrizitäts-Gesellschaft (German General Electric, AEG) one of the world's largest electrical firms.[12]

A less famous but arguably more important aspect of the Paris exhibition was the gathering of engineers working on a different aspect of electrification, what might be called the Achilles heel of the Edison system. The high cost and general difficulty of transmitting direct-current electricity long distances was one of the major "reverse salients" that limited the early growth of Edison's system. The problem, as Edison quickly recognized, was that the considerable current required to transmit direct current over longer distances would require such great amounts of expensive copper wiring that investment in such a system no longer made financial sense. Indeed, even in his boldest statements concerning the effectiveness of his central stations in the early 1880s, Edison spoke only of a half-mile radius.[13]

To many, this was not simply an abstraction, but a question of economic survival. If transmission could not be made economical, then electrification would depend on having energy sources located nearby. So far, that had meant setting up large and noisy steam engines in the urban areas where electrification was being tested. Vast natural flows of energy, like white coal, could not be made available outside the remote mountain regions where it resided. Determining whether electricity could be helpful in this respect, then, was a primary goal of several exposition attendees, including Miller.[14]

Miller recounted his immediate impressions of the exhibition in a report nearly fifty pages long that he filed with the Bavarian Interior Ministry. For his study of the new field of electrical engineering, Miller took advantage of the plentiful opportunities offered by the exhibition. In the exhibit's well-equipped reading room, he availed himself of international electrical journals and works on electricity. He was most enthusiastic about the fair's congress proceedings, where "the most famous scholars and electricians of the world" traded their viewpoints and experiences. There were also discussions and scientific presentations where professors and technicians demonstrated the apparatuses that populated the exhibits. Of course, there was the exhibition itself, where Miller could view the practical application of the things he had read and heard about. Finally, the hoopla was not limited to the fairgrounds. Numerous experiments with illumination, power transmission, and telecommunication took place at the exhibition's galas and operas, as well as throughout the city.[15]

Soaking in all this information, the young engineer summarized those achievements in electrical technologies that were closely related to civil engineering and waterpower, in order to justify his leave from state service. The engineer reported that hydraulic motors drove dynamos just as well as either gas or steam engines.[16] More importantly, Miller could also confirm to his superiors that the medium of electricity also held the promise of transmitting flows of natural energy—like water—over long distances.

As evidence, the young engineer could point to some of the first concerted attempts to transport power over longer distances. In a pioneering demonstration, the Frenchman Marcel Deprez (1843–1918) proved that high-voltage electricity could perform useful work 1,800 meters from its point of generation. Deprez's bold predictions about the future of electric power transmission probably raised more excitement than his actual demonstration. In one of the fair's technical presentations Deprez claimed it would be possible to transmit a useful amount of power a distance of fifty kilometers with high-voltage direct current. The assertion encountered hefty skepticism, if not downright antagonism.[17]

Miller, for one, was convinced. He reported that electric current was well-suited to transmit the energy of "large, often worthless natural forces" like waterfalls, tides, and wind in cases when the costs of generation and transmission were lower than those necessary to provide equivalent steam power. In practice, he explained, this meant that the source of natural power would have to be twice as large, since roughly half the energy would be lost in the transmission. Nevertheless, Miller was confident that this calculus would soon change in favor of water. In the conclusion of his report, he predicted that the immediate future would certainly bring progress, "if not in the almost perfected electric illumination, at least in the field of power transmission and the accumulation (*Ansammlung*) of electric current." The final sentence of his report Miller devoted to a plea: "May this new area of technology also be supported in every manner in Bavaria, so that the people may enjoy the advantages that the application of electrical current offers, so that the state can benefit from the capital that it possesses in its unused water power."[18]

Miller's report was an early summation of the potential of electricity to radically alter the value of water. It represented perhaps the first formulation of what his biographer refers to as the engineer's concept of "social electricity." While many of his colleagues were interested in electricity as a technological problem, Miller went a step further to consider how it could be employed towards certain social goals. Eventually he came to believe that by making electricity available to everyone—not just well-to-do urbanites—disadvantaged social groups could draw economic benefits. Miller argued that especially farmers, artisans, and tradesmen might once again find prosperity in the industrial-age economic order with access to electricity.[19]

The year following the Paris fair, Miller continued his efforts to unlock the power of the mountains. In 1882, the Bavarian organized his own international electrical exhibition in his hometown of Munich. Though Munich was not on the short list of Europe's foremost metropoles, the choice of venue vaulted Bavaria to the vanguard of world electrical development. In contrast to Paris, Miller was at pains to show not only the marvels of electricity, but also what electricity could mean for the use of renewable energy sources like waterpower. This went beyond merely showing that waterpower and other "natural forces" could be used to generate electricity. Electricity's role as a carrier of natural energy was also highlighted. As Miller later put it:

> the mission of electrical engineering was not at all fulfilled by the manifold applications of electricity; it was necessary to exploit the powers utilized to generate electricity more economically and above all to strive, with the help of electricity, to employ new energy sources, that until now have remained either incompletely exploited or not at all.[20]

In the historiography on electrification, the focus has tended to be on early uses of electricity, illumination in particular. The connection between renewable energy resources like waterpower and electricity as a carrier has received less attention. This, however, was arguably the real revolutionary potential of the new technology. From Miller's perspective, enormous amounts of waterpower were going unused in the Alps to the detriment of the masses.

Miller used the Munich exposition to highlight how electricity might function to transport energy. Five kilometers away from the exhibition grounds, the Bavarian connected an electric generator to a turbine powered by one of the city's numerous canals in order to light a number of bulbs at the fair. For the time, this was quite a novelty. An observer described the uncanny feeling of knowing that "in these simple copper wires flows raw, elemental water power, tamed by electricity, to be sent wherever one wants, and there shine bright as the sun before her tamers." He opined that anyone who could not sense that they were witnessing one of the most "wonderful and far-reaching" achievements of a century that was full of them, "was just completely incapable of appreciating any meaningful moment."[21]

As the high point of the exhibit, Miller offered his collaborator Marcel Deprez the opportunity to prove his controversial claims about long-distance power transmission. With financing from Baron Alfons Rothschild, Deprez and Miller set up a transmission line running some sixty kilometers from the village of Miesbach in the foothills of the Bavarian Alps to Munich. The high-voltage direct current flowing through the line was supposed to power a small pump that would create a two-meter-high waterfall at the fairgrounds. Set amidst a smattering of conifers on the edge of a rocky cascade, the waterfall (Figure 2.1) symbolized the role that electricity would play in making useful the remote energy of mountain waterpower (even though the generator was actually coupled to a steam engine). The Miesbach transmission pushed the outer limits of existing technology, and neither Miller nor Deprez were certain of success. When at the appointed time the waterfall indeed began to flow, Deprez, who had traveled from France to attend, joyously hugged Miller. Nevertheless, the experiment was not a complete triumph. Various components broke down, meaning the exhibit operated only for a very short time. The efficiency of transmission, moreover, was abysmal. While the experiment demonstrated that power transmission was theoretically possible, much work remained before the technology could be used to transport power economically.²²

Expert opinion on the significance of Miller and Deprez's experiment was correspondingly ambiguous. Among the impressed, however, was Friedrich Engels, who interpreted the event as nothing less than epochal. Immediately after the exhibition, Engels penned his colleague Karl Marx to inform him of the great significance of the transmission. Several months later in a letter to the German socialist Eduard Bernstein, Engels described the feat as an "electrical engineering revolution." He speculated

FIGURE 2.1   Artificial waterfall at Munich exhibition, 1882.
Courtesy of the Deutsches Museum Archive, Munich, BN 15419.

that electric power transmission would liberate industry from its present geographical confines and enable the harnessing of even the "remotest water power," leading to the abolition of the gulf between the city and the country.[23] One of Switzerland's foremost electrical engineers, Walter Wyssling, also held up the Miesbach demonstration as the moment when his compatriots realized the potential of electricity to tap into their country's considerable Alpine waterpower.[24]

Several years after the Munich exhibition, the Frenchman Deprez retired to another part of the Alps to continue work on the transmission problem. At the behest of Aristide Bergès, the mayor of Grenoble invited the engineer to come to the Dauphiné and experiment with the transmission of the abundant waterpower in the valleys surrounding the capital. Here Deprez managed to deliver nearly 70 percent of the 10 horsepower generated by a water turbine a distance of 14 kilometers. It marked the beginning of collaborations between the *hautes chutes* pioneer and one of the earliest proponents of long-distance transmission.[25] After Munich, the quest to tap into Alpine waterpower began in earnest.

Transforming Italy

On the southern slope of the Alps, a number of Italians took note of the energy developments in Paris and Munich. The Kingdom of Italy too took a keen interest, sending a delegation from the Ministry of Agriculture, Industry, and Commerce to Paris to report on the state of electrification. This report made clear the Italian state's awareness of electricity's potential to transport waterpower. As happened for Germany, the Paris meeting also resulted in the creation of an Italian Edison company, one that would come to dominate the country's electricity supply until the sector's nationalization in 1963. The Italian branch's striking divergence from the technology of its parent company makes clear the central role played by white coal in driving electrical innovation.

The Italian delegation to Paris was headed by Galileo Ferraris, professor of technical physics at the Museo Industriale in Turin and the foremost expert on the new field in all of Italy. Ferraris too held high hopes for electricity and its potential for transmitting Italy's abundant Alpine waterpower overland to industrial centers in northern Italy. In January 1882, he delivered a comprehensive report on the exhibition. Like Oskar von Miller, Ferraris was a civil engineer, and he too was fascinated by Edison's generator, particularly its dimensions. A device of its considerable size could produce electricity on a scale that opened the possibility of electricity serving as a "carrier" of motive power over long distances. Beholding Marcel Deprez's original transmission demonstration, Ferraris became convinced that "some of the most grandiose applications of electricity, such as those that can be carried out with regard to lighting, the transportation and distribution of mechanical energy, and some metallurgical works, can become in the near future actually practical and economical."[26]

A few months after Ferraris delivered his report, a man who would prove even more consequential for the future of the electrical industry in Italy followed up on these thoughts on electricity and white coal. This was Giuseppe Colombo (1836–1921), professor of mechanical engineering in Milan. Colombo would go on to become one of Italy's most important authorities on industrial technology and an active politician as well. Ideologically, Colombo belonged to that group of Italians who believed that Italy needed to modernize through technology and industrialization. He was a municipal councilor in Milan for most of the 1880s, a member of the Right Party in parliament from 1886 until 1900, and thereafter a senator. For two brief periods the engineer also held cabinet posts—first as minister of finance in 1891 and then at the Treasury Ministry in 1896.

Colombo is best known for his role in launching and helming the Società Generale Italiana di Elettricità Sistema Edison (Società Edison), the Italian franchise of the Edison company. Italian Edison would go on to become the country's most important electric utility. Colombo's business partner Thomas Edison gave him what ranks as high praise from the American inventor when he insisted that "he could not possibly be an Italian and he thought that if he [Colombo] looked up his ancestry it would prove that he was a Down East Yankee."[27] How Colombo felt about this comparison is unknown. More clear is his motivation to put Italian white coal to work in the *bel paese*'s industrial north. Colombo played a leading role in promoting the very technology that made this possible. It proved to be the technology of Edison's fiercest competitors.

Colombo's entrance into the quest for white coal began in March 1882. The occasion was a much-anticipated lecture he gave on electricity at Milan's famous La Scala theater, capital of Italian opera. The theater had been the site of some of the first efforts to promote Edison's lighting system in Italy. The first came in March 1881, when the finale of a ballet performance was punctuated by dancers decorated with small light bulbs that could be switched on and off. On the strength of that demonstration, additional electric lights were set up in the theater's foyer in February 1882. Such was the reception that soon thereafter a syndicate formed in Milan to explore the possibility of importing the Edison technology. The group included three of the city's most powerful banks backed by powerful Lombard industrialists. Its goal was to erect an Edison-style central station in the heart of Milan and challenge the city's longstanding provisioner of illumination, the French company Union des Gaz.[28]

Colombo's talk weighed in on the implementation of Edison bulbs at La Scala. In addition to praising the bulbs' performance—a stance that likely endeared him to the Italian Edison Company that he would soon join—the engineer also waxed prophetic about the future of the technology.

> A day will perhaps come when the power of water falling from our Alps will be driven to the plain and distributed to houses like drinking water or gas. What I am saying is not lyric or utopian. . . . Perhaps the plan is not ripe enough to become successful now, but I have no doubt that sooner or later it will be possible to do it. It is just a matter of time.[29]

A matter of time and his encouragement. In the summer of 1882, Colombo formally joined the syndicate pushing Edison technology in Italy. As he ascended to the top of company leadership, he held fast to his vision of distributing Alpine energy throughout the country. In the meantime, Italian Edison grew. By the close of 1884, the company operated Europe's largest central station in Milan. Colombo and his associates also equipped dozens of northern Italian textile mills with their own electric illumination plants. The short range of these facilities limited their markets, leading to operating losses. Though it represented a violation of Edison company guidelines, Colombo began looking for other potential technological solutions.

That he found them in his own backyard was not entirely a coincidence. As Colombo dreamed of distributing Alpine power throughout the Po basin, the Italian government also became active in promoting the tapping of mountain energy. Both the central state and municipalities sought to encourage the technological innovation that would enable long-distance electric transmission. They used the increasingly pivotal venue of international fairs to offer cash prizes to engineers working in this area. In so doing they created the setting for an early breakthrough. It proved to be a duo of foreigners that made the initial development upon which the global spread of electrical transmission networks has been based. At the Esposizione Generale Italiana in 1884, French engineer Lucien Gaulard and his British business associate John Dixon Gibbs were the first to demonstrate that by using a different form of electricity than Edison's direct current—so-called alternating current (AC)—in conjunction with transformers, power could be feasibly transmitted long distances.

Alternating current is another type of electricity that, in contrast to direct current, periodically reverses its flow. It is the product of a different type of generator, one in which the magnet used to create electricity rotates around a fixed coil. It is this continual changing of the magnet's polarity that causes the flow of current to change directions. Due to its physical properties, it is a comparatively simpler matter to change the voltage of alternating current either up or down. This seemingly small difference possessed enormous implications. Early in the history of electrification it was recognized that transmitting high-voltage electricity (regardless of form) resulted in fewer losses.

Gaulard and Gibbs's achievement was to incorporate the insight about high-voltage electricity with a new energy converter: the transformer. These contraptions, in the words of one energy historian, should probably win a contest for a device that is as important to the modern world as it is unknown.[30] By employing them to "step" AC voltage either up or down, they enable the functioning of the modern world's gigantic electricity grids. Subsequent inventors and companies would improve upon Gaulard and Gibbs's initial designs. These included the Austro-Hungarian firm Ganz & Company, which would go on to be a leading early supplier of technology for hydroelectric systems in the Alps, and American inventor William Stanley, who developed transformers for the Westinghouse Company and helped turn that firm into Edison's major competitor. In the United States, the commercial struggle between the two forms of electricity—Edison and DC and Westinghouse and AC—came to be known as the "battle of the systems." By the early 1890s, AC had triumphed, and to this day electricity coming out of wall sockets into peoples' homes is alternating current. The technology behind global electrification then owes much to the effort to tap into Alpine white coal.[31]

Gaulard and Gibbs had been engaged with the problem of transmission for some time. They introduced their first novel approach to the puzzle at an exhibition at the Westminster Aquarium in London in early 1883. There they used a high-voltage Siemens alternating-current generator connected to one of their transformers that in turn fed electricity to twenty-six lamps at the aquarium. The transformer succeeded in stepping-down the voltage of the current so that the bulbs could be lit without being destroyed by the high tension. The team submitted the apparatus as a solution to the problem "of the further industrial development of absolute distribution, that is to say, a system of distribution limited neither by the distance of the central factory from the point of consumption, nor by the number of consumers to be supplied." The leading British technical journal understood the potential of the new system, connecting it to recent pronouncements by Marcel Deprez on the prospect of using high-voltage electricity to convey "the wasted force of waterfalls and tides." Gaulard and Gibbs's demonstration suggested that the high voltage necessary for long-distance transmission might be made usable by lowering it just prior to its intended use.[32]

The main test of Gaulard and Gibbs's method, however, occurred the year after Westminster. The setting was the city of Turin, capital of the Italian Piedmont region. Turin had been selected to hold a large national fair in 1884, the Esposizione Generale Italiana, organized by the Società promotrice dell'industria nazionale.[33] On the banks of the Po, in addition to the usual topics and themes addressed by such fairs, the new field of electrical technology was well represented. In the runup to the fair, the Italian state intervened to ensure that the electrical category would include attention to the problem of transport. The Ministry of Agriculture, Industry, and Commerce proposed a special cash prize of 10,000 lira to the inventor who could come up with a practical solution to a problem of "special importance" in a country such as Italy lacking in fuel resources. This was the application of electricity to transport energy at distance. In a report prior to the fair, the ministry expressed great hope that the new technology would aid Italy in substituting the "inexhaustible motive force of its torrents and cascades" for the fossil fuels that it had to import for illumination and metallurgy. That this was viewed as a crucial problem in the country is confirmed by the release, in December 1883, of a royal decree from Umberto I formally establishing the prize solely for people working in this field.[34]

Further evidence that transmission loomed large in the minds of Italian political leadership is the fact that the Turin city council, independent of these upper-level deliberations, also began discussing offering a special prize along these very lines. For the Alpine city, gaining access to the power slumbering in the nearby Savoy Alps was paramount, so initial discussions centered around rewarding "the best electrical transmission to Turin of the motive force of our Alps" with a prize of 5,000 lira. Having learned of the royal decree, the council moved at a February 1884 meeting to scrap their separate prize and join with the central government to offer a grand international prize of 15,000 lira, a sum deemed worthy of the magnitude of the problem and the dignity of the Italian nation.[35]

The bounty had its desired effect. Attracted by the reward, Gaulard and Gibbs put their system to another test in Turin. Lacking the funds to purchase new equipment, they were forced to ship some of their devices to be able to participate in the international competition. After several preliminary trials, September 29, 1884, was selected as the date for the full demonstration. Installing their new transformers along a nearly 100-kilometer circuit, the pair succeeded in providing various types of illumination in the "Gallery of Electricity" (Figure 2.2) at the exhibition, the Turin rail station, and several other stations on the route into the Savoy Alps. The current was transmitted at just over 2,000 volts of AC power (considerable for the time) before being modulated by the transformers at the end point. Most witnesses and experts—including Giuseppe Colombo, who was in attendance—were convinced. He noted a "perfect result." Despite the fluctuations in current, all of the incandescent bulbs shone with a "very fixed light."[36] In the end, the jury awarded Gaulard and Gibbs 10,000 of the 15,000-lira prize. Out of a sense of national obligation, the remaining prize went to an Italian firm. Competing enterprises scrambled to incorporate transformers into their own systems. The pair's triumph at Turin, however, quickly faded. Thanks to their notoriety they found themselves embroiled in lawsuits concerning the provenance of their inventions and patents. Years of legal battles wore on Gaulard's nerves. Only four years later, he appeared at the Elysée asking to be conducted to the president

FIGURE 2.2 "Gallery of Electricity." To the left hang several chandeliers from the display of the Italian Edison Company. Located in the center aisle (but not visible in this image) would have been a Gaulard and Gibbs "secondary generator." The visitors give some hint as to which social groups were most interested in the new technology.

Source: Torino e l'Esposizione Italiana del 1884, *Cronaca Illustrate della Esposizione Nazionale-Industriale ed Artistica del 1884* (Turin: Roux E Favale & Fratelli Treves, 1884), 341.

of France with an urgent message: "I am God and God does not wait." He would die a broken man in Paris's Sainte-Anne Hospital for the insane only a few months later.[37]

But news of Gaulard's breakthrough spread. The spring following the exposition, the American inventor and entrepreneur George Westinghouse read about the demonstration in an engineering journal. Westinghouse was interested in entering the electrical industry and had been working on developing a DC system. Gaulard and Gibbs's AC system, however, appeared to Westinghouse to hold the potential for much greater returns. If electric power could be transmitted long distance, then power plants could take advantage of economies of scale. Westinghouse dispatched an associate to Europe to secure an option on the new system. He also ordered several of Gaulard and Gibbs's transformers to be sent to his factory in Pittsburgh.

Over the next few years, Westinghouse continued to develop the technology. Through turbulent times, he managed to secure the contract to provide electric lighting for the Chicago World's Fair in 1893. He also submitted a bid to provide the electrical equipment for the planned hydroelectric plant at Niagara Falls. Ultimately, Westinghouse beat out powerful rivals like Edison to win this prestigious contract. He was helped in this endeavor by a recent immigrant to the United States, Nikola Tesla. Since his boyhood in Croatia, Tesla had dreamed about harnessing the powers of nature. In his elementary classroom he played with models of waterwheels and newly invented turbines. Upon learning about Niagara Falls, he boasted to his uncle that he would one day move to America and use a wheel to capture its power. During his studies as a civil engineer at the technical university in Graz, the Croatian first began to dream about inventing a motor that could run on alternating current. He envisioned this motor as part of a larger electrical system, and throughout the 1880s he worked on his designs. Tesla saw his AC motor, and the systems that could supply it, as the future of electrification. He was also the decisive figure in convincing the Westinghouse Company to utilize polyphase AC in its bid to equip Niagara. This system got the nod, and when Westinghouse's power plant began transmitting alternating current to nearby Buffalo and beyond in 1896, it was registered as a historic triumph. Thereafter, new utilities in the United States overwhelmingly chose alternating current (and it remains the dominant form of electricity worldwide).[38]

For Giuseppe Colombo too the Turin demonstration pointed to the future, to the means to finally be able to utilize Italy's Alpine bounty. Though still leading the Italian branch of Edison's company, he steered his concern away from direct current technology and towards the alternating current trail that Gaulard and Gibbs had blazed. In 1889, the Società Edison took a first step by seeking a concession to utilize the power of the Adda. This river, with its source in the Alps on the border to Switzerland, was largely responsible for filling and draining Lake Como. Below the lake, the Adda snaked southward toward the Po to be carried into the Adriatic.

Some of northern Italy's largest textile factories used the river for mechanical power.[39] At the rapids near the town of Paderno, just to the north of Milan, the company determined to harness a falls that would make for the largest hydroelectric plant in Europe at the time. There, near an imposing high bridge, a bend in the stream made it possible to create a thirty-meter fall that would produce over 10,000 horsepower. Still, Paderno stood some thirty-five kilometers away from the metropolis. Despite the progress that had been made using high-voltage electricity and transformers for long-distance transmission, such a stretch remained a great technical challenge. It would take one more high-profile experiment to show not only that sending electricity overland was technically feasible but economical as well. Here again, Oskar von Miller entered the picture.

Alternating Currents

All of the various efforts to solve the problem of transmission converged on the German city of Frankfurt in 1891. In that year, the city decided to hold its own international electricity exhibition. While the expo was conceived with many of the usual motives, Frankfurt's city leadership had a particularly urgent bit of business they hoped to resolve. Ever since Oskar von Miller had organized the first electrical exhibition on German soil in 1882, the city also sought to construct its own municipal electric lighting system. By the mid-1880s, the battle over what form of electricity to use had reached a stalemate. Unable to make up its mind whether Edison's direct current or alternating current represented the wisest choice, the city moved to make this determination by bringing the brightest minds and best technology in the electrical engineering field to the exhibition.

Frankfurt chose none other than Oskar von Miller as technical director for the exhibition. In this capacity he had the greatest hand in designing the event and the power transmission that sealed its epoch-making reputation among his colleagues in the electrical world. As in Munich, Miller sought to make a bold transmission the highlight of the affair. For him, no other question came close to having the importance of electric power transmission and distribution. As usual Miller couched this view in social terms. In a presentation to the fair's organizers, he explained that transmission would be the central concern of the exposition:

> There is no time that would push so much to solve this task as ours; a time, in which the working day should be shortened, and one does not want to limit the necessities of life.⁴⁰

In activating renewable power sources and sending them overland, Miller hoped to defuse social tensions by shortening the workday while still maintaining modern quality of life.

To prove this was possible, he planned to transmit electricity along a route that nearly tripled the distance of the Munich and Turin demonstrations. The selection owed a great deal to coincidence. In 1890, Miller happened to have been engaged to construct a hydroelectric plant for the Württembergische Portland-Cement Werk in the town of Lauffen on the Neckar. The facility produced excess power that could be used for the purposes of the fair. Though the 170-kilometer distance between Lauffen and Frankfurt was larger than Miller's initial preferences, the remaining circumstances met the demonstration needs. Miller had equipped the hydroplant with alternating current generators, meaning the electricity could be stepped-up for transportation purposes. He calculated that a tension of 25,000 volts—over twelve times what had been used in Turin—would be necessary to send roughly 300 horsepower to the exhibition. These were truly unprecedented numbers, and they lent some risk to the enterprise. The route between Lauffen and Frankfurt crossed through four separate German states, and officials in each had to approve rights-of-way. To inform the public of the danger of the high-voltage electricity being carried, each mast along the way was affixed with a death's head symbol. The Badenese government made its permission contingent on the construction of enclosures for the transmission line at certain distances. Miller would later refer to them as "madhouses."⁴¹

After months of preparation, at 8 o'clock in the evening of August 24, 1891, Miller and his associates activated the transmission line for the first time. While several of the participants waited expectantly in Frankfurt, Miller remained along the route in Baden to intercept any bureaucrats with last-minute objections. After completing the circuit, observers at the exhibition jubilantly confirmed "the electricity is in Frankfurt." The day after, the Lauffen line lit bulbs at the exhibition for the first time. A few weeks later it also ran the pump for the ten-meter-high artificial waterfall—the symbol for the natural powers that electricity was unlocking—that Miller once again insisted upon including (Figure 2.3). In contrast to Munich, the power transmission functioned nearly flawlessly and managed to transport hundreds of horsepower while losing only 30 percent effectiveness along the way.[42]

While historians today recognize the Frankfurt demonstration as the outcome of continual technical development, contemporaries almost unanimously interpreted it as a crucial turning point in the history of electricity supply. Carried by this enthusiasm, Frankfurt chose alternating current for their municipal power plant. Lauffen-Frankfurt marked the end of a nearly decade-long development. In its wake, European interventions into Alpine hydrology increased dramatically. In place of small-scale modifications of waterways to generate power for isolated plants came a willingness to more drastically rearrange the Alpine hydrosphere to liberate larger amounts of energy. Electric power transmission changed attitudes on Alpine waterpower. With the ability to transmit hydroelectricity

FIGURE 2.3   Lauffen-Frankfurt transmission with waterfall, 1891.
Courtesy of the Deutsches Museum Archive, Munich, BN 37064.

beyond the waterways where it was produced, it now made sense to find large waterpower sites and exploit them as completely as possible. Oskar von Miller had a hand in designing a fair share of such facilities, which he called "overland works," throughout Central Europe. First on the Isar River just south of Munich and then on the upper Etsch River between the cities of Meran and Bozen, the purpose of these facilities was to transport electricity for light and power from an advantageous location to cities and industrial zones some distance away.

In the immediate aftermath of Frankfurt, however, no hydroelectric plant in all of Europe would match the Italian Edison company's Paderno dam on the Adda River. While Giuseppe Colombo had hesitated in 1889 to commit to a system for Paderno, the Frankfurt demonstration convinced him to opt for a high-voltage alternating current and transformer setup. The choice marked a fundamental break with his concern's proprietary technology and is an example of how the new system contributed toward the transformation of previously remote Alpine rivers.

Armed with the capability of transporting the power of the Adda on a grand scale, Italian Edison replumbed the river's rapids for maximum energy production. With the aid of a small dam, the Adda was split into two. Departing from the main branch, a second river of forty-seven cubic meters per second (about a tenth of the size of the Hudson River as it drains into the Atlantic) was created that fed both the power plant and allowed navigation around the rapids. This diversion canal flowed sluggishly along the hillside until just above the facility where it plunged rapidly to push seven turbines. When construction began in 1895, it was Italy's first commercial hydroelectric plant. Completed in 1898, it provided 13,000 horsepower of electricity for an immediately adjacent industrial zone. The remainder traveled to Milan via six wires for lighting, power, and tramway service. Along the thirty-kilometer route, about 10 percent of the original power dissipated from each line. Despite these losses, the plant was a commercial success. Within six years, the metropole's electricity demands had grown so much that Italian Edison built an additional dam on the nearby Brembo River. The era of large dams on Alpine rivers had begun.[43]

Visions of White Coal

Alternating current and its ability to transport the waterpower of the Alps to centers of consumption marked the final step in making white coal a potentially lucrative new energy source. The turn of the twentieth century, then, is when we find some of the most enthusiastic speculations on the prospects of Alpine waterpower, often in the pages of trade journals devoted to electrical engineering or water management that proliferated at the time. Some of these pertained specifically to mountain waterpower, and some more generally to hydroelectricity. In all cases, these ideas would accompany and undergird hydropower expansion in the Alps throughout the twentieth century.

As H. G. Wells prophesied at the beginning of this chapter, many believed white coal had the potential to completely transform the economic situation of Europe. For the French economist Pierre Leroy-Beaulieu, his country's newly valuable waterpower would not only improve the economic situation of the Alpine lands, but also the nations connected to them. Leroy-Beaulieu became professor of finance at the newly founded École Libre des Sciences Politiques in 1872 and was among the country's leading political economists at the time. The Sciences Po intellectual argued that Alpine waterpower could save the nation in the new century from its recent energy woes. Chief among these was France's lack of coal, which many blamed for a humiliating defeat in battle at the hands of Bismarck's Prussia in 1871. Indeed, this paucity had been exacerbated when France lost even more of its supplies when the newly formed German empire annexed Alsace and Lorraine as part of the war's settlement. In light of these energy traumas, Leroy-Beaulieu pinned great hopes on the prospects of Alpine waterpower. "The richness of France," wrote Leroy-Beaulieu, "in chutes of water, in white coal, as this energy that descends from the glaciers is called, could happily compensate in the twentieth century for her poverty in black coal, which harmed her so greatly during the nineteenth."[44]

Similar sentiments abounded in German-speaking Central Europe. Here such viewpoints were often also tied up with peculiar, German-nationalist demographic concerns. For the Baron von Liebig in turn-of-the-century Germany, Alpine waterpower offered a potential lifeline to the southern part of his country. Like many German nationalists, von Liebig worried about emigration of ethnic Germans from the empire. A major reason for this population drain, he believed, was the lack of modern industry. While regions like the Rhineland and Saarland possessed economic activity paralleled only in England and the United States, this action dropped off as one ascended the south German uplands. For von Liebig, the reason was as clear as day. "The industrial operations that require and consume coal here in Bavaria are too far away from the site of coal production in the Ruhr. Transport costs increase so substantially with the distance from the site of production, that one pays double the price for coal here in Munich as in the Rhineland."[45] Thanks to electricity, however, southern Germany could now access the type of cheap energy that would create towns and cities to house its excess population. Jakob Zinßmeister, editor of a Munich trade journal entitled *Weisse Kohle,* placed similar hopes in hydroelectricity. For him white coal, an energetic "knight in shining armor," would enable south German industry to continue to compete on the world market.[46]

Many saw the economic situation in the Alpine provinces of Austria as equally dire. The German engineer Wilhelm Pressel, one of the primary minds behind the Berlin-Baghdad railway, wrote about the "decline, impoverishment, and emigration" out of the Alps that he perceived at the turn of the century. Humanitarian considerations and state interests demanded an intervention, and fortunately the Alps possessed a natural resource that could reverse the downward economic trend: "In the abundantly available, considerable, and what's more perennial watercourses the Austrian Alpine lands have been conferred treasure of great value." It remained only for Austrians to make the commitment necessary to activate this power.[47] In a play on a famous formulation by Kaiser Wilhelm II, a Salzburg industrialist summed up the viewpoint of his country's business elite: "the future of our industry, but especially of Alpine industry, lies in the intensive use of our waterpower."[48]

One of the most prominent arguments in favor of waterpower was its renewability, though this term was not yet used. When reflecting on this characteristic of hydropower, the explicit comparison was of course nonrenewable fossil fuel. In a 1908 article entitled "The Mightiness of White Coal," an engineer from Saxony characterized the peculiar characteristics of hydropower in terms common for the time. He began by noting that while coal supplies seemed unending to many, their supply was in fact limited—and in danger of depletion with continuing industrial growth. White coal, by contrast, did not suffer from this problem. "As long as the sun continues to radiate its power onto the earth," the engineer explained, "it constantly raises water from the endless seas to the peaks of the mountains, where the water flows once again to the sea in an eternal cycle." Whereas coal seams all came with a more or less calculable expiration date, white coal did not. White coal, the engineer pronounced, "does not lie buried in the mysterious depths of the earth's womb, it renews itself daily before our eyes, and we can observe the processes that drive this cycle and learn to predict their effects."[49] Inexhaustible, product of eternal natural cycles: this was the language often used to describe what we would now call renewable energy. In a sense, the Saxon's depiction was less misleading than our present discourse because it at least acknowledged a situation—the cessation of the hydrological cycle—where the energy would no longer exist. The engineer concluded that using waterpower was economically preferable to burning coal, because the former was a case of adding value to natural resources while the latter represented their destruction.

In a unique form of this argument, the German chemist Wilhelm Ostwald contended that using hydroelectricity instead of coal represented not just economic prudence, but a more advanced stage of cultural development. Ostwald, the founder of physical chemistry and later a Nobel Prize recipient, proposed an "energetic theory of culture." This system conceived all societal endeavors as fundamentally attempts to increase the amount of usable energy, either by finding new sources of power or increasing energy efficiency (*Güteverhältnis*). The "character" of the dominant energy source, then, exercised a strong influence on the spirit of a corresponding time period.

For the most part, Ostwald was not very impressed with the moral qualities of his own era. While the intensive use of fossil fuel resources had permitted unprecedented development in his lifetime, Ostwald was doubtful of the wisdom of carbon-based economies. For him, exploiting fossil fuels was akin to stumbling into an "unexpected inheritance" that had allowed humanity to turn away from a "sustainable economy" (*dauerhafte Wirtschaft*) and live profligately. He regarded increases in energy efficiency on the other hand as both cultural progress and morally preferable. Squeezing more out of existing power sources was a means to avoid the struggle and competition to find new ones. The move from petroleum to gas lamps was an improvement in efficiency and therefore a cultural step forward. The same applied to using waterpower to generate electricity. Ostwald also did not refer to waterpower as renewable but emphasized it was the result of limitless solar energy raising water to the heights. In this manner it was part of the energetic "annual revenues" available to humanity. Ostwald's attempt to explain cultural developments from a natural science perspective met with disapproval from many corners, including the sociologist Max Weber, who believed making sense of historical processes was the purview of the nascent social sciences.[50]

In addition to its supposed limitlessness, contemporaries touted the hygienic and socially advantageous properties of hydroelectricity. By this time there was a widespread discourse about the ill-effects of the smoke and soot issued by steam engines. In the German-speaking world, one even spoke of a "plague of smoke" (*Rauchplage*). Visually it was possible to discern the superiority of hydropower to coal because dams were more aesthetically pleasing than the "black desert of a coal area."[51] In a speech comparing thermal and hydropower plants, Theodor Rehbock—rector of the University of Karlsruhe near the Black Forest—naturally found the latter to be preferable. Waterpower facilities, he explained in a speech, had the virtue of "greater cleanliness, the absence of the pollution of the atmosphere." He also added that unlike a coal-fired station, hydro plants were less dependent on the "skill and good will of the operating staff" and free from the "regular delivery of fuels."[52]

A newspaper supplement to the *Münchner Allgemeine Zeitung* in 1899 echoed this point on the "independence" of waterpower, explaining how hydroelectricity favored industrialists:

> In our time of strikes and trusts, there is substantial value in the independence of a source of power. According to experience, coal prices and correspondingly the production costs of steam power are subject to quite considerable fluctuations. The industrialist who works with steam eats his bread from the hands of the mine owners and the railway administrations. Every misfortune which befalls them, befalls him as well. If the miners strike, or if the mine operation is interrupted, so might the factory shortly

close. If the miners win a wage increase, if the freight prices are pushed up, if the owners of the coal mines want higher profits—it is always the industrialist who pays in the end.[53]

Factory owners using waterpower would no longer be beholden to the whims of coal syndicates, militant workers, or the railways and navigation companies responsible for transportation (and their workers' unions too). From this point of view, hydroelectricity was one tool to dismantle the growing political power of both the coal trusts and the labor movement.[54]

Others believed hydroelectricity would have important social impacts by bolstering the diminished prospects of artisans and craftsmen in the age of industrialization.[55] This was a particularly important concern for Oskar von Miller, proponent of "social electricity." In an 1890 letter to his wife Marie, Miller spelled out a vision for how electricity could help the "little man" regain his economic footing in the new industrial era. Miller recounted to his wife a meeting with Wilhelm Peter von Finck, financier of the "Isar Works"—the first hydropower plant that Miller erected in the wake of the Frankfurt exhibition on the Isar River south of Munich. Miller explained to Finck that he wished for the emergence of a company that would loan electric motors and other necessary equipment to craftsmen for a rental fee or by sharing in the worker's increased profits. Not only would such a system reverse the declining prospects of artisanal industry, but it would dampen some of the social strife that he saw as a consequence of German industrialization. Miller described to Finck "how good and timely it would be, if the factories were disaggregated and installed in the craftsmen domiciles again, where no quarrels about eight-hour working time and employment of women and children could take place." According to Miller, Finck appeared enthusiastic about the idea, if only because it seemed to be a good business as well. Still Miller did not have much hope that his plan would soon be put in action. As he wrote his wife, "A lot of water will flow down the Isar until this, one of my favorite plans, is realized."[56]

The increased value of waterpower also raised questions about who should have the power over its disposal. In all Alpine countries, the question of whether the state should develop waterpower resources or leave this field to the private sector (often categorized simply as "industry") generated serious debate. As early as the 1880s, movements to nationalize waterpower resources emerged in the Alps, particularly in Switzerland, Austria, France, and the German state of Baden. The discussions reveal much about Europeans' attitudes towards water, energy, and the role of the state in economic life.[57]

One of the first strong movements to nationalize waterpower emerged in Switzerland. In 1888, the formation of an association in favor of nationalizing Swiss waterpower, the "Frei-Land" Society, marked a beginning of the societal discussions about nationalization there. In 1891, the society petitioned the Swiss Federal Council to codify the nationalization of waterpower into the federal constitution. They called for federal ownership of all unutilized waterpower in Switzerland and federal regulation of all transmission of this power via electricity or compressed air. According to their petition, the federation should also have the authority to dispense with its profits as it saw fit.

Groups like Frei-Land justified their cause on a number of grounds. The transfer of natural sources of energy from private to public property, it argued, would redress social imbalances, reduce economic crises, and lead to a more just distribution of the benefits of industry. Nationalization was all the more justified in the case of water resources, which were already in most instances in public hands (except where private enterprise had obtained concessions). In nationalization, the Frei-Land Society also saw the best means of overcoming the negative effects of Switzerland's byzantine water law landscape. The Swiss confederation's twenty-two cantons (and three half-cantons) had a hodgepodge of different forms of water legislation that complicated the harnessing of waterpower, especially in the (quite frequent) cases where a particular waterpower project involved more than one canton.

The main impetus throughout West-Central Europe for making white coal a public good, however, stemmed from the strategic importance of waterpower in a region that mostly lacked fossil fuels. The Frei-Land Society emphasized the decisive importance for waterpower in Switzerland, a country with no coal supplies, and the risks entailed by allowing it to fall in the hands of "profit-addicted private speculators, tribute-demanding high-finance, or the stock exchange." Switzerland's waterpower wealth should instead "remain preserved for all time for all Swiss." The proponents of nationalization believed that only the intervention of the federal state could ensure that Swiss natural prosperity—by which they meant the valuable gradients of its Alpine waterways—would be properly utilized. Without state intervention, Frei-Land predicted the continued fleecing of cantons and municipalities by the private sector, which would gain piecemeal control of the country's waterfalls. In the place of uniform, well-thought-out waterpower projects, the country would reap nothing more than the "atrophied crumbling" of the natural falls. In the end, it would be the industrious Swiss folk (*gewerbetreibende Volk*) that would suffer the most, paying "taxes to private owners for every motor load, every electric light bulb."[58]

In 1895 the federal council followed the urgings of the private sector and rejected the Frei-Land Society's petition. The council maintained that since waterpower use always had to adapt itself to peculiar local circumstances, the private utilization of waterpower was actually a boon for many areas. From its perspective, only the "energy and prudence of the entrepreneur" could create the necessary market for the goods produced by waterpower. It also advanced the somewhat confusing arguments of one report that warned that state development of remaining waterpower was a potentially risky business proposition before doubting that the government would be able to raise the capital necessary to exploit the waterpower of all Swiss waterways anyway. This report also departed from the dubious notion that the state would develop its millions of horsepower at once, and so artificially expand national industrial production with no hope of finding a market. Another expert opinion advised that the complicated field of waterpower use did not lend itself to a state monopoly in the manner of the postal service and telecommunication. Even Switzerland's recent nationalization of the railways took place only after that sector had outlived its initial "growing pains." In the final instance, the federal council stood by the principle "the state should not speculate," as the monopolization of waterpower would mean "a constraint on the healthy development of native industry."[59]

Nationalization efforts also surfaced in the Austrian half of the Dual Monarchy in the 1890s. There the premier professional association for engineers and architects came out against the nationalization of waterpower in 1897. In response to a legislative initiative in the Upper Austrian provincial diet, the Austrian Engineers and Architects Association composed a report in which they argued against state monopolization as an obstacle to hydropower development. Tapping into this energy, the report stated, was merely a means to an end—namely the promotion of the commonwealth through industrial development and job creation. In contrast to those seeking nationalization, they did not argue that certain uses of waterpower were more beneficial to the public than others; rather, simply putting water to work was an advantage. Similar to the Swiss Federal Council, the association believed in the centrality of entrepreneurs in this process and rejected saddling them with any tax or fee that might hamper their activities. "If we want to raise labor and industry," the statement insisted, "we ought not tax tools, or the steam and waterpower that drives the motor." While the group did not want public ownership, its members welcomed state support through tax breaks, subsidies, and help with expropriating property necessary to construct a hydropower project if necessary. But the Austrian engineers and architects sought to preserve the freedom of entrepreneurs and uphold the sanctity of private property enshrined in the granting of waterpower concessions. Up through the fin-de-siècle, their viewpoint carried the day. No meaningful legislative challenge to private development succeeded.[60]

Conclusion

Thanks to the development of the turbine in the nineteenth century, the Alps became a landscape of potentially lucrative energy resources due to their abundant, high-energy waterways. Still, many believed that this energy could only reach its highest potential if it could be available outside of the remote valleys where it was produced. Trying to solve this very problem, a handful of engineers and entrepreneurs like Oskar von Miller, Gaulard and Gibbs, and Giuseppe Colombo focused on modifying the emerging technology of electricity. They succeeded in developing a system whereby alternating current, in conjunction with transformers, could be stepped-up and stepped-down in voltage to enable economical transportation and consumption. From here on, electricity was not only something that provided incandescent illumination: it was a potential carrier of energy.

It is in this latter form, as energy carrier, that electricity has spread across the globe. Acknowledging the connection between hydropower and electricity, then, revises our understanding of some of the most momentous energy transformations of the modern period. Though the early history of electrification is often told as a story of Thomas Edison and lighting in cities, the developments in the rural uplands have arguably been more consequential in their impacts. Thus, if we recognize the crucial role played by hydropower in the development of the technology that underlies our modern grids, we also glimpse a different facet of the history of coal. A technology developed largely to enable the harnessing of remote renewable resources like Alpine white coal ended up facilitating fossil-fueled electrification as well.

With the invention of turbines and alternating current, the possibility of tapping into every drop of the mountains' potential power—and finding a use for it—now existed. It remained, however, to be determined how this Alpine waterpower should be utilized. Here the question of ownership—always looming over the issue—reemerged with particular urgency. In the next chapter we will see how the dawning realization of the truly unique nature of Alpine waterpower led central state governments to see the value of white coal in a very specific way—and in some cases to be proponents of public control.

THREE

Exploiting Nature's Gifts

ACCORDING TO AN OLD LEGEND, the depths of the Lake Walchensee are prowled by a giant monster with the ability to unleash frightful power. Located some one hundred kilometers to the south of Munich, the Walchensee is one of Bavaria's many beautiful mountain lakes. Surrounded on all sides by the limestone peaks of the Kochel mountains—the last northern outliers of the eastern Alps—the lake sits perched 200 meters above the hilly pre-Alps, held back only by a kilometer-thick wall of rock known as the Kesselberg. As the legend has it, should the wickedness and sinfulness of the capital city reach a tipping point, the monster would smash through the Kesselberg with its giant tail, draining the lake and sending a flood surge to devastate the immoral capital. This nightmare preoccupied not a few of Munich's inhabitants through the ages. Up until the year 1783, a mass held in hope of averting such a disaster was given every three days in the city.[1]

Just after the dawn of the twentieth century, many Bavarians reversed their stance on this matter. They began to fervently hope for a chute of water cascading down from the Walchensee, albeit in a slightly more controlled manner. Starting in the summer of 1904, a swelling chorus of voices began to call for the conversion of the lake into a reservoir of high-pressure hydropower. The plan entailed taking water from the Walchensee and dropping it 200 meters to the plains below where its force would spin electric generators. Such an imaginative stroke did not come out of nowhere. In the years around 1900, energy boosters had begun to think more systematically about the peculiarity of the Alpine landscape and its value for hydroelectric generation. A growing consensus held that the seasonally fluctuating availability of energy derived from earth's natural cycles—solar, wind, and hydrologic—was a problem. A rational supply of electricity, it was argued, required constant and predictable inputs. This was not only dictated by the needs of the commercial systems that provisioned energy. Electricity possessed the physical characteristic that if it was not utilized at the moment it was generated, it was lost for the ages. But natural flows of energy defied easy control. In the Alps, water was most abundant during the summer melt and regularly dwindled in the winter. Mountain hydropower diminished precisely at the time when more electricity was needed for lighting. Here humans had to intervene.

An additional problem led engineers to concentrate on storage. As Europeans considered how abundant white coal should be put to use, their gaze fell upon one of the most significant consumers of fossil energy. This was the provisioning of traction—motive power—for the foremost land transportation device of the era: the locomotive. Here was an existing demand for energy that figured to continue in the future. What's more, if a way could be found to substitute waterpower for coal, the largely state-owned rail companies in the Alpine area stood to save significantly on fossil imports. Unfortunately, trains too required energy on a schedule vastly different than the flow of power in Alpine watercourses.

This problem of timing, inherent in all human energy use, could be countered in a number of ways. First, a means could be found to store electricity. This existed primarily in the form of batteries that found use in the electricity supply from the earliest days. But these were and remain expensive.[2] Alternatively, planners could have accepted the need for auxiliary power in the form of steam reserves to make up for shortfalls. This occurred frequently in practice, but engineers in the Alpine region showed a marked preference toward dispensing with these. A third solution, barely considered, would have been to adjust energy use to reflect the availability of white coal. Seasonal adaptation had been the rule for much of human history. It is conceivable that Europeans might have attempted to modify their societies to better match the flows of available power. For 10,000 years farmers had conformed the rhythms of their work years and days to the unsteady rhythms of energy delivered by the sun. This, however, was improbable in a world where fossil fuels like coal offered independence from earlier geographic and seasonal restrictions on energy use. By the late nineteenth century, Europe had grown accustomed to the convenience of coal. Instead of changing structures to fit available energy, energy planners resolved to adapt the environment to fit their needs. Renewable sources of energy like water would have to behave more like coal if they were to be used.

For this reason, the Alps began to loom particularly large as a landscape of potential energy storage around the turn of the century. The mountains seemed to possess more opportunities to store natural flows of energy than anywhere else. These occurrences were usually the product of glaciation, whether in the form of existing lakes or ice-carved valleys that could be transformed into lakes with some help from mortals.[3] Lakes and valleys in the Alps also enjoyed the general advantage of higher relief, another gift of geology. The epochal forces of nature had performed work that humans would otherwise have to do; in this sense they provided an energy subsidy from the past. If a way could be found to create falls from these locations, the energy bounty would be multiplied. From 1900 onward, energy experts began scouring the mountains for first lakes and later valleys that could be employed to make hydropower more like coal.

This chapter zooms in on the history of the effort to convert one Alpine lake, the Walchensee, into a reservoir. Until the mid-1920s, the incredible potential energy of the Walchensee—and its implications for Bavaria—made it perhaps the best-known lake in the entire range. Conceptually, lakes have received far less attention from historians than rivers or oceans. But as this chapter shows, they too have a distinct history. Through the lens of this one lake, we can see how the Alps came to be viewed as a unique energy landscape with the ability to counteract the perceived disadvantages of renewables. The case of the Walchensee reveals a state that began to "see" its rivers and mountains in terms of cubic meters per second of runoff and square kilometers of catchment area. Crucially, though, it was forces outside of the state administrative apparatus that roused the kingdom out of a reflexively conservative approach to the project.[4] By concentrating on lakes like the Walchensee, an alternative facet of electrification comes into view: namely, that of rail electrification. In the Alps, lakes and electric trains were inextricably linked.[5]

Finally fixing our gaze upon a single episode illustrates that the process of hydroelectric development in the Alps was always marked by bitter struggles. Questions surrounding the wisdom of modifying nature, who would have control over the resulting energy, and which social groups would bear the burdens of change hounded the project from its inception (and do so to this day). This was especially true of attempts to impound water for the purpose of energy storage. Creating reservoirs brought such difficulties that one scholar argues it was a major impetus for the fossil fuel transition in England. Setting up a steam engine to power a factory was a much simpler proposition than using the common good of water, and for this reason industrialists flocked to it.[6]

A King Among Alpine Lakes

The Alps are home to around 4,000 lakes. These take various forms such as cirque lakes, kettle lakes, finger lakes, and lakes formed by debris dams. Among the most striking of mountain landscapes, cirques are bowl-shaped valleys formed near the top of mountains by glacial erosion. Several neighboring cirques huddled near the top of a peak have created some of most stunning Alpine forms like the Matterhorn. As cirque glaciers retreat, many leave behind a threshold that impounds water. Finger lakes, by contrast, materialize at the edge of mountain ranges in the area where large glaciers once stretched out toward the plains below at the end of the last ice age. Upon receding, these sheets of ice left behind terminal moraines that act as dams. Lakes Constance and Geneva are examples of finger lakes, and they are by far the largest lakes in the Alps.[7]

All lakes fulfill important environmental and social functions. Though high-altitude ones possess lesser productivity due to their typical acidity and colder temperatures, lower-lying ones are critical biotopes. For this reason they are often utilized as fisheries and for fish farming. Larger lakes impact microclimates, above all by moderating temperature extremes. As increasingly overbuilt shores attest, the aesthetic appeal of Alpine lakes has become very important for tourism in the last few centuries. All Alpine lakes also play crucial roles in the hydrological cycle. Their function as retention spaces stands out as the most important in this regard. In fact, Alpine lakes store roughly twice as much water as do glaciers.

Beginning around 1900, lakes across the Alps fell increasingly under the gaze of engineers who spied in them a means to counteract the drawback of water as an electricity source. In a sense, lakes had always and everywhere been important for hydropower. They regulate downstream flows quite independently of any human interventions, creating conditions seen as ideal for waterpower exploitation. Many Alpine lakes had also undergone extensive "rectifications" during the nineteenth century, bringing their water levels increasingly under human control. It was thus not an accident that most of the earliest dams on principal Alpine rivers were located in the downstream vicinity of some of the larger Alpine finger lakes such as Geneva, Constance, and Como. The latter was the source of the Adda River that supplied the Italian Edison Company's Paderno plant.

Even the earliest efforts at harnessing white coal also included plans to incorporate reservoirs, usually in the form of lakes. Already in 1887, for example, Aristide Bergès petitioned the Isère prefecture in Grenoble for a permit to equip his paper mill at Lancey with a high-pressure reservoir in the form of the Lac du Crozet.[8] By the turn of the century, at least one French engineer was trying to popularize the idea of using lakes as power reservoirs by referring to them as *houille bleu*—in reference to their often stunning blue coloration.[9] Shortly thereafter the existing waterpower plants at Lac de Caillaouas and Lac de la Girotte in Savoy underwent extensive modifications to improve their efficiency.[10] The trend extended to Switzerland and Italy. The Brusio power plants in southern Switzerland harnessed the storage power of a handful of lakes to generate 50,000 horsepower of electricity as of 1910—a very substantial amount for this time period. Much of this energy provided traction for two Swiss railways, and about 20,000 kilowatts were reserved for export to the thriving cotton industry to the north of Milan.[11] Directly to the southeast in Italy, a high-pressure work connected to the Lago d'Arno in the Adamello range exploited two stages of around 1,400 meters to generate 20,000 horsepower. The Adamello Works also fed their power to Milan and its surroundings.

In terms of its importance for the surrounding polity—as well as its general dimensions—few measure up to Bavaria's Lake Walchensee. The Walchensee is Bavaria's largest Alpine lake. It owes its existence to the same processes that created the Alps. As the tectonic plate in this region shifted, it created folds and synclines—furrows between the peaks. Some, like the Walchensee, eventually filled up with water. Up until the turn of the twentieth century, the Walchensee was relatively undistinguished. Linguists believe the lake takes its name from *Wälsche*, denoting a Ladin-speaking population (Rhaeto-Romanic dialect) that inhabited the lake's shores for some 700 years, from the sixth to the thirteenth century CE. The lake remained relatively undeveloped until the nineteenth century. Until that time, only a nearby monastery used its waters as a fish hatchery. In the same year that Columbus made landfall in the Americas, the first road leading up to the Walchensee was constructed as part of a shortcut on the route between Munich and Innsbruck. The road was continually expanded and improved, and it remained the main transportation development in the area until the close of the nineteenth century. At that time the lake was integrated into Germany's burgeoning networks of railways. The proliferation of these railways would eventually influence the lake's fate.

In 1898, the initiation of regular train service to Kochel—the town just below Lake Walchensee to the north and situated plausibly enough on Lake Kochelsee—improved access to the area. With the trains came tourists and nature enthusiasts, drawn by the lake's fjord-like qualities and stunning color. Artists were especially attracted to the landscapes of the area, which they viewed as the antithesis of urban industrial life. Indeed, it became something of a headquarters for one of Germany's most significant modern art networks, Der Blaue Reiter. The expressionist painter Franz Marc had a house in Kochel and painted the Walchensee and its surroundings. In 1918 the impressionist Lovis Corinth moved to the shores of the Walchensee and produced a series of portraits of the lake.[12] One Bavarian professor spoke for many when he referred to the Walchensee as a king among lakes.[13] By the turn of the twentieth century, Lake Walchensee was home to several small towns and villas. Most locals earned their living either in the tourist industry, the timber trade, or fishing. The lake made an excellent home for whitefish and char in particular, but also pike, perch, and other varieties that laid their eggs along the shore.[14]

At the turn of the twentieth century, the Walchensee belonged to the south German kingdom of Bavaria—a territory that could look back on over a millennium of existence. The historic polity had emerged out of the feudal reorganization of Central Europe in the wake of the fall of the Roman empire. The word "Bavarian" (Baiuvarii, people from the land of Baiu) first appeared in the middle of the sixth century CE and seems to refer to an ethnic group from the region of Bohemia in the modern Czech Republic. They settled territory in the southern German linguistic area, from the upper reaches of the Danube basin in the north and west down through southern Tyrol and over to Enns River in the east. Over the centuries they began to distinguish themselves linguistically and politically from their neighbors, such as the Alemanians. The medieval period saw Bavaria rise to the status of a powerful duchy within the Holy Roman Empire.[15]

A decisive moment came in the late twelfth century when the Wittelsbachs rose to the ducal seat. They would remain the Bavarian dynasts until collapse in 1918 and exercise a critical influence on the history of the territory. One particularly important effect stemmed from their adherence to the Catholic faith even as the battles of the Reformation raged around Germany. Their eventual choice of Munich, a city founded by the Bavarian Heinrich der Löwe, as their residence would also raise that city to the status of a European metropole. Bavaria benefited greatly from the grand geopolitical struggles in early modern Europe. Following the Thirty Years' War, the Wittelsbachs became electoral princes in the Holy Roman Empire, and kings after the wars of the French Revolution. In the nineteenth century Bavaria stood after Prussia and Austria as the most powerful of German states. It remained a kingdom and junior partner to Prussia after it joined the Bismarck-forged German empire in 1871.

In terms of territory, the kingdom of Bavaria around 1900 had reached its high point. It was almost as large as the present-day country of Austria and just smaller than the US state of Maine. At this time, over half of the population still earned their living from the land. But as with much of Germany, pockets of industrialization had sprung up in urban centers. Augsburg had long been a textile production center, and Würzburg, Nuremberg, and Munich supported mechanical engineering hubs. Ludwigshafen on the Rhine was one of the original cores of the global chemical industry. Most of this economic activity was dependent on steam energy; Bavaria's spotty industrialization stemmed in large part on a lack of domestic coal. Despite being the single biggest polity with holdings in the Alps, Bavaria's share of the mountains was quite small. It was all contained within the province of Upper Bavaria, which occupied a tiny piece of the northern East Alps and the Bavarian plateau (Figure 3.1). This sliver contained Germany's highest mountain, the Zugspitze (2,962 meters) and was drained primarily by four Alpine tributaries of the Danube: the Lech, Loisach, Isar, and Inn.

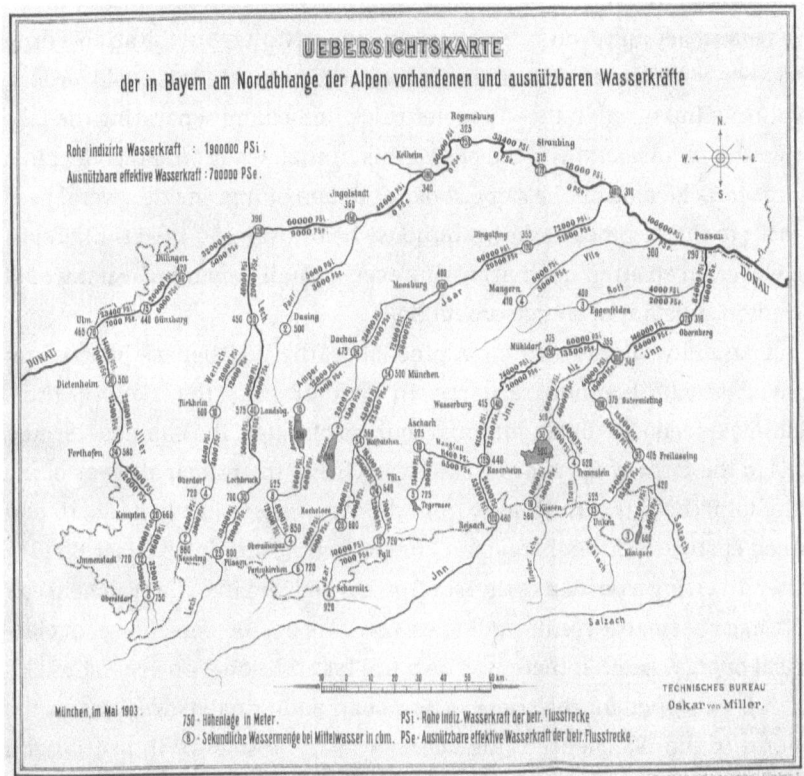

FIGURE 3.1 "Overview Map of the Available and Usable Water Power on the Northern Slope of the Alps in Bavaria," 1903. Oskar von Miller created this map as part of an effort to clarify the waterpower potential of Bavaria's Alpine waterways. Miller considered only run-of-river dams. The Walchensee in the upper Isar basin is not even depicted on the map. This omission demonstrates that at this point Bavaria's foremost hydro expert had not considered the possibility of using the lake as a reservoir for the energy of the upper Isar.

Source: Oskar von Miller, *Die Wasserkräfte am Nordabhange der Alpen* (Berlin: VDI Verlag, 1903).

The idea to convert the Walchensee into a power reservoir first emerged in the summer of 1904. In July of that year, the Upper Bavarian district government received a request for a concession to build a hydroelectric power station on the shores of Lake Kochelsee. The author was Rudolf Schmick, a hydraulic engineer from the southern German state of Hesse. Schmick had spent a considerable amount of time siting hydroelectric plants in Switzerland and aimed to put that experience to work in Bavaria. In the first place, Schmick was interested in taking advantage of the sizable drop between the Walchensee and the Kochelsee to its north. This could be done by tunneling through the kilometer-thick Kesselberg separating the lake from the plains below. At the end of this tunnel, water would collect in a surge tank located on the slope above and then plunge inside several penstocks to the turbines in the powerhouse below. The 200-meter-difference in elevation ensured that harnessing even a small amount of water would result in substantial energy production.

Like many higher altitude Alpine lakes, the Walchensee by itself received very little annual recharge. To avoid draining the lake over time, Schmick needed to find additional sources of water. The engineer located this in the form of the nearby Isar River. One of the principal rivers of the kingdom, the Isar turned away from the Walchensee just to its south and arced eastward through a region known as the Isarwinkel ("Isar Bend"). After diverting a portion of the Isar from its bed and into the Walchensee's drainage basin, the result would be a constant 20,000 horsepower of electrical energy. Later, if there was demand for additional power, Schmick's plan provisioned for the construction of an additional hydroplant on the shores of the Walchensee. The energy was to be sold for industrial and lighting purposes to consumers in the plant's immediate vicinity.

From an environmental standpoint, Schmick's proposal was radical. He intended to completely replumb the upper Isar basin using the Walchensee as a high-lying reservoir and tapping into a portion of the power of the Isar to fill it. In so doing, he established a template whose basic outlines underlay all future considerations of the Walchensee project. The scheme required massive interventions into the ecology of the Isarwinkel. In the first place it was necessary to modify the lake's outflow so that it would drain downwards toward the turbines in Kochel. This meant installing a floodgate at the source of the Jachen, the small creek that had traditionally drained the basin eastward into the Isar. Going forward, the Jachen Valley would of necessity receive less water and at times run dry.

Charging the reservoir with the Isar River also carried its own consequences. To produce the constant power Schmick envisioned, the engineer calculated that during the part of the year when the Isar ran lowest—the winter months—almost the entire flow of the river needed to be diverted into the Walchensee. For roughly four months, a section of the Isarwinkel would run more or less dry. Throughout the whole year, the river's flow would be substantially weakened. Additionally, to ensure sufficient power production in the dry winter months, the Walchensee would be called upon to store a portion of the Isar's summer melt, raising the lake's level considerably. And though Bavarians had long modified the Walchensee's level, henceforth the ebbs and flows would be determined by the schedule of electric power generation. Moreover, all of that additional Isar water being diverted from the Isarwinkel would now be flowing into the Kochelsee and through another river valley before it finally rejoined the Isar to the north of Munich. One watershed's loss was another's gain.

The project prompted varying reactions from the Bavarian state. Local authorities were initially critical of the environmental and social disruptions such a transformation would cause.[16] The central state government immediately recognized the magnitude of the project. Officials in the Ministry of the Interior, the department that dealt with hydraulic engineering projects, were far more positive. In his evaluation of Schmick's project, the head of the Bavarian Supreme Construction Authority (*Oberste Baubehörde*) noted its unprecedented scope and scale for Bavaria. "From the first glance it is clear that its realization would be of outstanding economic importance—not only for the city of Munich, but the entire land," he wrote. For the bureaucrat, the key comparison was the Bavarian city of Augsburg. There, intensive use of much smaller amounts of traditional waterpower from the Lech River since the medieval period had made the city—the seat of the famous Fugger banking family—a thriving textile production center by the nineteenth century. The possibility of "exploiting 20,000 horsepower from the Isar with the help of the Walchensee," he imagined, "opens the prospect of a previously unthinkable industrial development for Munich, in which the broadest circles would share." Based on these considerations alone, he concluded that the project should not be rejected out of hand.[17]

The chief of the newly established Bavarian hydrographic service agreed, exclaiming that the project "is so obviously grand, that further words about it are not to be wasted." "In all of Bavaria," he continued "we will probably not find one single site which nature has so favorably created for the exploitation of waterpower on a grand scale." Above all he referred to the Walchensee's unique environmental configuration. It not only permitted the generation of large amounts of electricity extremely cheaply, but its proximity to Munich promised markets for this energy as well. For the hydrographic agency, the Walchensee project was like a "coal mine" that could produce the equivalent of 200,000 tons annually. The crucial difference, however, was that this coal mine was "inexhaustible as long as the Isar and the Walchensee have water" while actual fossil fuel supplies would be exploited in the foreseeable future. In an early indication of the Bavarian state's attitude toward the project, the civil servant remarked that nowadays no one would consider allowing such a "gold mine" to fall into private hands. Rather, it was "self-evident" that the Bavarian state would build and operate the Walchensee facility.[18]

Despite positive reactions from the state's key agencies, all recommended a wait-and-see approach. This was because it was clear to all involved that the Transport Ministry, which had recently begun considering using hydroelectricity to power the state-run railways, needed to weigh in. In the fall after Schmick submitted his concession, the Transport Ministry responded to the magistrate of the city of Munich that it would need to study the question before any decision could be made about the future of the project.[19]

Storage Gaining Traction

It took the intervention of another outsider to finally convince the Bavarian government of the uniqueness of the resource that Bavaria held in the Walchensee. The catalyst was another German engineer living in Switzerland, Ludwig Fischer-Reinau. Fischer-Reinau had been born in the lower Alsatian town of Keffenach a few years after the region had been annexed from France into the German Reich. He studied mechanical engineering in Saxony before matriculating at the University of Zurich in his early thirties.[20] Fischer-Reinau's experience in Switzerland, a country at the forefront of white coal development, had a huge impact on his views. In the summer of 1905, Fischer-Reinau determined to spread the cutting-edge ideas about hydroelectric development he was witnessing in Switzerland to his German fatherland. His focus fell on Bavaria, the German state that most closely resembled mountainous Switzerland. First in an article published in a major German engineering journal and then later at public presentations in Bavaria during the fall, Fischer-Reinau made a case that Bavaria should follow Switzerland's lead in utilizing mountain lakes as reservoirs. His rhetoric gives insight into how engineers saw their work underpinning nationalist and even civilizational missions.

For Ludwig Fischer-Reinau, harnessing the power of mountain lakes was an international task of epochal importance. He viewed the Alps as a mighty barrier (*Grenzwall*) to progress in the heart of Europe. Due to the high mountains and rugged relief, Fischer-Reinau believed that the people living in and at the edge of the Alps lagged behind those of the plains in both intellectual and economic terms. This state of affairs, he argued, had naturally created a constant push for equilibrium between the peoples of the Alps and the surrounding regions. Only in the modern period, with the help of "monstrous sums of technical work and jingling coins," had Europeans succeeded in piercing through the mountain barrier from both sides with the construction of Alpine railroads. This creation of a new means of accessing previously isolated territory, according to Fischer-Reinau, had led to the recognition of a new task of equal importance: "the task of exploiting the waterpower of the Alps." In mountain lakes, Fischer-Reinau argued, lay the key both to reversing the Alps's arrested development and putting the power of its water to work.[21]

This was because lakes appeared to be the solution to a twofold problem—part natural, part technological. On the one side were the hosts of problems that came with using water to generate electricity. On the other side was the question of how to use this power. The key, in Fischer-Reinau's mind, would lie in using nature in a new way to overcome these issues. On the natural side of the ledger, Alpine water had its disadvantages as an energy source. Though the energy slumbering within Alpine streams was abundant, it was not evenly distributed throughout the year. Instead, its availability was determined seasonally. During the cold season, when the sun's direct rays drifted away from northern latitudes, much of the precipitation and runoff otherwise destined for Alpine waterways remained stored in snow and icepack. With the more direct sunlight and increased temperatures of the spring and summer months, rain and snowmelt swelled the proportions of Alpine streams. Even neglecting extraordinary precipitation events, the flood stages could be over one hundred times greater than minimum flows. For Fischer-Reinau this was evidence of the unruliness of Alpine water. In a common refrain, he noted that seasonal and even daily ebbs of waterpower availability happened precisely when electrical energy was needed most in Central Europe at the turn of the twentieth century. On cold, dark winter days, electricity for illumination and in some cases heat was at a premium. On dry summer days, after the summer melt was over, people flocked to the trains for day trips to the mountains. These problems were exacerbated by the peculiarities of electricity as an energy carrier. Above all, it had to be used at the instant of generation or it would dissipate. It also defied easy and inexpensive storage.[22]

Ludwig Fischer-Reinau was the first to raise in Bavaria these issues that engineers across Europe had been discussing since just after the turn of the century. It was a familiar one within the organic energy regime: namely, that the availability of many of the prevailing energy sources fluctuated seasonally. Generally speaking, the spring and summer months in the northern hemisphere were ones of energy surplus, as solar radiation shone more directly on these latitudes. The winter was the opposite. Up until the eighteenth century or so, this remained a vexing problem but was simply a fact of human economic and social life. The transition to intensive fossil fuel use, first in England and then elsewhere, was in part so revolutionary because it offered a way past these limitations. Coal could be burned at any time, theoretically allowing economic activity and transportation to take place all year round.

Increasingly, Central European engineers began to make the case that without special action, hydroelectricity would continue to be inferior to coal as an energy source. In the pages of technical journals, they carried out intense debates over the preferability of one source over another and the accompanying social and political consequences. Artur Budau, engineering professor at the technical university in Vienna and one of the world's foremost experts on turbine construction, submitted that, from the point of view of a hypothetical factory owner, coal trumped traditional waterpower. For an industrialist, a heap of coal in the factory yard was a reassuring guarantee of uninterrupted energy supply for a plant using thermal power. Those dependent on waterpower were not so lucky.

> Here the factory owner is dependent on the mood of the water. The temporary surplus that accompanies the flood is completely worthless for him. The times of water shortage, on the other hand, are oppressively felt. It is mainly this circumstance that so unfavorably influences the profitability of waterpower plants today, as one usually builds his facility for that water volume which is safely available the whole year through. The potential bonus is foregone, since one would have no means to dispose of it.[23]

Flow variability and the inability to site a single hydroplant to efficiently take advantage of all available waterpower were economic problems that negatively influenced a plant's profitability. But Budau also saw more important reasons for counteracting this flaw. The professor worried about the dangers of Europeans burning non-renewable coal. Budau correctly understood that coal was essentially irreplaceable over any human time scale and perceived Europe's coal supplies as finite. When they were used up, he foresaw an economically disastrous exodus of coal-dependent industries to parts of the earth with still bountiful resources. Conserving coal was necessary for preserving European industry and by extensions its *Kultur*—a word often used by German nationalists who juxtaposed their culture to the abstract "civilization" of the French Revolution. The best way to do this in his view was by rationally exploiting waterpower through the use of "storage facilities" (*Akkumulierungsanlagen*). Here he meant the creation of reservoirs. The point of utilizing reservoirs was to make waterpower a better substitute for coal wherever possible, slowing what he viewed as a profligate wasting of precious carbon resources.[24]

To combat the inconstancy of Alpine water, Fischer-Reinau agreed with his colleague about the necessity of finding ways to store water. By using mountain lakes as reservoirs, Fischer-Reinau explained, "a large portion of Bavaria's Alpine waterpower will receive effective accumulators that can even out the differences between water flow and consumption through relatively small variations in their water levels."[25] Since high-lying water gained power by falling greater distances, it required comparatively small amounts to create an energy impact. The engineer, however, was not satisfied with simply demonstrating the usefulness of reservoirs for Bavaria. Fischer-Reinau believed they had a specific duty to perform: to allow the kingdom to transition from steam to hydroelectric propulsion on its mainline trains.

For almost as long as rail travel has existed, there have been efforts to move trains using electricity. In fact, some of the earliest applications of electric motors were used in this attempt. In 1839, Robert Davidson used an onboard battery to move a five-ton locomotive at a speed of six and a half kilometers per hour along a flat stretch of the Glasgow-Edinburgh line using electric current from a battery. Forty years later at a Berlin exhibition, Georg von Siemens powered a streetcar with electricity delivered to it from a cable set up near the track. In the wake of this demonstration, the 1880s saw the rapid expansion of electric tramways and narrow-gauge railways. Electric traction quickly proved superior to steam locomotives not only on streetcar lines, but also on suburban passenger trains with heavy traffic. Subways only became possible with the advent of electrification. Especially in the United States, electrified "interurban" lines connecting cities proliferated.[26] In the Alps, narrow-gauge mountain railways also found widespread use.[27] The success of these enterprises encouraged further experimentation with using hydroelectricity to drive main lines. The electrification of the Burgdorf-Thun railroad in Switzerland using waterpower in 1898 gave hope to those who sought to apply white coal to this end. These hopes were improved in 1902 when a previously steam-driven light railroad was switched to hydroelectricity just across the Swiss border in Italy around Lakes Lugano and Como north of Milan.[28]

But it was not until the outset of the twentieth century that efforts to expand electric traction targeted the large-gauge, mainline railways that carried the bulk of passengers. The Alps were one center of this activity (another was Scandinavia), thanks to the widespread desire to use hydro to power state railways. Switzerland took the first step, spurred on by interest from the private sector. In July 1902, a group of Switzerland's most powerful interests in the engineering, manufacturing, and banking fields reached out to the Swiss railway ministry and federal railway company suggesting the need for a study group on the question of electrifying mainline railways. These interests were motivated by a desire not to fall behind in a potentially lucrative new market. But they also couched their proposal in terms of Swiss national interest. While other countries might see electric traction as the best way to achieve superlatively fast rail connections between two distant transportation centers, Switzerland's interest lay elsewhere: in the "reduction of coal imports," which they characterized as a "tribute abroad."[29] In October of that year, the Swiss legislative assemblies (*eidgenössische Räte*) formally requested that the Swiss Federal Council explore the practicality of introducing electric traction. From the point of view of the railway department, the effort to use Switzerland's "domestic powers of nature" for the country's railroads was "a great and beautiful goal, that could bring our land extraordinary advantages." In 1903, the ministry formally agreed to membership in the Study Commission for Electrical Railway Operation (*Studienkommission für den elektrischen Bahnbetrieb*), along with members of the Swiss Electrical Engineering Association and the Swiss Federal Railways. The railway department also secured a federal annual subsidy of 10,000 Swiss francs to finance the work of the commission.[30]

Elsewhere in the Alps, other governments also began exploring the possibility of hydroelectrification. In Bavaria, the state's engagement with alternative power for its trains began a year after Switzerland's when its transport minister asked Oskar von Miller to study the possibility of electrifying several routes that ran along the Alps in the southern portion of the kingdom.[31] In 1905 industrial interest groups in the Austrian half of the Dual Monarchy also admonished their government to investigate the suitability of using waterpower for moving trains, particularly those crossing the Alps. The imperial government responded by creating a study group and commissioning a hydropower cadaster in 1906. While the cadaster was to catalogue all potential waterpower sites, one of its explicit goals was to determine whether hydroelectricity could be used to power any of Austria's Alpine railroads. That same year an Italian law stipulated that by 1911, 320 kilometers of rail had to be electrified, with the energy supplied almost completely from waterpower.[32]

Electrifying the mainline railways in the Alps could be justified on several counts. Technologically, electric traction offered advantages over steam, some that were particularly significant in mountainous areas. Oskar von Miller noted that one advantage of electric traction in regions with steep gradients was that it enabled downhill-traveling trains to generate electricity that could then directly be used to power the train uphill.[33] But the greatest motivation for the states was for an improvement in both their immediate financial situation and their long-term economic security. As the secretary of the Swiss exploratory commission emphasized, "The main importance for Switzerland lies much more on the economic side, in the application of its waterpower instead of coal imports from abroad."[34]

At the time, railroads were one of the primary consumers of coal, and Alpine states made considerable outlays to procure the fuel. The expense by itself might have been bearable, but many state governments—and military authorities—bristled that their transportation networks operated at the mercy of coal suppliers. Worries about dependence on syndicates abounded, whether or not the coal came from abroad (Figure 3.2). One German economist, for instance, called the cartelization of the German coal industry "the sword of Damocles hanging over our economic life" and lamented that "the price politics of the coal syndicate drives the coal prices ever higher."[35] Other Alpine districts in countries without coal stocks experienced perhaps even more acute pressures. The desire for economic independence from coal imports would increase dramatically after the First World War's disruptions to fossil fuel supplies and the general coal crisis they unleashed (see the next chapter).

Engineers analyzing the feasibility of using waterpower to provide electric traction quickly arrived at the necessity of storage. The problem lay primarily in the unique daily pattern of energy demand for main-gauge railways, which differed considerably from other types of loads. The typical daily demand for electricity around 1900 was driven by the need for illumination. Portrayed graphically, this curve hovered at a low point throughout the early morning hours, rose slightly as the workday began, and then returned to the minimum until the early evening. At this point, the demand rapidly reached a peak that lasted several hours before descending. Mainline railways posted a much different curve. During the day, with myriad trains stopping and starting continuously, the demand for energy oscillated constantly between minimum and maximum. With a series of jagged peaks, its energy curve looked in some ways like a stylized

FIGURE 3.2 "On the Coal Emergency: A deputation of the Trade Association of German Thieves and Burglars presents the nomination for honorary membership to a coal baron." This illustration from October 1900 on the cover of the leading German satirical weekly, headquartered in Munich, hints at the social awareness of regular coal supply crises.

Source: *Simplicissimus* (October 15, 1900).

representation of the Alps themselves (Figure 3.3). As with the mountains, the spikes in energy use for trains could also reach tremendous heights. The Swiss study commission calculated that hydropower plants would have to be able to handle "formidable fluctuations" in demand, with maxima up to five or ten times greater than the average load.[36] Waterpower, as it had been harnessed and financed up until this point, was not capable of efficiently meeting these spikes in demand by itself.

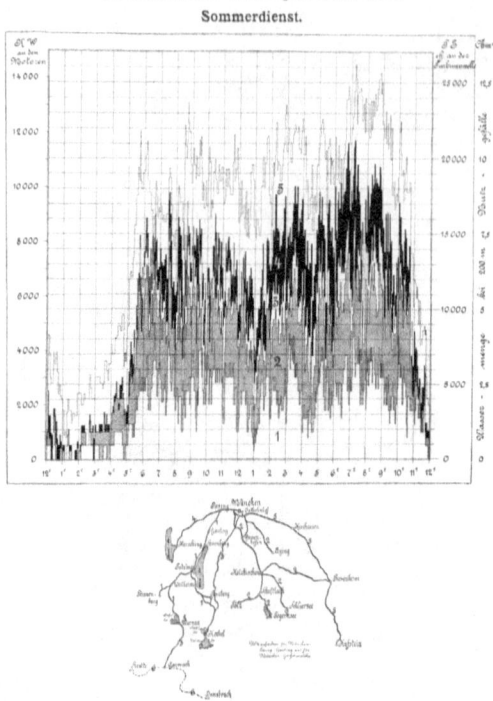

FIGURE 3.3 "Load for Electric Operation on Routes 1–5 on the Map Below—Summer Service." This diagram comes from a Bavarian state publication on how its public waterpower should be utilized. Engineers arguing for the use of lakes as reservoirs consistently used graphics such as these as evidence. They pointed to the multiple, rapid spikes in energy demand as a supply challenge that only lakes could address. On the small insert map, five railways leading to the capital of Munich are illustrated, as well as the Bavarian lakes that would enable their electrification.

Source: K. Oberste Baubehörde, *Die Wasserkräfte Bayerns* (Munich: Piloty & Loehle, 1907), 53.

Storing high-pressure waterpower in lakes emerged as the early solution to the problem of using white coal to power trains. When train traffic ceased during the nighttime hours, water could be stored in a lake until it was needed to cover daytime peaks. The Swiss commission charged with studying waterpower for electric traction concluded: "Only such waterpower can be considered that is permitted by accumulation in lakes."[37] The commission advocated that some of Switzerland's most promising lakes "that due to their nature are especially suitable for electric traction, must be cautiously set aside and safeguarded for their timely use."[38] The group's findings proved influential. Governments across the Alps would follow Switzerland's lead in arguing that exploiting the peculiar Alpine landscape for its storage possibilities would be critical for the success of any electrification projects.[39] A scramble to reserve potential reservoirs for state designs broke out.

It was this knowledge that Ludwig Fischer-Reinau brought to Bavaria in 1905. While powering trains was the most difficult task for an electric plant, he reasoned that "precisely the hydraulic power of the Bavarian Alps is capable of meeting these challenges, because the incorporation of dammed-lakes provides never-failing batteries."[40] Using Bavaria's hydropower for traction also made sense because there was still skepticism about the future of electricity. Traction loads represented an already existing market for energy that would no doubt persist into the future.

Fischer-Reinau, like Artur Budau and Oskar von Miller, understood his profession in political terms. Perhaps in part because of his upbringing in a contested border region, he frequently framed his ideas as a means to strengthen the German nation. Nationalist rhetoric appeared repeatedly in his pronouncements. His arguments combined a rationalist faith in technology and its ability to promote economic growth that would likely have appealed to other nationalist liberals of the time. The engineer saw water as a part of "national wealth" (*Nationalvermögen*) and argued that hydroelectric development would become a "lasting" (*Bleibendes*) component of it. This was in contrast to fossil-fueled steam plants, which were "impermanent" (*Vergängliches*). Fischer-Reinau, moreover, viewed rational hydropower development as the duty of all Germans to create the economic conditions under which coming generations could prosper. In a clear criticism of Germany's intensifying imperial policies, the engineer argued that harnessing white coal was a "*domestic* task, whose successful resolution will strengthen the roots of our people and create a bequest for which our children will thank us" (emphasis added). For all these reasons, Bavaria's mountain lakes—geomorphological features created over eons by the forces of nature—were "a precious gift of a benevolent nature." He admonished Bavarians to do their part so that they would be used "for the benefit of the people."[41]

The focus on lakes as power reservoirs coming out of Switzerland completely altered the Bavarian state's thinking about the Walchensee project. In October 1905, the Royal Bavarian Railway concluded its study of the feasibility of using Bavarian waterpower for electric traction. The railway administration that only a year earlier had been searching for conventional, run-of-the-river sites on Bavarian watercourses was now singing a different tune. It declared it "obvious" that in all plans to utilize Bavarian hydroelectricity to power the railways that "attention must above all rest on the mountain lakes as natural batteries (*Akkumulatoren*), and it is only a question of special studies to determine in what manner and scope the mountain rivers and lakes can be drawn upon to produce electric energy."[42]

The railway agency did not require further investigation of the importance of the Walchensee. In a special evaluation of the project, it reiterated the significance of rail electrification for the Bavarian state and the decisive role that the Walchensee had to play in this endeavor. The railway authority explained that conserving coal supplies was one of the most important duties and believed that posterity would be thankful for sparing these stores of energy and warmth. It noted that coal consumption at the present time was anything but economical. The smoke and soot caused by ovens, factories, and locomotives symbolized the inefficiency of burning coal. To arrest the wasteful consumption of this precious resource, the report declared, was the "calling" of flowing water:

> If any land, then Bavaria—so poor in stocks of mineral coal—has the greatest interest to make use of the energy in its mountain waterways, and place electric power instead of coal, hereby minimizing the very substantial outlays for the purchase of foreign coal and improving both the industrial competitiveness and the financial performance of its railways, which suffer from their significant distance from coal stocks.[43]

In the wake of Fischer-Reinau's intervention, Bavaria's transport leadership determined that it could not give its approval for a concession to exploit any large waterpower site in state possession until it was completely clear that the energy was not needed for electric traction. This stance applied especially to the Walchensee, which promised large amounts of particularly cheap electricity. The railway administration further declared that given its significance, the *"Walchenseewerk"* must be carried out by the Bavarian state. A few months later, the Bavarian government formally adopted this policy. It resolved that the broader question of Alpine lakes would be systematically investigated by the state. Following Fischer-Reinau's suggestion, a memorandum would be compiled on how these and all other waterpower resources within the kingdom should be utilized. State engineers retreated to draw up their own plans for the Walchensee.[44]

Resistance

As with all hydraulic engineering projects, exploiting mountain lakes to create reservoirs of power required environmental change and therefore social disruption. This guaranteed that the new idea would be challenged. The earliest major resistance to the state's plans, however, took a somewhat surprising shape. An active group of enthusiasts actually clamored to make the Walchensee part of an even greater intervention into local ecosystems. The existence of a movement to make the lake part of an unprecedented technological utopia reveals something about the attitudes of the broader public toward energy in this era. Its ability to pressure Bavarian authorities to augment their own plans also suggests that it is not simply states whose vision matters in formulating large-scale schemes to improve the human condition.[45]

The first public movement related to the Walchensee project crystalized around the person of Fedor Maria von Donat. A retired major in the Prussian army, Donat had long made a name for himself traveling around Central Europe, agitating for large-scale hydraulic engineering projects. He claimed, with some justification, to have been one of the first to put forward a modern plan for the draining of the Pontine Marshes in Italy (a project eventually executed under Mussolini).[46] In the summer of 1905, Donat also approached the Bavarian government with a proposal to exploit the hydropower of the Walchensee.

The genesis of the idea shows how much imagination goes into envisioning a hydraulic energy landscape. As Donat told it, the idea came to him during a sojourn just across the Bavarian border in the Austro-Hungarian crownland of Tyrol. During a stay on the Achensee, another

Alpine lake that actually lay within the same catchment as the Walchensee, Donat had an epiphany. The Achensee, he realized, could be utilized as a hydroelectric reservoir.[47] While studying local topographic maps, he noticed the striking relief of the Walchensee and approached the Bavarian government with the idea to build a power plant there as well. Donat sought to generate power on a scale far greater than Schmick. His scheme would produce a minimum of 92,000 continuous horsepower, with a temporary maximum output of 150,000. The major boasted that the result would be a power plant with no peer in Europe and that bested even the superlative Niagara Falls project.[48] He too suggested that some of the electricity could be used to power Bavaria's state-owned railways.

Donat's scheme was similar to Schmick's, but with several notable differences. Donat, like Schmick, sought to exploit the sizable difference between the Walchensee and the Kochelsee by tunneling through the mountain separating them. Moreover, Donat had also resolved to divert water from the Isar River basin to increase the capacity of his plant. Donat, however, sought to capture not just a portion of the river's flow, but the entire runoff of the upper Isar basin. To do this, he advocated constructing a dam wall across the Isar Valley. Behind the barrage, a new lake would emerge—the Isarsee, as Donat dubbed it. The major also planned to impound the Riss—the upper Isar's largest tributary—within the Isarsee. Despite the state's dismissive stance towards Donat's extreme project, he persisted in trying to acquire a concession.

To win support for his Isar dam, Donat took his ideas public. Both in the newspaper and presentations, he criticized Schmick's Walchensee project as too cowardly and conservative. The Bavarian press gleefully characterized the disagreement among Donat, Schmick, and the Bavarian authorities as a "battle for the Isar" (*Kampf um die Isar*). In a published version of his lecture, entitled "The Power of the Isar," Donat laid out how the river could solve most of Bavaria's energy problems, from the powering of its railways, to the fixation of nitrogen for synthetic fertilizer, to the provision of cheap electricity for the state's artisans. "Generally, we have no idea how inhumanly rich Bavaria is, and that thanks only to the exploitation of the Isar. It is said, that the Isar carries golden sand, and that it is just somewhat costly to fish it out: Here you have the gold in thick—thick—really thick clumps!"[49] From Donat's point of view, lesser Walchensee schemes were in fact wasteful: "one must always remember that each cubic meter that flows [unused down the Isar] equals a loss of 2,825 horsepower . . . such a crippling of the glorious waterpower of the Kingdom would be an inexcusable crime against the Bavarian state and commonwealth."[50]

That Donat managed to garner considerable attention and assemble a substantial following hints at a broader popular enthusiasm for the fantastic promises of the new energy source. It also suggests that Europeans were every bit as susceptible to the siren song of the "technological sublime" as supposedly more modern Americans.[51] We can get an impression of the level of public interest as well as the types of people who may have been excited by the major's utopic energy scenarios from newspaper articles at the time. One of Munich's leading papers, the *Münchner Neueste Nachrichten*, took time to describe the crowd gathered at a presentation Donat gave before the local chapter of the German League of Land Reformers (*Bund deutscher Bodenreformer*). This group saw itself as part of the international movement led by the American reformer Henry George. The crowd is depicted as "extraordinarily large" and boisterous. In particular, its shouts and applause increased during the second half of Donat's talk, when the speaker peppered his lecture with quotations from Otto von Bismarck, the former chancellor of the German empire. The audience also seemed to take special pleasure in Donat's frequent attacks on public figures. Both the club venue and the disposition toward Bismarck suggest that large-scale energy development appealed to German nationalists with an interest in addressing the social disparities caused by industrialization.[52]

In the process of working out its plans for the Walchensee, the Bavarian government rejected Donat's idea of damming the upper Isar. State objections were of a primarily technical nature. In early January 1905, the Bavarian state construction authority's evaluation of Donat's Isar dam concluded the plan was "not worth discussion." The report questioned Donat's understanding of the morphology of mountain rivers, stating that while the major lived in the mountains "he seems not to have acquainted himself with the main characteristic of all mountain watercourses, namely their excessive sediment loads." The high sediment load of the upper Isar, according to the report, meant any dam there would fill up in a relatively short time period. With a touch of defensiveness, the author of the report noted that Donat had acquainted himself with the publications of Otto

Intze. Intze, the grand master of German dam building, had been instrumental in impounding water in northern industrial districts.[53] Intze's monumental works in Silesia and Westphalia, however, had not been built on waterways with heavy sediment loads. The report declared that if not for the sedimentation problem, "then numerous dams for the purpose of water storage would have been built long ago."[54] In February 1907 Donat took the extreme step of appealing directly to Bavaria's Prince Luitpold to intervene on his behalf in the Walchensee affair. The regent rejected the Prussian's pleas after a brief consultation with his cabinet.[55]

While Donat's Isar dam failed to gain support, his activities nevertheless had important impacts on planning for the Walchensee project. The Prussian major's success with the Bavarian public pressured state engineers to increase the dimensions of their Walchensee project. Although they had admired the Schmick project from the start, their final plan far exceeded it in scope and scale. The new conception was expressed in a magisterial memorandum entitled *The Waterpower of Bavaria* that was published in October 1907. Across five hundred pages and two volumes of historical background, international surveys, maps, and tabulated hydrological data for Bavaria's waterways, the state made its case for what should be considered waterpower and how it should be used. This included its plans for Bavaria's Alpine lakes and the Walchensee in particular. The introduction to the tome signaled the recognition that a new era had dawned. The "moment of waterpower exploitation" had arrived on the world historical stage, and its advent heralded a new chapter in the age-old relationship between humans and water. Rising coal prices gave water a new value, and waterpower could no longer be used according to custom. The signs of this new perspective were to be found not just in Europe but across the globe. In Egypt, for example, the new economic circumstances necessitated the destruction of the Temples of Philae that were to be drowned behind the first Aswan dam.[56] Accordingly, the state Walchensee project went far beyond the scale of Schmick's project. Seizing on Donat's idea to fill the lake with as much water as possible, the state project was slated to produce a maximum of 56,000 horsepower—nearly triple the capacity of Schmick's project. Furthermore, the state planned to make much greater use of the Walchensee's storage capabilities. Over the course of a year, it would cause the level of the Walchensee to sink by up to twenty meters.[57]

As the state's superlative plans took concrete shape, additional resistance surfaced. In Germany at this time, as in other parts of Europe, movements aimed at protecting nature from urban and industrial expansion emerged. Some groups imagined natural landscapes as constitutive of a German national identity and therefore deserving of protection.[58] Others understood landscapes as the anchor of regional distinctiveness.[59] Both currents flowed into sentiment to protect the Walchensee. Historians have already documented how the Walchenseewerk gave impetus to burgeoning nature conservation movements in Germany. Thus far, scholarly attention has mostly been devoted to the significance of the Walchensee opposition for the institutional history of the German environmental movement.[60] Less interest has been paid to the actual nature that conservationists sought to protect.

In the case of the Walchenseewerk, the principal questions related to the value and function of Alpine lakes. Conservationist objections to the project reveal that they were concerned above all with aesthetics. One aspect they homed in on was the lake's color. Bavarian nature protectors argued that the lake's unique turquoise color would be ruined by sediment carried into the lake by the Isar. Their most stringent objection, however, was to the enormous alterations of the lake's level necessitated by its reservoir function. Professor Schmidt, a member of the state's newly created conservation bureau, lamented that the Walchensee would be degraded to a dam (*Talsperre*, literally "valley closer"). Instead of a daily ebb and flow, its level would rise only with the floods of the warmer months and be drained the rest of the year. For him, the worst part of this situation was the exposure of the muddy banks of the lake. Perhaps trying to garner support from the pious in heavily Catholic Upper Bavaria, Schmidt declared that converting the Walchensee into a reservoir of these dimensions would be an "irresponsible assault against a unique memorial to the Creation."[61]

Opponents frequently framed the issue as one of the preservation of natural beauty versus economic gain. Typical of these viewpoints was Professor Franz Kreuter of the technical university in Munich. Called to provide an expert opinion on the state's plans, Kreuter questioned blithely sacrificing what to his mind was the finest lake in all of the Bavarian Alps for the sake of profiteering. Kreuter also saw an irony in disfiguring the very natural beauty of the area that had recently brought a certain prosperity to the region by attracting thousands of tourists. The professor noted that even the "practical Americans" had made the sacrifice of preserving one of the most beautiful areas in their country—"as big as a small kingdom"—in all of its elemental glory. Here he referred to the creation of Yellowstone National Park, a conservation measure made so that the "people's enjoyment of nature would not entirely waste away." The fact that it would be the state reaping the profits from a transformed Walchensee changed nothing about the situation for Kreuter, but he acknowledged that the potential financial gain probably made the decision to build an easier one.[62]

The idea of using lakes as reservoirs prompted resistance throughout the Alps. In Italy, at the same time the Walchensee controversy was still raging to the north, the country's major tourist and cycling club publicized protests over the lowering of the level of the Lago di Antillone in the Piedmont. In an article entitled "A Blue Eye Snuffed Out" in the club's magazine, they portrayed the lake as a local cultural touchstone threatened by energy development. The article's author painted a vivid picture of the lake as the place where women would congregate after church. There the elder women would gather by the shore to read mystic omens from cross-shaped lilies while the younger divined their romantic prospects from the buds of the same flowers. The publicity proved effective in rallying a number of prominent organizations to the lake's defense. Eventually lawsuits forced the electric company to divert additional water into the lake to maintain its level.[63]

The arguments of conservationists did not persuade most Bavarians of the plan's folly. But for many conservationists, this had not been the point. They did not reject the Walchensee hydropower out of hand. As with the case of the Piedmontese lake, objections focused on the dimensions of specific project components. They often centered on the extent of water level fluctuations or the amount of water to be diverted into the reservoir. Though this strategy did not prevent hydropower development from taking place, nature protectors succeeded in injecting questions about both the intrinsic and spiritual value of nature into a discussion that had primarily revolved around technological and economical considerations.

Their protests helped pull the state back from its plans to execute a larger scale project. First, the state promised that the Walchensee would not be drained any lower than six meters below its current level. Conservationists' critiques of the official state project were also one major reason that in 1907 the Bavarian government decided to hold a contest to elicit new Walchensee ideas. Internal discussions make clear that state officials viewed a public competition as a means of relieving some of the critical pressure it was feeling in response to its plans. The contest, some hoped, would deflect criticism about the government's "one-sidedness" and perhaps even result in a few useful new ideas. Others anticipated a certain "reassurance" for the state engineers who had drawn up the project, who hoped to have their ideas confirmed by a broad swath of their colleagues. Both the appearance of impartiality and competence seemed especially valuable "for a government that must always be concerned about cover."[64] As a nod to the nature protection movement, the contest instructions insisted that all entries appropriately consider the landscape. An international jury met in July 1909 to crown the winner. Among the over thirty proposals received, the titles of some submissions such as "No Dam" and "Power and Natural Beauty" make clear that the message about nature protection had been received by some. The eventual winner, entitled "Simple and Safe," was a far more modest proposal than the original state project, and the jury especially appreciated its less intense use of the lake as a reservoir.[65] The state incorporated many of the suggestions of "Simple and Safe" into an adapted Walchensee project. From this point forward, nature protection ceased to play a decisive role in the development of the Walchensee plan.

Which is not to say that concerns about environmental impacts disappeared. Instead, they shifted to a different arena. As the case of the Walchensee shows, reservoir schemes also prompted conflicts over their social and economic impacts. In Bavaria, as throughout the Alps, resistance hardened against the same major hydrological changes that drew conservationist ire. Raising and lowering lake levels not only created eyesores but disrupted fisheries. The "interbasin transfers," as hydraulic engineers call them, that diverted water into the reservoirs and augmented power production also created stark new hydrological circumstances. Moving water in this manner meant that one catchment area would be dealing with a surplus while another one was deprived.

Feeding the Walchenseewerk necessitated the diversion of considerable amounts of water from the upper Isar valley. For a stretch of over fifty kilometers, a substantial portion of the Isar would be removed from its bed. From autumn until the spring melt, sections of the river here would run completely dry, upsetting the situation of local flora and fauna. The removal of this water would also affect livelihoods. The state itself recognized that one Isar segment, ten kilometers in length, would be "completely devalued" for fishing, using language betraying a utilitarian perspective. On the other hand, the damming of the Walchensee's previous outlet, the Jachen, would not impact fisheries there because experts surmised they had already been ruined by timber rafting.[66] While agriculture did not have an important role in these upper reaches of the valley, the timber transport trade did. Prior to industrialization, mountain rivers were the best way to transport the most important source of energy and building material: wood.[67] The Isar and its tributaries were the only transport route for the valley's considerable wood resources. In the years before the emergence of the Walchensee plan, an average of 77,000 cubic meters of timber had been floated annually down the Isar to Munich. A study found that any disruption of upper-Isar rafting decreased the price the region's forest owners could gain for their goods.[68]

In the conflict around the project, a "raft master" (*Floßmeister*) Georg Willibald from the Isarwinkel became the personification of his guild. Willibald led a peaceful demonstration against the government project in May 1910. He argued that the timber transport required exactly 200 days annually where the upper Isar flowed with at least as much water as it carried during the winter low. For this reason, timber transporters like himself could only agree to the project if the state agreed to divert no more than the surplus floodwaters of the Isar into the Walchensee. Willibald enjoyed reminding audiences that after 1,400 years of floating timber on the Isar down to Munich, his guild expected that their lifeline would remain open.[69]

The river diversion also worried the inhabitants of riparian communities along the upper Isar. Chief among them was the spa town of Tölz. Hotel owners and other participants in the tourist industry declared that a diminishing of the Isar would seriously endanger the town's position as a popular tourist destination. In addition to concerns about the aesthetic impacts to the Isar, townspeople wondered whether the lack of water could lead to hygienic problems. Tölz, like most river communities, had charged the Isar with removing its wastewater. The diversion also raised legal questions. As the Bavarian state conservation agency put it, "important settlements, that in any event owe their emergence mostly to their respective waterways have a right to the same. One cannot take the Isar from Munich, the Main from Wurzburg, the Rhine from Cologne."[70]

Ultimately, neither those engaged in the timber transport nor the majority of the inhabitants of the Isarwinkel outright rejected the state's Walchensee project. Like the nature protectors, both groups accepted the justification for such a facility—or, at the very least, they saw little hope of standing in the way of what was viewed throughout all of Germany as a prime example of progress. The Walchensee scheme was so grand that it made the newspapers throughout the empire, including the front pages in commercial centers like Frankfurt. Though the editors of the *Frankfurter Zeitung* expressed some sympathy for those who would be impacted, they probably spoke for most Germans when they argued that changes to the region were only a matter of time. "Some day," they opined, "the great sacrifice will have to be made if it promises such great profit for the entire land and its future."[71] In the face of such momentum, residents of the Isarwinkel focused on ensuring an adequate water supply in their valley. They demanded a reduction in the amount of Isar water that would be diverted and insisted that additional tributaries be left alone. At the margins, they were able to win some concessions.

The Politics of Reservoirs

In the final instance, the decision whether to transform the upland for energy storage purposes was a political question. Making generalizations about the politics of reservoir building and nationalization of waterpower resources in the Alps is difficult. As the Bavarian case shows, regional peculiarities and happenstance played a role. People across the political spectrum supported hydroelectric development. The question of whether it should be carried out by the private sector or the state, however, often split society along more traditional lines of Left and Right. But even this equation broke down at times. Many Bavarian liberals and conservative-minded politicians, for instance, could support the progressive politics of public power when it came to giving their historical polity a competitive leg up. Their identity as Bavarians outweighed allegiance to laissez-faire economics. This seems to have been the animating principle of Bavaria's last king, who supported state development as long as it succeeded in finally beginning work on the project. Conversely, the overwhelming majority of Bavarians who decried the transformation of the landscape did so from a conservative critique of the materialism of modern civilization.

Nominally speaking, the three most powerful political factions in the kingdom—the Liberals, the Catholic Center, and the Social Democrats—all supported the state's Walchensee project. The willingness of the Liberals—who previously had advocated for laissez-faire economic policies—to push government intervention dated to a perceived "coal crisis" that gripped much of Germany in 1900.[72] Thereafter, it was a member of Bavaria's Liberal government—the transport minister—who actively pushed to have the Walchensee power reserved for the state-owned railways. Catholic Center politicians sought state electrification as a means to aid the Catholic middle class and farmers they viewed as their most important constituents. A constant in the political calculus was the support for state-led electrification by Bavaria's Social Democrats. The vast majority of Bavaria's Socialists—with their bases in the cities of Munich and Nuremberg—toed the more moderate, revisionist line of the national party. While eschewing calls for revolution, moderate Socialists did call for the state to intervene more decisively in the economy. By the fin-de-siècle, the third largest faction in the diet, the Social Democrats, always saw themselves as the vanguard of state-led electrification.[73]

This coalition led the bicameral diet to approve two rounds of funding for the Walchensee project in 1908 and 1910. Nevertheless, headway remained completely in the realm of words and not deeds. While a project of such scale certainly required lengthy negotiations with interested parties, the impression existed that state's Liberal leadership was dragging its feet in the hopes that the private sector, too, could eventually be folded into the venture.[74]

The fallout from a political earthquake in 1912 also suggests that the battle lines were not so cut-and-dried. In that year, the Bavarian regent dissolved the government, ending decades of support by the crown for liberal government. As replacement, the regent appointed Georg von

Hertling, chairman of the national Catholic Center Party in Germany and future *Reichskanzler*. The move has traditionally been interpreted as an attempt by the aging monarch—he would die before the end of the year and be replaced by his son—to circle the wagons against the rising tide of social democracy. There is evidence that the state's progressive electricity politics played a role in this regard. Already in August 1912, however, the new monarch proclaimed his support for a state-led Walchenseewerk. He justified his stance by declaring that his kingdom was in the "predicament of possessing no hard coal," and arguing that only by activating its waterpower resources could it promote industry and prevent falling further behind. He criticized the lack of progress on the Walchensee and reasoned that the risk for the state would not be so great.[75]

Despite the support of the king and the dominance of the Center Party in the diet and government, however, momentum completely stalled. How to explain the lack of progress? Later, Bavarian Social Democrats would accuse the new prime minister of purposely dragging his feet in the matter in order to prevent realization of a public project. A group of patricians in the upper chamber of the Bavarian diet sent the government scrambling when they demanded to know whether the state still intended to carry out the project. The immediate fallout was a report by the new government that the financial benefits of rail electrification did not look as positive as initially believed.[76] In response, the government shifted responsibility for the plan from the Transport Ministry to the Interior Ministry. The move received the blessing of the diet in the spring of 1914, but the future of the Walchensee remained very much in doubt. The events of the coming summer, however, would decisively change the calculus behind reservoir building in both Bavaria and the rest of the Alps.

Conclusion

After the turn of the twentieth century, the desire to put white coal to work electrifying the era's most important mode of transportation provoked an energy reckoning of sorts. It necessitated a direct comparison between Alpine hydropower and coal, the extraordinarily versatile fuel that enabled steam travel in the first place. Engineers dominated these discussions and they contended that only by finding a way to store waterpower could it compete with coal. In the engineering imaginary, high Alpine lakes revealed themselves to be the ideal solution. Eons of natural forces had already embedded these stores of great potential energy in the landscape. The argument that they would therefore make comparatively inexpensive reservoirs seemed plausible, to some even ordained by Creation.

Focusing on the case of the Walchensee reminds us that the making of a new Alpine energy landscape was contested at every turn. Creating high-pressure reservoirs out of lakes necessitated unprecedented interventions into regional environmental and social circumstances. Not surprisingly it prompted resistance from municipalities and social groups whose livelihood would be impacted by the changes. The consequences of hydropower development for natural landscapes even spurred some opposition from elite circles. Somewhat surprisingly, some Bavarians opposed the state's plans on the basis that it was not going far enough in transforming its Alpine waterways into energy sources. From the Bavarian case we can surmise that throughout the Alps, there would have been similar popular enthusiasm for the new energy bounty. This broad support, along with a strong regional identity, combined to make the politics of state intervention in reservoir construction accepted even among parts of the conservative camp.

By the early 1900s, then, the general blueprint for harnessing white coal had emerged. Initially, individual waterfalls had been tapped by entrepreneurs for small-scale industry. They were then supplemented by larger run-of-river dams (dams without water storage) on rivers in Alpine valleys. Finally, a move toward economic rationalization led Europeans to see that the highest calling of the Alps was to make waterpower behave more like coal. Thanks to the centrality of storage in energy systems and the far-reaching environmental disruptions necessary to achieve it, states across the Alps began to take a more active role than ever before. From this point forward, the question of the role of the state in energy development would never disappear. By the summer of 1914, Europe was hurtling toward a crisis that would make energy a question of national survival.

FOUR

Emergency Power

> The influence of mountains on the conduct of war is very great; the subject, therefore, is very important for theory. As this influence introduces into action a retarding principle, it belongs chiefly to the defensive.
>
> —*Carl von Clausewitz,* On War *(1832)*

IN THE EARLY MONTHS OF 1918, Louis-Jules Arrigon was traveling in the southeast of France. As the First World War approached what many anticipated to be its decisive months, the French journalist had gone to the region to observe an extraordinary transformation of the countryside. Writing about his experiences after the war, Arrigon noted that the word "miracle" had often been used in reference to France's war effort.

> Miracle, the marvelous strategic reestablishment of its armies at the Marne; miracle, the amazing defense and inviolability of Verdun, which Europe witnessed under the formidable German assault. Miracle, the reconstitution of its industries, their expansion and adaptation to the war, effected under the menace of the enemy and despite the invasion of its richest *départements*.

Arrigon believed that the term deserved to be used once more, to characterize the development "realized at the height of the war, by the French industries of the South-East, Savoy and Dauphiné, which drew from the inexhaustible reserves of the white coal of the Alps the power necessary to operate and achieve the results from 1914 to 1918." There he witnessed and later described the results of a rapid expansion in all sectors of hydropower usage: electro-metallurgy, electro-chemistry, paper manufacture, and power transmission.[1]

Directly after the signing of the armistice ending the First World War, future British Foreign Secretary Lord Curzon declared that the Allies "had floated to victory upon a wave of oil." He referred specifically to the "tremendous army of motor lorries" he had seen in France and Flanders.[2] When historians have thought about the connections between energy and the First World War, then they have increasingly thought about oil. But it is worth remembering that Lord Curzon uttered his famous lines at a dinner for the Inter-Allied Petroleum Conference. At such an event, one might expect a certain immodesty about the part that oil had played in the victory. While the First World War probably stands as the watershed moment in which oil ascended to become the key global energy resource of the twentieth century, it would be difficult to argue that hydrocarbons actually won the war. True, they powered many of the new weapons that made their debut at this time—the submarine, the airplane, the tank. In due course these machines would decide the outcomes of epic battles. But in the second decade of the twentieth century, they could not yet tip the scale. And from a quantitative standpoint, oil was a junior partner during the Great War.[3] Something left out of all of these reckonings, finally, is that petroleum-burning motorized vehicles owed their existence increasingly to electrified factories. The Allies may have in part rode a swell of oil to success, but they did so in contrivances likely built by the power of the electron.

In this regard, the Alps were home not only to a theater of war, but also an underappreciated economic front from 1914 to 1918. Though it has fallen from memory, white coal also played a vital role in the conflict. The Great War prompted a flurry of energy development on waterways throughout the French Alps. This energy was instrumental for the French economy, both during the war and beyond. To a degree, a similar story played out throughout the Alps, albeit with slightly different timing. Combatant countries like France and Italy that experienced the greatest disruptions of their prewar coal supply found it necessary to seek additional energy in the mountains during the conflict. Moving north and east across the chain, the calculus changed. For Switzerland, in the words of one of its foremost electrical engineers, the First World War began long before 1914, at least in an energy sense. Without coal of their own and dependent on imports, the Swiss felt pressure to develop their Alpine waterpower long before the upheaval of 1914–1918. Despite war-induced coal shortages, the lack of available labor and other raw materials stymied energy expansion for the course of the war.[4] For the German empire and Austria-Hungary, domestic coal resources more or less sufficed to meet the demands of their war economies. In the former Central Powers, the decisive moment came after the war. At that time demobilization and the loss of coal supplies due to peace settlements combined to focus attention on putting Alpine waterpower to work.

Historians have recently begun writing the environmental history of warfare. The effort has been motivated by the belief that military conflict is a distinct and often decisive contributor to environmental change.[5] Though research has considered the First World War, much work remains to be done.[6] This is particularly true of the war in the Alps. Compared to the historiography of the main theaters of war, scholarship on the Alps has been sparse. Until very recently, even the academic study of the Alpine front depicted the combat in this region as a holdover from a more traditional era of warfare, where individual abilities and extraordinary physical and mental power still made a difference. The shorthand for these viewpoints resides in the widespread icon of the hardy mountain warrior.[7] Despite this literature's fascination with the mountains, they serve mostly as a backdrop for human heroism. More recent research on the war in the Alps has explored different avenues of inquiry, emphasizing the impacts of the war for civilians, as well the conflict's influence on ideologies and mentalities. Environmental historical analysis, for the most part, has not been one of the new axes of exploration.[8]

This chapter adds to our understanding of the environmental history of the Great War by showing that the conflict was a watershed not only for global oil consumption but for the Alps as an energy landscape. Due to the disruption of coal supplies and the increased demands of fighting total wars, the states of West-Central Europe sought to meet energy needs with an alternative source. As would remain the case until the advent of nuclear power in the 1960s, the most abundant and feasible substitute in the region appeared to be coursing down from the peaks of the Alps. The war and its immediate aftermath led to a veritable boom in dam building. This was in part because the stakes of the conflict also helped remove some of the societal obstacles to larger projects, and to projects that intervened more drastically into local hydrology.

What follows is a look at two cases emblematic for the wider history of hydroelectric development in the mountains during the First World War. In certain regions—particularly the French Alps—the emergency power harnessed in the mountains was crucial to the war effort. Returning to the Bavarian Alps, the completion of the Walchenseewerk in the postwar period illustrates patterns noticeable throughout the Alps after the end of hostilities. A combination of disruption to the prewar coal markets in Europe and heightened nationalism made developing Alpine waterpower irresistible. It is not a coincidence that the plan to convert the Walchensee into a reservoir, the subject of the last chapter, finally moved forward during this period after nearly two decades of deliberation.[9] Nor was it an accident that when the project came online, it was owned and operated by the Bavarian state to provide electricity for one of the globe's most advanced electrical grids—and one in public hands no less. Particularly after the emergency of the Great War, governments throughout the chain seized control of their water resources to drive postwar economic recovery. These actions complicate widespread views of mountain polities as fundamentally conservative or tradition bound. Bavaria and other governments solved the problem of white coal development by and large through what was viewed as a progressive policy: public ownership.[10]

The Miracle in the French Alps

France's need for a miracle was set up by the first few months of the conflict. In the hopes of swiftly knocking France out of the war, the Germans launched a surprise offensive in August 1914 through neutral Belgium and into northern France. While France managed to stave off defeat at the battle of the Marne, the ensuing crystallization of the Western Front severely disrupted the country's economy. From October 1914 until the end of the war, some three-quarters of the rich coal seams of the *départements* of Nord and Pas-de-Calais lay either in enemy hands or within striking distance of its artillery. This had disastrous consequences for French coal production. Before the war France produced on average some 40 million tons of coal annually. At the same time, the country required over 60 million tons to satisfy its domestic and industrial demands. France made up the shortfall through imports, primarily from Great Britain, but also Belgium and Germany. The German invasion made the deficit even greater, removing at a stroke two-thirds of France's annual production in addition to blocking key sources of imports.[11] The incursion into the north also resulted in the loss of almost all of this region's metallurgical industries. When added to the losses suffered in the eastern part of the country, French heavy industrial output decreased by almost three-quarters. All of this happened precisely at a time when France desperately needed to ramp up wartime production to satisfy matériel shortages, particularly of ordnance.

The waterpower of the French Alps emerged as the country's greatest source of supplementary power.[12] As we have seen, in the decades before the war, the French Alps had been the birthplace of white coal. The new form of hydraulic energy spurred the growth of the region's paper industry—spearheaded by Aristide Bergès at Lancey. With the advent of hydroelectricity, the nascent electro-metallurgical and electro-chemical industries also established themselves in this same area. The majority of this development had concentrated itself in the northern French Alps, where the climate, relief, and economic conditions had proven more conducive to early hydro exploitation (Figure 4.1). Within the northern Alps, the main activity took place in the Sillon Alpin—the large depression between the pre-Alps and the central massifs running roughly from the Arve River to the region around Gap in the south—and the large valleys of the inner Alps. Indeed, in these large interior valleys (those drained by the upper Isère and Arc Rivers), white coal supported a density of industry and energy exploitation greater than anywhere else in France, or anywhere in the Alps for that matter.

During the winter of 1914–1915, many of these existing facilities switched to war production to help allay the country's most pressing needs. The paper industry supplied cotton for munitions manufacture, the electro-chemical industry numerous chemical compounds, and the electro-metallurgical sector provided aluminum, iron alloys, steel, and cast iron. But as the war progressed, it became clear that France needed

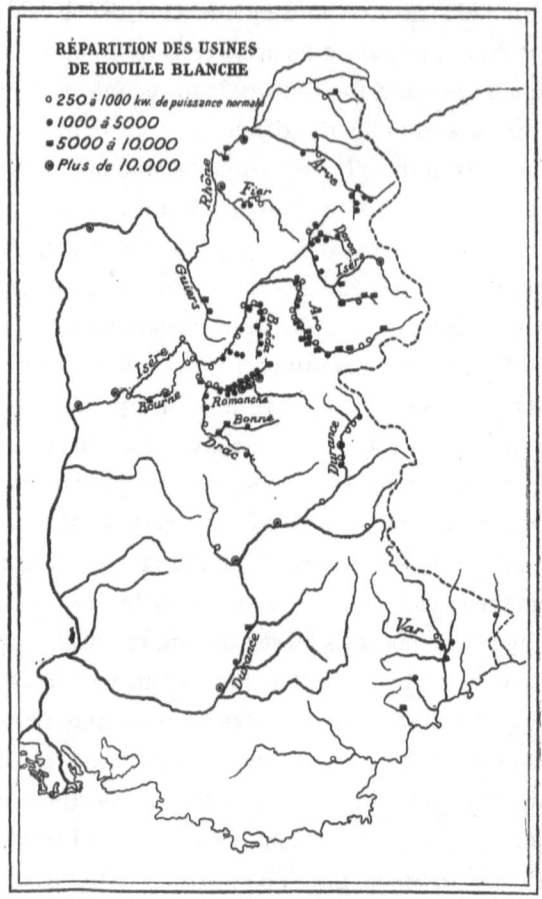

FIGURE 4.1   Distribution of white coal plants. This map, made for an interwar study of the contribution of hydroelectricity to the French war effort 1914–1918, shows the density of hydroelectric plants in the French Alps.

Source: Raoul Blanchard, *Les Forces Hydro-Électriques Pendant la Guerre* (New Haven: Yale University Press, 1924),17.

to develop additional sources of power. For one thing, coal was becoming ever scarcer. An increase in domestic production had been negated by the loss of imports caused by submarine warfare. Moreover, new material programs to supply France and its allies demanded new sources of production.

A movement to increase the exploitation of France's white coal emerged. In part, this movement was motivated by a desire to solve the problem of the coal crisis by substituting white coal for its fossil cousin. The harnessing of chutes was seen as a means to rid the French of the humiliating necessity of importing their energy from abroad. After the new facilities had done their part for the defense of the nation, they would then provide the power for French industry, light for both city and country, and traction on the country's railways. In order to pay for this critical program, the French state intervened. In addition to building some facilities entirely with public money, the state advanced generous credit to entrepreneurs, to be paid back within ten years after the cessation of hostilities. The state also assumed responsibility for critical groundwork, literally so in the form of the geological studies necessary to site future dams.[13]

Wartime hydro development in the Alps proceeded in two broad phases. The first phase unfolded under the motto *Faire vite!* From 1915 to 1916, French industry focused on quickly expanding capacity by augmenting the exploitation of white coal in the places where it already existed. Doing so saved time by avoiding the necessity of building transmission lines that cost both time and money. The centers of activity during this phase centered on the Arve, Arc, and the Grésivaudan. In the latter valley, the Bréda, a tributary of the Isère, was further put to work. With fifteen existing hydro facilities before the war, the Bréda was one of the most heavily exploited waterways in all of France. The war witnessed the erection of three additional power plants, more than doubling the energy wrung from the river basin to over 15,000 kilowatts. Nearby, the Grésivaudan paper industry availed itself of the last remnants of power remaining in the left-bank torrents rushing down to the Isère.

The journalist Arrigon provides a window into this transformation through his reporting on his visit to the paper factory of Aristide Bergès. As part of their wartime expansion, the Papeteries Bergès created an additional 1,000 kilowatts of power by equipping another chute on the upper stretches of a nearby torrent. The additional energy enabled the Papeteries to expand its traditional production and venture into new areas as well. The plant began manufacturing cellulose and nitrocellulose (sometimes known as guncotton), the latter at the behest of the Service des Poudres. A contingent of 150 workers at the factory even produced bombs. The wartime activity sparked growth throughout the establishment. The number of workers at the mill nearly tripled, from 1,000 in the summer of 1914 to 2,800 in 1918. The new production necessitated the expansion of the facilities outside of the gorge it had always called home. Transportation to the plant also improved after the connection of its private rail station to the nearby Paris-Lyon-Marseille railway. This boom came at a propitious moment for the factories, as the entire French paper industry had been suffering prior to the war. Increasingly, the branch that had previously produced the pulp necessary to manufacture paper or cardboard found itself importing the material from abroad (particularly Germany and Austria-Hungary). The imports cost some 100 million francs, prompting the general director of the Papeteries Bergès to complain about how his industry's tributary status to foreigners hampered its domestic and global aspirations. By 1918, the future of the Papeteries looked considerably brighter.[14]

Besides modifying local waterways, the mill's expansion wrought other environmental changes. One of the more far-reaching was the impact of pulp production on the region's forests. Whereas before the war, the factory consumed some 160 cubic meters of wood every day (to produce 40 tons of pulp), by 1918 this number had climbed to 380 (for 150 tons of pulp). In anticipation of the operation's expansion, Bergès's mills acquired vast forest holdings at the outset of 1915. These were located not only in the factory's vicinity in the Dauphiné, but places farther afield in the Savoy and Jura. By 1918, the company held forests valued at some 5 million francs and worked by 800 Spanish loggers. To get the timber to the factory in Lancey, the firm commanded a fleet of 200 horses, 15 trucks, and 8 motorized tractors.[15]

In this early phase, France also put her Alpine waterpower to an unprecedented use: producing chemicals for use on the battlefield. To hear Monsieur Arrigon tell it, Germany forced France to make use of these more "murderous chemical resources." From his perspective, since the outbreak of hostilities, the German combatants had acted barbarically, violating the articles of the Convention of the Hague repeatedly. This flouting of the rules of warfare culminated in spring 1915, when the Germans used poison gas against their French counterparts at the second battle of Ypres. This succeeded in momentarily buckling the French line. But the troops and the home front quickly bounced back. Industrialists set about establishing facilities to produce chlorine gas. By May 1916 they were churning out huge quantities of the substance. At the end of the war, seven such plants existed in France, almost all of them in the Alpine region. Arrigon visited one of the most important of these liquid chlorine factories located within sight of the Belledonne massif in a suburb of Grenoble. The factory had become operational six months from the beginning of construction in September 1915. After the war, the factory switched from producing poison gas to several other chlorine byproducts, including sodium bicarbonate, hydrogen gas (for aviation), calcium chloride (*chlorure de chaux*), and sodium hydroxide (*soude caustique*). Small wonder then, that "the odor of chlorine reigned both within the factory and all around."[16]

The second phase of hydro development in the Alps took place in the years 1917 and 1918. Under the specter of the failure of France's Nivelle offensive, and the intensification of submarine warfare, the "captains of white coal" launched endeavors of unprecedented scale in France. Though some of these dams were still completed during the war, their size raises the question of whether their proprietors only had the country's short-term interest in mind. Activity during this period tended to concentrate on those areas previously neglected. The main theaters of this new spirit were in the south. French industrialists tapped into the mightiest river of the southern Alps, the Durance, with the creation of two large run-of-river dams. In the exterior valleys, the French engaged the slower rivers whose power sprang from their sheer quantity of water. The lower stretches of the Fier and the Isère were made to generate tens of thousands of kilowatts, respectively. French industry even tackled the mighty upper Rhône.[17]

The ultimate destination of this waterpower was also new. Increasingly, the companies harnessing the power of the French Alps transmitted this electricity outside of the mountain region. The new power generated by rivers at the foot of the Alps often found its way to nearby industrial centers such as Saint Étienne and even Lyon. But these cities also increasingly became the destination even of energy produced in the new facilities of the mountainous interior. Alpine hydroplants also found new customers in Marseilles and the Mediterranean coast.[18] As we will see in Chapter 5, from 1914 onward the trend of using white coal outside of the mountains spread throughout the region.

In the conclusion to his account of the wartime developments in the Alps, Monsieur Arrigon attempted to summarize the significance of what he had seen in the mountain valleys. "Under the influence of war," he wrote "destructive, sower of death and ruin, here, but there, terrible creator of energy and wondrous stimulating power, the *départements* of the Alps, most particularly Savoy and the Dauphiné, have given in a few years the spectacle of an unheard of industrial boom."[19] There is much to support his reading. In 1914 Alpine water produced some 302,000 kilowatts of electric power, accounting for nearly two-thirds of all French hydroelectricity. During the conflict, an additional 233,000 had been wrested from the waters of the French Alps, equaling an increase of 77 percent in the span of four years. Since its emergence in the late nineteenth century, France's white coal industry had never experienced such an enormous period of growth. The increase in Alpine white coal accounted for over half of the country's new hydro development during this time period (although hydro development elsewhere in France slightly reduced its proportion of national production). From a quantitative standpoint, Alpine waterpower was the single most consequential addition to France's wartime energy supply (Figure 4.2).[20]

In light of the sheer scale of the changes enacted, it is understandable that many could perceive the business in the French Alps as an example of a marvelous effort for the national cause. But the wartime determination to bend the waters of the French Alps was also attended by bitter social conflicts as well. Raoul Blanchard briefly detailed some of these difficulties in his introduction to a study on French hydro development during the war. Hydro development provoked lively debates within the engineering community about the proper technological approaches to the problem. Even thornier was the issue of the infamous *barreurs de chutes*. These were riparian owners less impressed with the importance of white coal for the nation. The *barreurs* took advantage of the great demand for hydroelectricity by speculating in the real estate necessary to build dams and diversions and selling to power companies at elevated prices. But the greatest problems had to do with labor. The same environmental conditions that made the mountains the ideal place to build waterpower plants also made them one of the least pleasant for a construction worker to live. Remote valleys offered few possibilities for recreation, and mostly poor lodging (if at all). The same seasonal changes in hydrology that plagued rational hydro exploitation also made for miserable work conditions on the construction

FIGURE 4.2 "White Coal Coming to the Aid of Black Coal." The snake represents the German invaders.

Courtesy of the Maison Bergès–Département de l'Isère (France).

sites themselves. Cold and snow in the winter prevented building just as assuredly as the summer floods. Construction workers on Alpine hydroplants, therefore, had to subject themselves to living in austerity and irregular seasonal employment. Even under normal conditions, it would not have been a simple matter to muster the workers necessary to rearrange France's Alpine rivers. The marshaling of most able-bodied men for the war effort made the situation acute. To skirt the labor issue, some industrialists imported colonial contingents and foreigners from countries such as Spain, Greece, Italy, and Armenia. In some cases, prisoners of war were also put to work. In doing so, they enjoyed the full support of the French government, particularly the minister of armaments.[21]

Lancey employed workers hailing from Italy, Spain, Switzerland, Serbia, Montenegro, Belgium, Armenia Greece, China, and Vietnam. Arrigon likened the scene he beheld there to "a kind of Babel where all of the nationalities, all of the races, all of the colors mix in picturesque variety." He noted that special barracks had been constructed for the East Asian contingent, and their doors were covered in chalk characters "of the Far East." Despite its upbeat tone, Arrigon's account could not help but imply the existence of some social tensions. The writer noted that among this group, the Italians, who had a history of migrating to the Dauphiné for work, were held in great esteem and would be later welcomed to return in large numbers. Other nationalities, he observed, had not experienced a "full success with the crucial test imposed on them," but he believed that a core of good workers nevertheless existed among this group. Once the bad and mediocre laborers had been removed, he asserted, "the better ones, increasingly trained in their task and increasingly specialized, would furnish excellent service."[22]

The entrance of large numbers of women into the labor force was another result of the wartime white coal expansion. In many parts of the French Alps, women made up an important part of the workforce. In the *département* of Isère, for example, 4,000 women worked in the machine construction industry by 1918, compared to 200 only four years earlier. While this development surely offended some sensibilities, the renowned geographer of the French Alps, Raoul Blanchard, predicted a future for female labor, while at the same time repeating the widespread opinion of the physical limitations of their sex. "The women are quite appreciated, they are a group full of vigor (*pleine de mordant*); all of the bosses expect to keep them after the war, as long as they specialize in those tasks that do not require great physical effort."[23]

The impact of the First World War on the waters of the French Alps lasted beyond the cessation of hostilities in November 1918. A slew of projects conceived or proposed during the conflict would only reach completion later. Some of these undertakings took longer because they incorporated new technological or economic elements in their designs. Above all, these newer plans sought to regularize the flows of Alpine energy through the use of reservoirs. Indeed, this movement to rationalize waterpower exploitation first took hold in wartime France. As described in the previous chapter, elsewhere in the Alps this movement initially focused on Alpine lakes as potential reservoirs, and in France it was no different. At the time, Lac de la Girotte was the most important lake-turned-reservoir in the French Alps. This small body of water, 1,736 meters above sea level, sat perched above the Dorinet de Hauteluce valley in the Arly basin. The lake itself measured only fifty-seven hectares in area, and it received such scarce recharge that it could provide but little water for energy purposes. However, Paul Girod, the owner of one of France's most important steel mills in the nearby town of Ugine, had the idea to transform the lake into a reservoir that would not rely on natural inflows to be filled. Girod outfitted the lake to be part of a pumped-storage hydroplant. During the day, water drawn from the lake plunged in a penstock to a power plant in the valley below to produce electricity to help meet the demands of the workday. At night, electricity from elsewhere operated a pump that sent water back up into the lake. In this manner, the plant obtained reserves of water it could use to cover the peaks in demand.[24]

Plans to achieve regularization of flows on a far grander scale also emerged. Large dam projects materialized in France—and throughout most of the Alps—in the course of the war. Indeed many of France's largest Alpine dams, realized in the interwar and postwar years, were children of the war. Schemes to erect dams on the swift-flowing Romanche and Drac Rivers for example, which would only be completed in the 1930s, first surfaced in 1918. Large-scale support for dams on the Durance and upper Rhône rivers finished after the Second World War—superlative projects that would rival some of the largest anywhere in Europe—also first gained momentum during the Great War. The latter scheme, symbolized by the large dam at Génissiat, was inextricably connected to the greater question of the development of the upper Rhône that would occupy all of France for several decades.[25]

In response to the war effort, industrialists in the French Alps intensified and expanded their modifications of Alpine hydrology. The war accelerated trends that had already existed and even extended the geography of dams to the southern French Alps that had been largely untouched prior to 1914. After the war, boosters for the hydroelectric industry like the Bergès family sought to depict these developments—not without some justification—as having contributed mightily to the defense of the country. Progressive-minded people like Louis-Jules Arrigon, who believed in the power of technology to improve society, welcomed these changes. There is good evidence that a majority of the French thought this way as well.[26]

The Central Powers and the Eastern Alps

While the Allied war effort depended on harnessing new flows of energy in the western Alps, a different situation obtained further east. In the eastern Alps, largely controlled by the Central Powers, new dam construction was largely absent. Neither Germany nor Austria-Hungary (or Switzerland for that matter) felt the same pressure to develop new energy sources as France or Italy. Which is not to say that all was quiet on this front. For the states of the eastern Alps, the First World War did impress the need for additional energy development. But they spent the war years planning rather than executing new projects in the mountains. Groundbreaking would wait until the cessation of hostilities, when governments anticipated the need for jobs for demobilizing soldiers.

This pattern can be seen in the case of the largest energy project to be realized in the region during the war years and their immediate wake. As detailed in the previous chapter, by the spring of 1914 the fate of Bavaria's Walchensee was up in the air. Already by the fall of that year, however, the outbreak of war created a new urgency to exploit this power. The remainder of this chapter follows the Walchensee project through to its conclusion in the mid-1920s. It shows how the war and its aftermath created consensus of the need to tap into additional flows of white coal in the remainder of the Alps.

At the outset of hostilities, the white coal pioneer Oskar von Miller volunteered the services of his engineering bureau to the state for the duration of the war, to ensure that the Walchensee project did not fall by the wayside. When the war continued past autumn 1914, and it became clear that there was no decision in sight, Miller conceived a new function for the lake. In October 1915, Miller submitted a new plan to use the Walchenseewerk in a scheme to supply all of Bavaria with electricity. Miller argued that the cheap, readily disposable Walchensee power should be made accessible throughout Bavaria. He proposed building a statewide high-voltage electricity grid—the *Bayernwerk*—to interconnect existing utilities. The

Bayernwerk would purchase electricity from Bavaria's power stations and distribute it to utilities throughout the kingdom.[27]

The key to the Bayernwerk was the Walchenseewerk, which would function as reserve power for the entire system. The Walchensee would encourage the development of Bavaria's remaining hydropower resources because its stored energy could—more rapidly than any other source—be mobilized to cover any power shortfalls due to seasonal changes in streamflow. Miller envisioned the Bayernwerk as a public-private partnership, composed of its power plants, municipal utilities, and the firms contracted to construct the grid. Pooling the power of all electric plants in the kingdom, Miller calculated, would both expand the electricity supply and lower generation costs, allowing the Bayernwerk to service previously disadvantaged regions such as eastern Bavaria and enabling farmers and artisans to benefit as well. The state, he argued, should be interested in the Bayernwerk because it would be making a "major contribution toward securing significant advantages for nascent industry and agriculture after the war."[28]

Miller's Bayernwerk pitch landed perfectly in the changing political climate in Bavaria during the First World War. The Great War finally forged the political will for state intervention in the electricity supply. Concern that the war economy was leaving agrarian Bavaria even further behind Germany's more dynamic economic regions ultimately prompted this accord. Within Bavaria, the perception reigned that the lion's share of the imperial government's wartime investment had gone to already heavily industrialized German states. Bavarian leaders tried to secure subsidies to promote key war industries on their soil as well. In 1916 the Bavarian state government attempted to raise the interest of the imperial government in harnessing Bavarian waterpower to produce the nitrates so desperately needed for the war effort. Eventually, the Reich built new nitrate plants near the lignite mines of eastern Germany, as thermal power plants could be erected more quickly than dams.[29]

Bavaria's political class perceived this as a severe economic blow, and by 1917, groups that previously opposed state involvement in hydropower development changed their tune. A majority in the Bavarian diet supported the state's new policy of "industrial advancement" and its centerpiece, the Bayernwerk.[30] As the war raged on, the Bavarian Ministry of War dropped its earlier objections to railway electrification, insisting on the importance of the Walchensee power in the interests of "economic mobilization."[31] By 1918, the Upper Bavarian Chamber of Industry and Commerce in Munich, which publicly rejected the state Walchensee project in April 1914 as an example of "state socialism," now viewed the work's development as urgent.[32] The diet unanimously passed funding for Miller's Bayernwerk proposal in the summer of 1918 with the anticipation that construction would begin with the war's end.

This came quicker than envisaged, and the first phase of the peace—that is the cessation of hostilities and demobilization—ensured that this time the Bavarian diet's resolutions would actually be followed by deeds. The swift and unexpected end of the First World War made civilians out of millions of German soldiers overnight. German leaders worried about the social consequences of failing to find employment for its demobilizing warriors. Putting these men to work took on an even greater significance in light of the political turmoil in Bavaria and Germany. On November 8, a day after the ruling Wittelsbach dynasty withdrew from power, the Bavarian socialist Kurt Eisner announced the creation of the Free State of Bavaria. The next day, Germany underwent a political revolution when the Social Democratic Party of Germany declared the establishment of what would come to be called the Weimar Republic in the wake of the Kaiser's abdication. Even as centrist Social Democratic governments established themselves at the state and the national level, socialists and communists pushed for greater change. The situation, Oskar von Miller insisted, demanded "the fastest possible initiation of the preliminary construction work." Miller saw to it that construction began a month after the armistice, a full two months ahead of schedule.[33] Throughout the postwar period, Bavarian politicians continued to emphasize the value of the Walchenseewerk as a postwar job creation measure.

The postwar economic and political situation convinced many Germans of the necessity of substituting hydropower for coal. Strikes in Germany's great coalfields disrupted fuel provisioning and revived an old refrain about hydroelectricity's superiority as an energy source since its production and distribution was not dependent upon politically unreliable workers.[34] For other Germans, one of the war's most important lessons was the necessity of conserving limited coal resources. Critics deemed the government's wartime policy of constructing gigantic steam-powered stations to meet electricity demand as quickly as possible a "reckless overuse" (*Raubbau*) of its irreplaceable stocks of thermal energy.[35] For one writer, the "emergency of the war and the upheavals of the revolution" had finally convinced Germans that their modern state depended on coal and therefore its conservation. This was best achieved by developing German hydroelectricity: "Gigantic powers, for the most part unexploited, slumber in Germany's watercourses and lakes. White coal!"[36] The state of Bavaria's chief hydroelectric engineer agreed about the war's impact on energy perspectives, arguing that it "brought a complete change" in attitudes towards waterpower, by removing the previous prejudices against its costliness as an energy source.[37] In part because of the widespread conviction in the importance of protecting German coal supplies, the Weimar government would later help finance the Walchenseewerk in exchange for an investment stake in the enterprise. As one prominent Bavarian engineer put it, however, saving coal was not merely a means of making up for past sins. It was necessary in light of what the peace settlement would take from Germany.[38]

Germany anticipated that the peace would come with an energy price. Though the Treaty of Versailles is commonly remembered for the reparations it imposed upon the defeated country and for the infamous "war guilt clause" that provided the legal basis for those payments, it also had an intentionally profound impact on Germany's coal supply. Signed by the German delegation in Versailles on June 1919, many of the territorial losses and other conditions prescribed by the treaty focused on the former German empire's coal resources. In the east, the new Polish state's demands for the coalfields of ethnically mixed Upper Silesia separated Germany from a region that had supplied about a quarter of its prewar production. In the west, Germany barely avoided greater losses when France's plans to detach the whole of the coal-rich Rhineland and create a new neutral state were thwarted. France's efforts to improve its dismal fuel situation—the retreating German armies largely destroyed the mines under its control—did result in ownership of the considerable supplies in Germany's Saar region to the north of Alsace-Lorraine. Taken together, Germany lost 40 percent of its coalfields due to Versailles.[39] The peace also required that the country make substantial coal deliveries to its neighbors.

These stipulations had significant impacts on Bavaria's coal supply. The loss of control over the Saar mines stung in particular, as the state's only significant domestic production occurred in a small piece of the region called the Saarpfalz. To procure its fuel, then, Bavaria remained completely reliant on the two primary sources of its prewar coal: the Ruhr and Bohemia. Ruhr coal production was hampered by the economic situation and by strikes, however, and high demand for its product existed throughout Germany. To aid in rationing fuel, the Weimar government set up the Reich Coal Commission. The Bohemian coal Bavaria had previously imported, on the other hand, now belonged to a new Czechoslovak state with different priorities.

The anticipated and ultimate loss of German coalfields made the completion of the Walchensee reservoir a necessity. Confronted with skyrocketing coal prices and eventually a crippling "coal crisis" (*Kohlennot*), Bavarian policymakers moved to accelerate progress on a Walchensee project modified slightly to reflect the new energy circumstances. In January 1919, the government appointed Oskar von Miller the commissar for the Walchenseewerk and the Bayernwerk and granted him special executive powers to push construction as quickly as possible.[40] Soon thereafter, the state also altered its plans for the Walchensee for a final time by reestablishing rail electrification—which had fallen by the wayside in the wartime reservoir plans—as a critical component of the scheme.

In the midst of this energy crisis, inflation in early 1920 forced construction prices drastically higher. The cost of the Walchenseewerk had been steadily rising since the end of the war. Already by early 1919, the reservoir's cost had tripled from the initial projection of 25.5 million marks in 1917. In 1920, however, the plant's price tag increased by 50 percent compared to the end of 1919. At this point the bill for the project stood at 185 million marks, roughly eight times the original estimate. Prices for the Bayernwerk experienced a similar trend, and the Bavarian state confronted the necessity of raising a further 200 million marks (in addition to the 250 million already spent) simply to bring both projects within a year of completion. Still, neither the diet nor the state flagged in its support for these initiatives. The astonishing rise in coal prices caused by Versailles meant that even under these circumstances, supplying Bavaria's industrial center in Nuremberg with waterpower generated from the Walchensee and transported by the Bayernwerk was 10–20 percent cheaper than coal. Moreover, Bavarian government officials believed that German coal prices would remain high for at least a decade, because the necessity for massive investments in the mining sector, combined with the coal deliveries to the Entente, would prevent an expansion of capacity for some time.[41] The only real debate provoked by the inflation was how to raise the necessary funds to continue construction.

In 1921, both the Bayernwerk and the Walchenseewerk were incorporated, the former as a quasi-public enterprise (Figure 4.3). The Walchenseewerk corporation, on the other hand, was fully owned by the state. A month after the formation of the company, it marketed its public bond issue with the explanation that the purpose of the Walchensee was the "rashest possible elimination of the grave economic consequences of the coal crisis for the transportation system and Bavarian industry."[42]

Under the circumstances, the construction of the Walchenseewerk and the Bayernwerk proceeded relatively quickly. Both projects continued to be disrupted by the upheavals that rocked Bavaria and Germany in the postwar period. Relations between workers, construction firms and the state became tense at times—as when armed communists interrupted negotiations demanding wage increases.[43] Another disruption occurred in

FIGURE 4.3   Cross-section of the Walchenseewerk. In the background is the Isar River, whose water was diverted to fill the reservoir. The lower-right inset shows the Bayernwerk grid.

Courtesy of the Deutsches Museum Archive, Munich, CD 59212.

January 1923 when France occupied the Ruhr—in part to force Germany to pay its coal reparations in a timely fashion. The ensuing strikes made critical construction materials scarce and exacerbated a hyperinflation that made the price increases of the early 1920s seem small. To complete the final half-year of construction, the Bavarian government required the help of the Reich to supplement its recent payments of 37 billion marks with an additional 107 billion.[44] One Munich satirical magazine quipped that the state was so eager to complete the Walchenseewerk because it needed the electricity to run its money-printing machines.[45] Nevertheless, both undertakings were completed on schedule. On an afternoon in late-January 1924, the Walchenseewerk began feeding electricity into the Bayernwerk. A Munich newspaper reported on the "historic moment" in the "winterly still of our mountains" where the Bavarian state executed its visionary plan for the regional electricity supply, making the Bavarian economy "increasingly independent from coal imports and securing its competitiveness for the future." The author declared the Walchenseewerk a "masterpiece of reconstruction" (Figure 4.4).[46]

FIGURE 4.4 Painting of the Walchenseewerk. This art was commissioned for display in Germany's foremost science and technology museum, the Deutsches Museum in Munich. It demonstrates engineers' desire to have their works perceived as art too.

Courtesy of the Deutsches Museum Archive, Munich, BN 23481.

Conclusion

The First World War and its immediate aftermath marked a watershed moment in the history of the Alps as an energy landscape. With the coming of war and its disruption of European coal supplies, the Alps began to loom large as a landscape of alternative energy. In the combatant countries of Italy, and especially France, the demand for increased energy in wartime resulted in a spike of hydropower development. In France, Alpine white coal became the most important substitute for the coal it lost to Germany in the opening months of battle. Throughout the rest of the mountain region, the conflict drew increased attention to the development of Alpine waterpower and provoked societal discussions about the necessity of augmenting its exploitation after the war's conclusion.

Elsewhere it was not the early months of war that sparked a flurry of hydroelectric development, but the first few months of peace. In the postwar period, the newly formed German and Austrian states joined Switzerland, France, and Italy as white coal countries. Austria, especially, would view the development of its Alpine waterpower as a question of state survival. The new state had little coal and little in the way of industrial energy sources generally. Alpine water appeared to be the best solution to this situation. After the First World War, moreover, the future of the Alps as an energy landscape became a matter of international importance. The mountains were the focus of competing visions about how to develop Europe's energy sources—with international cooperation on one side and autarky on the other.

FIVE

Between Cooperation and Autarky

> While an artificially impounded lake can balance out the difference in yield of several months, the glacier is a natural equalizer (*Ausgleicher*) for the weather characteristics of the changing years and decades. The correct recognition of and prediction of all these phenomena will increasingly be of immediate practical significance for us. The glacier must step in for the lack of coal. While depleted coal does not replenish itself in the ground, heavenly precipitation replaces the melting of the glaciers; therein lies the great advantage.
>
> —*Albert Heim, Glacier Commission of the Swiss Natural Scientist Society (1916)*

IN A 1922 ARTICLE, THE Frenchman Henri Cavaillés drew a metaphor-rich comparison between hydropower and coal. "One is sedentary, terrestrial, autochthon; it is, according to widespread opinion, white coal. The other is nomadic, maritime, cosmopolitan; it is the carbon of the mines."[1] Cavaillés was stating, in a more vivid way than was usual among engineers, some of the reasons that he and most of his contemporaries considered coal the superior energy source. Chief among these was the relative ease of transporting the fuel to wherever it might be needed, a characteristic that gave coal unparalleled flexibility of use. White coal, on the other hand, was bound to the earth, and as such, far more subject to earthly rhythms. From this point forward, Europeans would call upon the mountains in particular to redress these issues.

After the First World War, a consensus emerged advocating ever more drastic measures to redress white coal's perceived disadvantages. The first solution involved dams. At this time, Europeans began in earnest to impound the water of the Alps behind large dams. In the mountains themselves, such dams usually took advantage of glaciated valleys that served as ideal storage reservoirs. From this point forward, exploiting the mountains for their unique opportunities to store natural flows of energy became the dominating trend. Engineers, moreover, argued that in their quest to turn the mountains into a natural battery, they were simply doing what was necessary to make waterpower a match for coal. Alpine dam building took place against a backdrop in which hydroelectricity was beginning to be seen worldwide as perhaps the most modern form of energy, with the power to remake the world and arrest the exhaustion of precious coal resources. Utopian visions of hydroelectricity abounded. If the justifications for Alpine dams seem mundane in contrast, this largely has to do with the reality that they owed their existence primarily to one purpose—energy. Other dam projects counted flood control, irrigation, even navigation as major reasons for their execution. The discrepancy here derived from the exceptional nature of the Alps as a well-watered range in earth's temperate latitudes.[2]

Which is not to say that mountain water did not figure into increasingly ambitious plans for Europe's energy system. As the decade progressed, the idea of making white coal more nomadic and cosmopolitan by transporting it far beyond the borders of the mountains themselves also gained greater currency. In the 1920s, energy experts viewed transporting abundant Alpine hydroelectricity to centers of greater consumption as both economically and politically vital in the aftermath of the carnage of the First World War. In the economic sphere, many believed that activating unused energy resources would deflate some of the pressure from the social question and in so doing remove one of the primary causes of war. Politically, the hope was that interconnecting electrical utilities would also lead to deeper connections between countries, perhaps even to the creation of a truly united Europe.[3] The discourse about putting Alpine energy to work in the spirit of international cooperation is therefore part of a larger history of creating a Europe united by technological infrastructure.[4] As historians have shown, a cohort of international system-builders took inspiration from the internationalism of the League of Nations and the pan-European movement to propose constructing a European grid. They were joined by others who dreamed of networks of European motorways, railroads, and more.[5] As with these other missions, the idea for a European grid always competed with both national and regional interests and reached its apogee before the worldwide depression.[6] The examples from this chapter indicate that proponents believed they were employing technology to solve problems related to nature in addition to economic and political ones. The grid was another way to pool Alpine energy, and in so doing counter its vexing insistence on fluctuating with the seasons.

By the end of the 1920s, however, the internationalist sentiment had been joined by a competing desire to harness Alpine waterpower in the pursuit of economic autarky. These viewpoints emanated in particular from Central Europe's most powerful fascist states of Italy and Germany; some have suggested they were in fact a defining feature of transnational fascist movements.[7] Like the internationalists, autarkists recognized the centrality of resources for the economic future of states. They drew, however, fundamentally different conclusions about the logical consequences of this reality. Recent studies have underlined how the concept of autarky emerged as a reaction to the resource implications of the First World War, and particularly the fear of the social impacts of a blockade as had occurred during that conflict.[8] It is telling that despite the vast ideological gulf between these competing visions, they were based on more or less the same technological foundation of large Alpine dams feeding power into a large-scale grid. This convergence suggests that dams in particular offered political regimes of all stripes valuable political utility, from the dynamic image they suggested to the energy they provided. One way or another, the idea that the Alps must serve as a central component of a continental energy network—as Europe's battery—was here to stay.[9]

The close of this chapter, then, considers a key connection between Italian autarky plans and the Alpine environment in the border region of South Tyrol. In Italy, the dictate of land reclamation propelled the Fascist state's extractive relationship to nature. Autarky was simply reclamation by another name. Within Italy's self-sufficiency project, dams assumed a particularly important place. Most Fascists viewed the country as having been doomed by providence with a poor resource base that could only be overcome through electricity. In the *bel paese*, this meant above all Alpine water, which South Tyrol possessed in spades. What the territory did not have was a majority Italian-speaking population, so the Fascist authorities linked efforts to expand the country's energy base with a campaign of "Italianization." The case of South Tyrol in the interwar years reveals an understudied facet of Italian environmental history, and represents, perhaps, an example of imperialism at home to contrast with better-known efforts abroad.[10]

Ascension of Dams

Dam building in the Alps was based on technology that dates back to ancient times. The oldest physical traces of dams have been found in present-day Jordan and were built some 6,000 years ago. These structures diverted water to reservoirs as part of a sophisticated water supply system. By the close of the first millennium BCE, dams could also be found in China, Central America, and the Mediterranean basin. Most early dams consisted of embankments composed of rock and earth. Developments in Sri Lanka show that even early on, these structures could reach considerable proportions. A dam raised there in the fifth century CE remained the world's tallest for almost a thousand years. Roughly seven centuries later, Sri Lankans built the largest dam by volume for almost a millennium, a fifteen-meter-high structure that stretched almost fourteen kilometers in length.

In the early modern period, Spain continued the hydraulic tradition passed down by Roman and Islamic culture to set the global bar for tallest barrage. The 46-meter-high dam at Alicante, finished in 1594, remained the world's highest for some three centuries. In the nineteenth century, the center of dam building shifted to Great Britain, where almost 200 dams taller than 15 meters were erected. The majority of these structures retained drinking water for the country's rapidly expanding cities, but many also provided power for the Industrial Revolution. By the turn of the twentieth century, British dams nearly outnumbered the rest of the world combined. At this time, armed with applied sciences and industrial materials, societies began building dams to harness hydroelectric power. The Alps, flush with white coal, became one of the early centers of hydroelectric dam building.[11]

Dams come in different varieties with different purposes. The earliest dams overwhelmingly served to store water for domestic use and irrigation. Only later did the goal of energy generation emerge. Hydraulic experts distinguish between two broad types of dams. First there are *run-of-river dams* that create only small reservoirs and that cannot effectively regulate downstream flows. These dams are often created by lesser obstructions called weirs or barrages. *Storage dams*, on the other hand, are designed to create reservoirs to control the flow of water. These are dams in the sense of the German word *Talsperre* (literally "valley closer"). While the distinction is sometimes difficult to draw, in many cases it offers a useful way of understanding both the function of a dam and its impact on the environment. Up until the 1920s, the "dams" that proliferated on the Alpine landscape were generally of the run-of-river variety. With a few notable exceptions, these were smaller constructions whose primary purpose was not to store water but to divert it to a favorable location where a fall could be exploited.

As the twentieth century progressed, simply harnessing new energy sources no longer sufficed. As was often the case, Switzerland anticipated the new trends. On the occasion of the completion of a comprehensive study of the available waterpower in the country in 1916, the Swiss experts spelled out a vision for the energy future of the Alps that would triumph and hold sway until the 1970s. This Herculean task of surveying the land's watercourses required over twenty years to complete, since the Swiss Federal Council first ordered it in 1895. The results filled five volumes jammed with tables and illustrations. Though it could easily get lost amidst all of the data, the foreword, written by the director of the Department for Water Management, contained the new statement of principles. By way of justifying how the work had taken so long, the director also explained how opinions about white coal had decisively shifted:

> Of all the problems that today occupy the engineers of our country, the question of the creation of storage basins is without a doubt the most important due to its great importance for the national economy.[12]

Storage was the only means of coming close to utilizing all of the potential power streaming throughout the country. From the perspective of Swiss industry (and their government partners), allowing any amount of water to flow to the sea without generating electricity was a loss of national capital, a literal waste of power. In studying the question of how this "waste power" could be utilized, the director and his colleagues arrived at an ironclad conclusion. "Doubtless the solution of the storage basin problem in the high mountains (*im Hochgebirge*) provides the best guarantee for a complete utilization (*Nutzbarmachung*) of our white coal." Looking forward, the trick would be to find ways to store water that would otherwise flow unused past hydropower plants. Reservoirs would allow Swiss utilities to continue to provision hydroelectric power in winter (*Winterkraft/ force d'hiver*). Though stated in typically dry language, the bureaucrat spoke for most energy experts when he put his finger on the uniqueness of the Alps in permitting the creation of dams: "The topographic formation of our Alps lend themselves exceptionally." The mountains provided three major means to create reservoirs: existing lakes, alluvial valleys turned into artificial basins through damming, and canyons where water could also be impounded. Using a justification that would become common, the director noted that in many cases, Alpine dams would actually restore the landscape to a previous state of existence.[13]

Based on this logic of rationalization, the decade of the 1920s inaugurated the era of reservoir building in the Alps.[14] In the wake of the First World War—and from the mid-1920s in particular—Europeans began taking advantage of the Alpine environment in order to create hydropower reservoirs. The calculus behind building storage dams was much the same as that which prompted the initial movement toward water storage that had focused on natural lakes after the turn of the century (see Chapter 3). Notions of electricity supply dictated that the ability to control the timing of hydropower was critical to its continued expansion. Initially, expansion focused on using waterpower as a substitute for coal on mainline railways, and on more efficiently utilizing flows of Alpine waterpower. Early on, taking advantage of the storage capacity of lakes had appeared as a simple means of gaining some of this control. Using existing lakes as reservoirs entailed less severe disruptions of local hydrology than the creation of new lakes behind dam walls. But after the First World War, Europeans began closing off valleys behind walls of concrete in earnest. Many of the most suitable natural lakes had already been converted into reservoirs. A new hunger for domestic energy sources also no longer hesitated at massive incursions into the Alpine environment.

One of the foremost French historians on Alpine dams estimates that the interwar years witnessed the country's first true boom in dam construction (the second one would come after 1945). Before the Second World War, four major dams arose to impound the water of the French Alps. Of these, the Sautet Dam is perhaps the best example of the trend towards both energy storage and transport that characterizes the new chapter in tapping mountain energy. This dam, which would come to stand astride the Drac River southeast of Grenoble, also revealed the increasing role played by state governments in the damming of the Alps. For it was the state that ensured that in its final dimensions, the Sautet would be more substantial than might otherwise have been the case. A metallurgical company had submitted the original plan to build a dam, four meters in height, at Sautet in 1918. In its outlines, the scheme resembled the handful of smaller dams that had built on the lower stretches of the river up until 1914. However, in the wake of the First World War, a new national law gave the French state increased purview over licensing for dams. Armed with this new capability, the state rejected the plan, insisting that a more extensive project be considered.

Building a gigantic barrage that could make the Drac's flow more conducive to modern French society became the life's work of Ernest Dusaugey. Hired to consult on the project, Dusaugey imagined that the site had the potential to do more than supply electricity for one local firm. Properly equipped, the Sautet could stimulate economic activity and furnish electricity as far away as Paris. He envisioned a 126-meter-high obstruction and a reservoir of some 100 million cubic meters. At first, the chances that the larger Sautet Dam would materialize seemed very slim. A facility of this size required massive capital outlays. However, the support of the French state and local senator Leon Perrier—first president of the Compagnie National du Rhone—were instrumental in securing public funding. From the state's point of view, the remarkable natural conditions of the Sautet required a dam that could exploit them for energy production as well as storage. In 1923 the government gave its blessing to Dusaugey's plan, making it one of the first large dams in the French Alps. Thanks to the Great Depression, however, it would not be finished until the mid-1930s.[15]

Dusaugey's memoirs offer some insights into why Europeans moved to exploit the storage capabilities of the Alps. Written in the darkest hours of the Vichy regime in 1943, his remembrances bear the political scars of the period. As Dusaugey described it, the Sautet Dam was nothing less than physical proof of his profession's refusal to succumb to the paralysis that the social unrest of the twenties and thirties—"those disastrous years"—had wrought in France. As to the barrage's dimensions and purpose, Dusaugey indicated that it was simply a matter of vision. Whereas others had not thought of what might be possible in this area hidden by a bend in the river, engineers were rightfully more curious. In his mind "it was not possible to imagine a place that was more suitable for building a large dam." Dusaugey, like many others, claimed that geomorphology had practically destined the Alps to impound water. He did mention the classic byproducts of dam construction: sedimentation in the reservoir and the swifter Drac's scouring of its bed below the impediment. But these consequences were considered "inevitable reactions from nature under attack."[16]

Dams were also seen as the only means to harness certain types of Alpine waterpower. Especially after the First World War, energy experts advocated building high-Alpine dams in order to impound water at as high an elevation as possible to increase the energy bounty. While there was no concrete definition of what constituted "high-Alpine" hydro, the term generally referred to dams over 1,200 meters above sea level. High-altitude waterpower exploitation was thus limited primarily to France, Italy, Switzerland, and Austria. Nowhere in the mountains, however, was the difference between summer and winter drainage so stark as in the high Alps. Throughout the chain, summer floods regularly made up 80 to 90 percent of the total annual flow at high altitude. In glaciated areas, the ratio between summer and winter flows approached twenty to one. Such irregular flows were anathema to developing notions of the electricity supply, and the only way to artificially influence them was to capture and store this water. In 1932, one German expert estimated that high-Alpine waterpower represented about one-quarter of the possible annual hydroelectric power production in all of Central Europe—including the vast waterpower resources of the Scandinavian lands (Figure 5.1). At the time, this power alone would have been enough to cover three-quarters of Central Europe's demand.[17]

It was not merely the quantity of the high-Alpine waterpower that made it so attractive for the European electricity supply, but the quality as well. No energy source was better suited to provide valuable peak power than

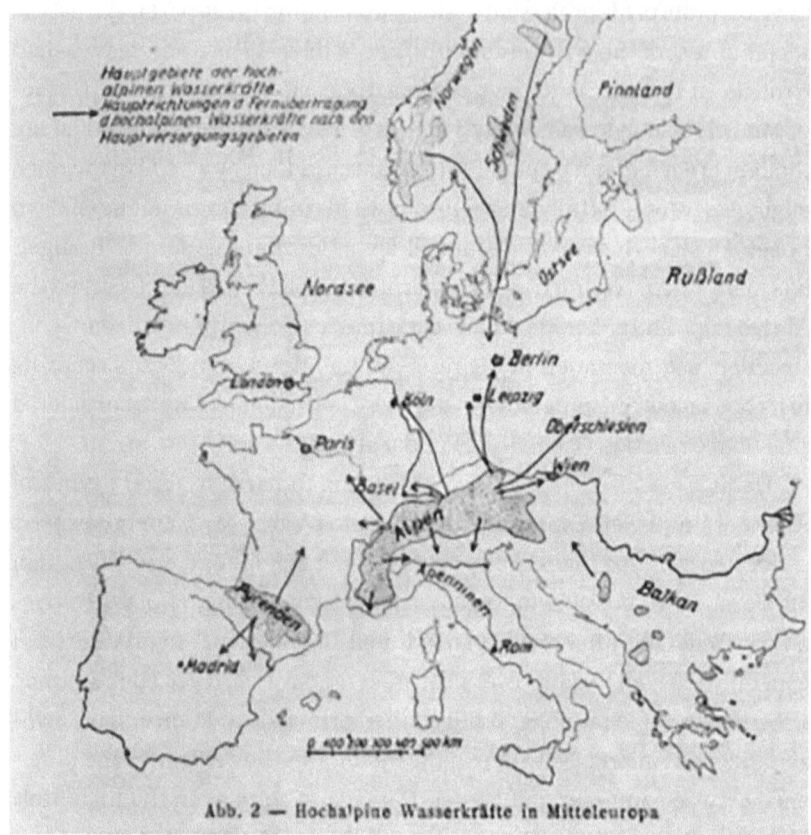

Abb. 2 — Hochalpine Wasserkräfte in Mitteleuropa

FIGURE 5.1   High-Alpine waterpower in Central Europe. Though not the sole location of high-Alpine energy in Central Europe, the Alps were the most important source. The diagram shows the prevalence of the idea of transporting white coal to Europe's urban and industrial centers. By the time this map had been drawn, the longest transmission line from western Austrian to Cologne (Köln) had already been built. Under the Nazi regime, Germany would attempt to send Austrian high-Alpine waterpower to the industrial zones around Leipzig and in Upper Silesia (Oberschlesien, where the extermination camp Auschwitz-Birkenau was located).

Source: Emil Mattern, "Die Hochalpinen Wasserkräfte in Mitteleuropa," *Die Wasserwirtschaft* (1932).

hydroelectricity. In contrast to thermal plants, which take some time to ramp up, hydroelectric facilities can increase production almost at a moment's notice. Even if they were not providing peak power, utilities had a much easier time fetching a decent price for their hydroelectricity if they had some control over its flow. Indeed, German electric engineers eventually came to refer to the hydroelectric current they could not regulate as "waste current" (*Abfallstrom*), and it presented a greater challenge for a utility to market.

Exporting White Coal for the Greatest Possible Advantage

Accompanying the rise of Alpine dams were new ideas about why and how this energy should be harnessed. During the Great War, the pervasive urgency of national defense had meant that white coal was understood in these terms as well. Dams were constructed for the national cause. In the interwar years, this nationalist motivation was joined by a different spirit, that of international cooperation. Influential voices within the electrotechnical community began to advocate large-scale exploitation of Alpine white coal as a means to reconstruct Europe's economy. Economic growth, they believed, would reduce the chances of a further conflict via prosperity for all. In advocating international solutions to problems in the electricity supply, interwar engineers only made explicit what is fundamentally always the case: that engineering is never free from political considerations. Those involved were part of a broader peace movement in the wake of the First World War.[18] This was a diverse sociopolitical movement born of revulsion at the carnage of 1914–1918. Related to these antiwar sentiments were "pan-European" ideas about forging European unity that also gained momentum in the 1920s.[19] In this era then, a number of proposals emerged to utilize Alpine waterpower in a cooperative manner. In some respects, they can be seen as intellectual forerunners to later arrangements like the European Coal and Steel Community and even the European Union. Cooperative use of energy resources appeared as a means to overcome the self-defeating logic of European nationalisms.

One particularly important forum that promoted the idea of sending Alpine waterpower far and wide was the World Power Conference (WPC). The WPC was the brainchild of the Scot D. N. Dunlop. Dunlop (1868–1935) wanted to provide a forum for engineers and scientists to discuss energy supply and technology. The founder's idealism nicely complemented his own business interests. Prior to the WPC, the Scot had worked for electrical trusts and acted as head of the British Electrical and Allied Manufacturers' Association—a trade association and primary sponsor of the conference. The World Power Conference—later renamed the World Energy Council—held principal meetings every six years beginning with the 1924 inaugural conference in London (Figure 5.2). In between these main meetings, smaller sectional conferences were held to discuss more circumscribed energy themes. At the World Power Conferences of the 1920s, those interested in energy discussed the field as a means of promoting international cooperation in the wake of the First World War.[20]

FIGURE 5.2 Cover of the World Power Conference brochure and badge from the inaugural meeting in 1924. The motto "Directing Forces Develops Resources" suggests the role that engineers and the power industry could play in promoting economic prosperity.

Source: World Power Conference.

The theme of the first meeting of the World Power Conference—held in London simultaneously with the British Empire Exhibition—was how to use the power and fuel resources of the world to the greatest possible advantage. Setting the template for later meetings, the conference was attended by national delegations comprised of representatives of the engineering profession and figures in the energy industry. Some forty countries and colonies sent committees; the US contingent alone numbered 300 members. One of the major goals for the conference was to gain a better conception of available world energy resources. Such an inventory was the prerequisite to developing these supplies. In this effort, each delegation composed and presented a report on the energy situation in their homeland. Additional participants contributed papers on other topics. Future US President Herbert Hoover, for instance, wrote one on the economic aspects of energy.

Many of the participants in London expressed the belief that as a group, their ability to shape the development of energy resources enabled them to make an apolitical contribution towards healing the wounds between nations. O. C. Merrill of the American delegation argued that they had to come to London to talk about power in a different way. "Not the power of territorial possession or economic aggression," he explained, "but that of mechanical energy and electricity, the greatest tool ever placed in the hands of man." Speaking at the conference, the Prince of Wales—and future King Edward VIII—suggested that the participants comprised a sort of technological League of Nations, with a responsibility no less important than the actual existing international organization.[21] Not all participants left London satisfied with the results of the meeting or impressed with its ostensibly cooperative spirit, however. A correspondent for the German trade journal *Die Wasserkraft* reported the French delegation's displeasure with the invitation of the former Central Powers to the meeting. He also decried the lack of true exchange of opinion and discussion.[22]

In London, the question of Alpine energy was—like most topics at this meeting—treated in a general way. On the one hand, the lands of the Alpine massif discussed the significance of white coal in their surveys of national power resources. But the Alps also figured prominently in the theme of international cooperation promoted by the conference. The power development of navigable rivers and the export of electricity emerged as significant topics of the conference: they were two main avenues of potential international cooperation and they both had Alpine connections. Many believed—as the conference's publicity brochure explained—that Europe's great rivers held the "secret of the industrial transformation of Europe."[23]

The navigable stretches of Europe's large rivers could generate considerable power—enough, some felt, to become the Continent's main source of electricity—if harnessed under international administration. But for the most part their power development had stalled due to the conflicting interests of navigation. Developing the power of the Danube in particular, was viewed as a means of rehabilitating some of those lands along its banks that had suffered the most from the previous war. The problem of developing the energy of Europe's navigable rivers extended to the upland stretches of the large rivers draining the Alps—the Rhône, Rhine, Po, and Danube—as well.[24] The Alps also stood out as a model of electricity export with great potential for growth. At the start of the 1920s the movement of electricity across national borders remained quite rare. One of the few countries that did export power was Switzerland, which fed hydroelectricity to many of its neighbors. At the WPC, exporting Switzerland's yet untapped hydropower seemed an ideal way to stimulate international economic cooperation by activating unused resources.[25]

A more focused discussion of the necessity of transporting Alpine waterpower out of the mountains took place at the next WPC meeting in Switzerland two years later. The follow-up to London was the first of the so-called sectional meetings, organized around the narrower topic of waterpower and inland navigation. There were many reasons to hold a conference on waterpower and shipping in the town of Basel. On the transport side, Basel was notable as the head of navigation on the Rhine. And as Dr. Tissot, the president of the Swiss national committee to the WPC, reminded the audience in his speech before the general assembly, Switzerland was also at the head of the class in matters hydroelectric. Tissot maintained that electrification in Switzerland was "the most developed and advanced" of any country. In Switzerland, Tissot noted, conference attendees would find some of the most modern hydroplants in the world.[26] These included one of Switzerland's new interwar dams, the Wägital, that had created one of the country's largest reservoirs to supply winter electricity for the city of Zurich.[27]

In Basel, the connection between Alpine waterpower and international cooperation was given symbolic representation in the conference's badge, which depicted a heavenly handshake before a scene of a mountain dam (Figure 5.3). In the words of a representative from the Basel city canton, the intertwined hands had two meanings. First, they represented the conference themes of navigation and waterpower exploitation. But they also signified an invitation "to the peoples, separated by the world war, to extend each other their hands in common on the neutral ground of our land."[28]

FIGURE 5.3 Cover of the badge for the World Power Conference held in Basel in 1926. The image visually connects Alpine dams, rail electrification, and international cooperation.

Source: *Transactions of the World Power Conference, Basle Sectional Meeting 1926* (Basel: E. Birkhaeuser, 1927).

In his speech during the opening session, President Dunlop also had words of admiration for Switzerland's waterpower accomplishments. But to his mind, Switzerland's success was merely one part of a broader regional success, born of international cooperation with the potential for more. Indeed, Dunlop saw the Alps as one of the foremost electrical landscapes on the planet.

> In the Alps especially, all the possibilities of complete co-operation and unification are present; the Alpine Power Block as spread over Switzerland, N. Italy, Austria, South-Eastern France and Bavaria, represents now generating plant [sic] aggregating over 4 million kilowatts, capable of producing 20 billion units, and as such has become one of the greatest power zones in the world.[29]

Dunlop believed the "Alpine Block" had resulted from the efforts of five countries working more or less cooperatively to utilize to the fullest the natural energy resources present in the region, and he judged that the bulk of this achievement happened in the postwar period. Moreover, he believed the Alpine model was one worth emulating. "To constitute power zones in every part of the world where nature supplies the necessary resources, and through those zones raise the economic and industrial standard of development, remains one of the greatest problems now confronting scientists, industrialists, and engineers, and there is no doubt that the path to greater economic prosperity lies in this direction." While he recognized many obstacles lay before the realization of such a goal, Dunlop avowed that "When man will the good [sic], then the tempest of national and international jealousies will be stilled and the order of the starry heavens will be reflected on earth."[30]

Following Dunlop's speech, one conference panel focused on the possibility of extending that Alpine power zone. In the section devoted to international "exchanges" of electricity, the subject of linking up Alpine white coal with other regional energy systems made up a fair portion of the dialogue. For one thing, as many of the panel's participants noted, the Alps were in fact one of the few places on the globe where international exchanges of electricity took place. In the report summarizing the essays, Jean Landry, president of the large Swiss utility Energie de l'Ouest-Suisse

(EOS) and professor at the University of Lausanne, noted that significant exchanges of electricity only occurred between Sweden and Denmark, Canada and the United States, and Switzerland with its neighbors Germany, France, and Italy. Indeed, Landry observed, in 1925 Switzerland had exported nearly 20 percent of its total electricity production. Nearly half of this energy went to France where it made up some 3 percent of that country's electricity supply. The common denominator in all of the global electricity exchanges, Landry explained, was that one of the partners possessed more hydraulic energy than the other.[31] Why this should be the case was obvious to the panelists. As Theodore Rich from Great Britain put it, "when the question of the export of electrical power first came up, it was primarily on the basis of hydraulic power, because it is impossible to carry, as we have heard, the products of a waterfall in a cask or box."[32]

The panel agreed that electricity exchanges should occur between areas with abundant and complementary energy resources. For Oskar von Miller, the units involved were not nations but regions with specific power resources. Exchange was not an international question, rather "a question between regions rich with waterpower and regions that control large amounts of coal or other fuels."[33] Robert Haas, director of one of the original large hydropower plants on the High Rhine, concurred, arguing that one of the most sensible long-distance exchanges in Europe would be between the Alps and Germany's lignite fields. Haas maintained that any geographic area where the lay of the land enabled the natural storage of water in lakes or artificial storage in reservoirs had the potential to generate large amounts of hydroelectricity. In this respect, the Alpine and pre-Alpine lands of Germany, Austria, and Switzerland were "richly blessed." Haas contended that these regions should take full advantage of that gift by exchanging surplus waterpower.

For these engineers, the necessity of exchange was obvious. As Landry explained,

> Nature has not distributed her wealth at all uniformly, and it is necessary therefore that what can be only imperfectly carried out within political boundaries should be obtained, wherever possible by the exchange of energy between countries which, from point of power resources, are supplementary.[34]

Another panelist marshaled a map demonstrating the vast geographic diversity of energy resources in Europe, and hence the necessity of exchange (Figure 5.4). For the engineers in Basel, this relatively uncomplicated state of affairs was spoiled by politics. Robert Haas complained of the difficulties of exchanging electricity between Germany and Switzerland and blamed them on the latter country's national law on waterpower exploitation. Passed during the First World War, the law made it necessary for utilities to acquire federal permits for electricity exports and thus exposed any

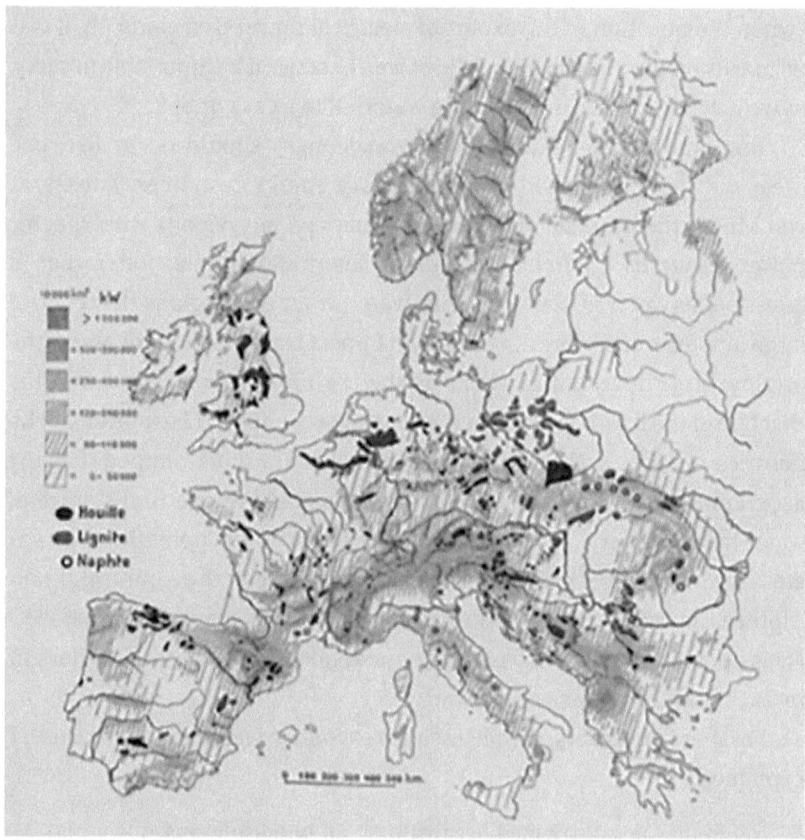

FIGURE 5.4 Geographic distribution of European energy sources in 1926. It is notable that the map only considered coal, hydropower, and petroleum.

Source: H. Niesz, "L'Echange D'Energie Electrique Entre Pays, Au Point de Vue Economique et Technique," in *Transactions of the World Power Conference, Basle Sectional Meeting 1926*, vol. 1 (Basel: E. Birkhaeuser, 1927), 1025- ?.

potential exchange to domestic political struggles. In the panel's closing discussion, Haas made a plea "that the statesmen, the lawmaker, the diplomat might not ruin, what the engineer, the businessman (*Grosskaufmann*) have created, by calling to life the possibility of the exchange of electrical energy." For this traffic in electricity represented "another important link in the peaceful cooperation of the nations."[35]

Germany's Aorta

Even as these engineers were discussing the necessity of electricity exchanges in Basel, one of Europe's most powerful utilities was busy establishing the most significant long-distance transmission of electricity on the Continent up to that point. That company was the Rheinisch Westfälisches Elektrizitätswerk AG (RWE), Germany's largest utility and supplier of electric current for the bulk of the West German industrial juggernaut.[36] RWE's expansion plans focused on tying into the high-altitude waterpower of the Alps. Beginning in the mid-1920s, the company commenced constructing a high-voltage link from West Germany's industrial heartland to the Alps of western Austria. When it was finished in 1930, the line represented the most substantial long-distance transmission in Europe and the long dreamed of extension of Alpine energy into the commercial areas of the lowlands. It also became a model that other European utilities would seek to emulate, a first step toward a continent-spanning grid.[37]

The RWE was a product of the fossil-energy riches of western Germany—particularly the Ruhr—and the efforts of its most important director Hugo Stinnes (1870–1924). In its early years, the company survived off the combination of coal mining, steelmaking, and the chemical production provided by the Ruhr valley's copious anthracite stocks. Eventually the existence of RWE would contribute to these industries' success as well. After acquiring the company in 1902, the mine owner Stinnes became one of Germany's most important industrial barons, thanks largely to the success of RWE. Stinnes concentrated his efforts on making RWE the main supplier of electricity in the region through acquisition of other utilities and development of its grid. Just before the start of the Great War, RWE constructed a gigantic lignite (brown coal) power plant in the vicinity of Cologne, which it then rapidly expanded in order to meet the rise in wartime demand.

The growth of the lignite-fired Goldenbergwerk continued in the postwar period, as the Treaty of Versailles and Germany's reparations agreements stripped the country of much of its anthracite supplies. By the end of the twenties, it was Germany's largest power plant with a capacity of half a million kilowatts.

Already in the early 1920s, RWE initiated a plan to expand its grid southward. After preparing the way with a series of acquisitions of south German utilities, RWE—inspired by the worldwide leader in long-distance power transmission of California—began construction of a high-voltage link to the Austrian Alps beginning in 1924. Starting in the lignite fields around Cologne, this 800-kilometer-long transmission line was to feed power from the Austrian province of Vorarlberg, which was in the process of completing a high-altitude dam in the Ill River basin. After six years of construction, transmission began in 1930 (Figure 5.5). The line represented a remarkable technical achievement as the first instance of a 220-kilovolt transmission in all of Europe. It would go on to form the backbone of Germany's burgeoning high-voltage grid. It also represented the linking of two of Europe's most important electricity-producing regions: the Rhineland-Westphalian powerhouse and the Alps were now wired in parallel on the same circuit, enabling the two regions to complement and supplement each other. The iconic heavy industrial district could now take advantage of the cheap electricity that only could be provided by hydropower; Alpine hydroplants on the other hand, could now fall back on the unparalleled reliability of fossil fuel energy.

Bringing Alpine waterpower to the Ruhr was part of RWE's corporate strategy of rationalization they called *Verbundbetrieb*.[38] RWE believed there were economic gains to be had by coordinating a grid of power plants spread out over a large area. In this system, the hydropower of Alpine reservoirs was to play for the distant Ruhr the role that it had been playing on a smaller scale since the turn of the century. This stored waterpower would act as a reserve for the network, to be employed whenever needed. With the Alps covering the grid's auxiliary power, RWE argued that it could use its other power sources—run-of-the-river hydro, lignite, and anthracite—more efficiently. The company's managers considered the west German switching station at the head of the link to the Alps to be the greatest concentration of electric power in the world.[39] A later observer likened the

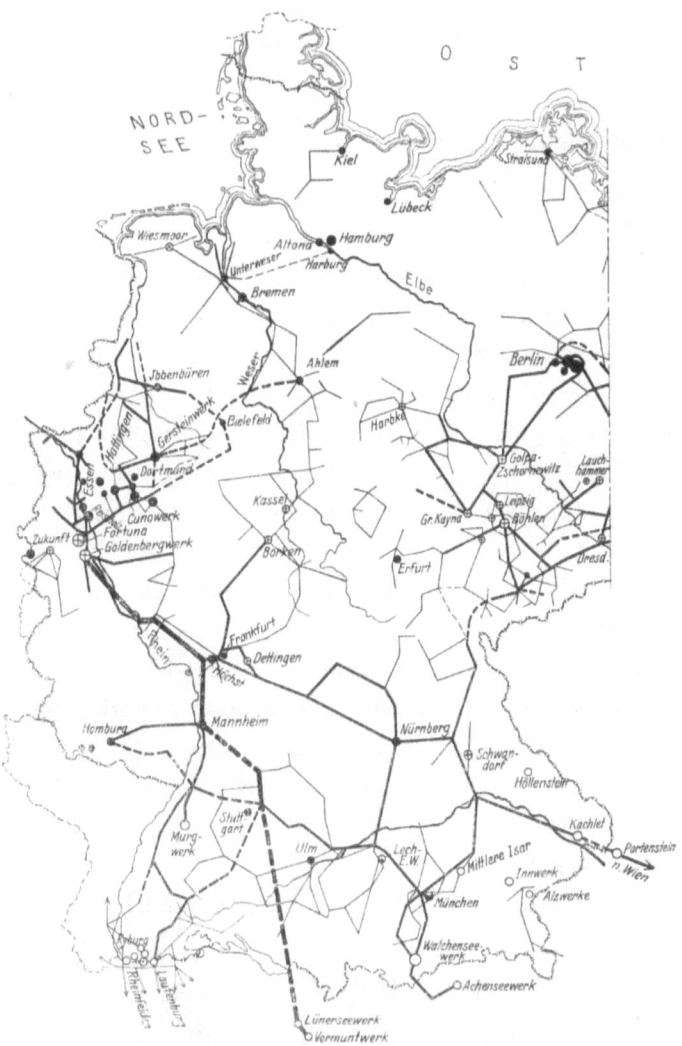

Abb. 70. Deutschlands Kraftwerke und Hochspannungsleitungen (Anfang 1928).

FIGURE 5.5   Germany's power plants and high-voltage lines. The dotted line in the lower left shows the 220 kV transmission line being built by RWE to interconnect western Austrian Alpine dams to industrial western Germany (the area around Essen and Dortmund). Stretching almost 800 kilometers in length, it would later be called "Germany's aorta." Black dots represent anthracite, the circles are hydropower, and the circles with crosses are bituminous electric plants.

Source: Gerhard Dehne, *Deutschlands Grosskraftversorgung* (Berlin: Springer, 1928), 132.

RWE transmission line into the Alps as "Germany's aorta."[40] Similar thinking inspired even grander visions for the role of white coal.

The Alps and Superpower for Europe

One of these schemes marked the highpoint of the 1930 WPC meeting in Berlin. At the second principal meeting, Oskar Oliven, an executive with a German electrical company, outlined his plan for the creation of a European "superpower grid" (*Großkraftnetz*). Oliven saw the historical expansion of Europe's electricity grids as too haphazard and illogical. Like most of his engineering colleagues, he particularly decried the lack of information exchange among nations and blamed it on short-sighted politics. Oliven believed the time had come to cooperate to create something big. "Now we stand before a task," he explained, "confronting all of the peoples of Europe, that only can be solved if we overcome all of the personal, material, and political difficulties and open the visible and invisible borders to electric energy."[41] Oliven recommended the creation of a high-voltage continental grid that would interconnect all of Europe's primary sources of electrical energy. He envisioned the construction of five primary transmission lines, two linking east and west, and three running north and south. His plan also saw for the further development of Europe's energy sources to feed into the grid. In his scheme, the Alps represented the fulcrum of the entire system (Figure 5.6).

The primary advantage of his plan, Oliven argued, lay in its ability to balance electricity production and consumption in Europe. In his conception, Oliven accorded the Alps a special role:

> An electric connection between the waterpower of the high Alps, which reaches its maximum in summer due to the melt, the still-to-be built hydraulic works of the Danube at the Iron Gate, and further the power of the Adriatic coast, where atmospheric precipitation falls down to the valleys in the form of rain mainly in the winter, creates a balance that will make it possible to avoid large dams, or at least postpone them for future.[42]

The same logic held for linking Alpine white coal with the waterpower of Central Europe's lower mountain ranges, which also reached its maximum during the rainier portion of the colder months. The possibilities of bal-

FIGURE 5.6 Oskar Oliven's "Proposal for a European Superpower Grid," presented at the World Power Conference in 1930. The Alps are depicted as the most central of a number of regional power sources that would be fed into the European grid.

> Source: Oskar Oliven, "Europas Großkraftlinien: Vorschlag eines europäischen Großkraftnetzes." In *Gesamtbericht, Zweite Weltkraftkonferenz/Transactions, Second World Power Conference, Compte rendu, Deuxième Conférence Mondiale de l'Énergie*, vol. XIX, edited by F. zur Nedden, 30–39 (Berlin: VDI-Verlag, 1930).

ancing the load were even grander as the scale involved increased. Oliven argued that

> on our continent, there exists a natural storage of water due to differences in climatic and atmospheric circumstances, and if we exploit this natural water storage at the right time and in the right places, and utilize our superpower network to deliver it to the necessary locations, we will be able to save capital on construction while using our works substantially more efficiently.[43]

The superpower grid, Oliven claimed, would also balance consumption. As an example of how the supraregional expansion of grids improved utility load factors, Oliven pointed to the recent success of RWE, Europe's largest utility. Thanks to its expansion into the Alps, RWE enjoyed the best load factor, meaning it most efficiently utilized its existing capacity. Oliven promised that a European grid would be even more efficient, as it would take advantage of various geographic and climatic conditions that RWE had not yet capitalized on. Harvests, vacations, construction: all took place on various schedules throughout Europe. If serviced by a single grid, aggregate capacity could be transferred to meet periodic demands. The European grid would also flatten out peak demands that hindered the economy of works. Interconnection would allow certain power plants to run at or near full capacity constantly, by allowing them to ship electricity to wherever it was needed in the grid. When it was dark in Rostov, Oliven noted, Bucharest still had an hour of light. With the existence of a European grid, Oliven optimistically argued, unused afternoon capacity in Romania could be dispatched to Russia to provide nighttime illumination.[44]

Oliven's thoughts marked the high point of interwar plans for cooperation in European electricity. Even as he presented his work at the World Power Conference in Berlin, the world was beginning to slide into economic depression. In Europe, the downturn extinguished much of the internationalism that had nourished plans like Oliven's and made financing them extremely difficult. By the 1930s international cooperation gave way to a different motivation for energy development.

The Alps and Autarky

After the First World War, a new justification for harnessing the power of the Alps emerged: the quest for national economic independence. In the interwar years, voices throughout West-Central Europe called for the development of white coal as a means of achieving national autarky. Neutral Switzerland, often an exception to European trends, emphasized the importance of hydro in bolstering its independence and neutrality. But it was especially the rise and consolidation of fascist political movements that inspired a significant portion of the hydroelectric development in the Alps during the final decade of general peace. One of the most consequential examples of Alpine hydropower development in the name of interwar autarky comes from the Italian region of South Tyrol. This Alpine territory on the south side of the Brenner Pass belonged to the Austro-Hungarian crownland of Tyrol until 1918, when Italy succeeded in acquiring it as compensation for its participation in the Allied war effort. Beginning in the early 1920s, the Italian Fascist state pursued waterpower development in South Tyrol for the benefit of Italian industry outside of the province.⁴⁵ Eventually, hydro projects were conceived as a means to "Italianize" the area, by providing an industrial base to employ ethnic Italian migrants. These efforts culminated with the creation of an "industrial zone" in the province's capital of Bolzano around the time of Mussolini's formal announcement of his country's autarky program in 1936. The transformation of South Tyrol's waters in the interests of autarky represent an important chapter in the ecological history of Italian Fascism. It also highlights that dam projects can be used to attract populations as well as displace them.

Located in the heart of the Alps, present-day South Tyrol (Südtirol/Alto Adige) is Italy's northernmost province.⁴⁶ The area is the historical cradle of a larger Tyrolean state, which evolved from a county in the early medieval period to become a possession of the House of Habsburg in 1363. It would remain one of the core holdings of the multinational monarchy until its dissolution in 1918. At that time Italy annexed the South Tyrol, while the northern remainder became part of the newly formed Austrian republic.⁴⁷ The territory lies immediately south of the main ridge of the Alps. In the eastern part of the province are the Dolomites, famous as one of the most aesthetically impressive mountain formations in the world. South Tyrol's nearly 8,000 square kilometers are drained primarily by one river, the Etsch/Adige. From its source in a tiny spit of land where Switzerland, Austria, and Italy all come together, it flows out of the province into the north Italian plain before entering the Adriatic north of Venice. The Adige's main tributary is the Eisack/Isarco, South Tyrol's second-longest river. The Isarco flows south from the Brenner Pass where it joins with the Adige to the south of Bozen/Bolzano. The confluence of these two large valleys and transportation arteries has helped Bolzano ascend to the status of the region's economic and political capital.

The fact that the Adige flowed to the Adriatic has long been seen by Italian nationalists as natural proof that all of the southern portion of the crownland of Tyrol that it drained should be part of Italy. This idea that the southern Tyrol belonged to Italy reaches back to the early nineteenth century, and it gained credibility thanks to the work of several Italian geographers who advanced a "watershed theory" of politics. Nationalist clamor for the annexation of South Tyrol intensified with the outbreak of the First World War. One of the conditions of Italy joining the Entente in 1915 had been the annexation of Tyrol up to the Brenner Pass. After the Allies frustrated Italian designs for territorial aggrandizement on the Adriatic coast in favor of the new state of Yugoslavia, they gave in to Italian insistence on annexing the southern Tyrol. While the border change did bring hundreds of thousands of Italian speakers from the southernmost portion of Tyrol (now Trentino) into the kingdom, it also placed 250,000 German-speaking Tyroleans inside the north Italian frontier. This region—with a large German-speaking majority—was given the name Alto Adige. The Italian state would soon go about putting the upper Adige to work.

Before 1918, South Tyrol was an overwhelmingly agrarian region. This state of affairs changed dramatically after its incorporation into Italy. Particularly after the rise of Mussolini in 1922, the Italian state sought to systematically develop the waterpower of its new territory. Mussolini's Fascist regime was especially interested in nationalist goals like expanding military industries and promoting autarky in strategic economic sectors. Electricity supply was one such arena—indeed, one of the only sectors where Italy could hope to approach something near self-sufficiency. Though the country possessed few of the strategic resources necessary to increase industrial and military might, it could boast a large amount of potential electricity in its waterpower—particularly in the Alps. Electricity supply was all the more important because vast quantities of cheap electricity could in part compensate for a lack of heavy industry—through the magic of electro-chemistry and electro-metallurgy. During the 1920s then, Mussolini's regime promoted white coal development throughout Italy's Alps.[48]

In this climate, the development of South Tyrol's almost untouched waterpower resources had the full support of the state. In 1923, a royal decree changed South Tyrol's water law to facilitate hydropower construction. The new electricity powered South Tyrol's first large industrial operation: a nitrogen plant run by Italy's leading chemical firm Montecatini. The factory provided work for large numbers of Italian-speaking immigrants from the city of Turin. Around the same time, the Società Idroelettrica dell'Isarco turned its attention to harnessing the power of its titular river. The

Kardaun plant, completed in 1929 with the help of a $5 million loan from the United States, became Europe's largest hydroelectric plant for a spell.

The proximity of the Isarco to Bolzano made its energy particularly useful for the centerpiece of Italian autarky plans in the region: the creation of the so-called Bolzano industry zone. Beginning in 1935, and with critical financial support from the Italian state, a number of major Italian industrial firms set up South Tyrolean branch operations in former orchards south of Bolzano. Suddenly, Bolzano was home to a burgeoning metallurgical industry, a magnesium plant, and automobile production. From an economic standpoint, the location of these industries in the Alps was questionable. The raw materials they required for operation often had to be transported from far afield. Since the advent of long-distance power transmission in the 1890s, most Italian companies had opted to transport electricity to the more accessible north Italian plain. But the various works in the Bolzano industrial zone—executed with massive state support—were not located there solely for economic purposes. They were also intended as a means of Italianizing the capital and its region.

Large dam projects are well known for their tendency to displace people. The textbook example of this phenomenon is now China's Three Gorges Dam, which has forced millions to abandon their homes. But the example of South Tyrol during the Fascist era shows that such projects can shift populations in different ways too. Since constructing hydropower plants requires considerable numbers of workers, and since the newly converted energy creates opportunities for new industries, the effect can be one of attraction as well. In the opinion of one of the foremost historians of South Tyrol in the twentieth century, hydroelectric construction projects from 1920 to 1940 were responsible for the first wave of Italian immigration to the region. This was because—as a part of the effort to Italianize South Tyrol—they used exclusively Italian manpower. The Italian consortium that built the Kardaun plant, for example, employed 5,000 Italian workers recruited from the Veneto, Emilia, Abruzzia, and Lombardy. A veritable village was created with its own Carabinieri station and priest. The Bolzano industrial zone that drew its power from plants like Kardaun anchored even more new souls to the region. Where in 1935 an apple orchard had stood was by 1942 a thriving business district employing 7,000 workers. This worked out to more than 10 percent of the city's population. By 1947 an additional 12,000 laborers had joined them. Nearly all of these workers came from neighboring northern Italian regions.[49]

While the migrants who helped construct dams and run factories were not necessarily forced to settle in South Tyrol, their circumstances often strongly encouraged the move. Most were farmers impacted by the worldwide depression who now had to survive by selling their labor. The Italian state incentivized the industrial firms that hired them by providing 800 lira for each "real"—meaning from outside the area—Italian brought to South Tyrol. Free apartments in newly constructed high-rises acted as a further enticement to the workers. The luckiest ones secured spots in the so-called Semirurali. These were buildings designed in a pastoral style with apartments for four separate families. Architects modeled them after Mussolini's birth house in Predappio. Taken together, the tapping of new energy sources and the various inducements had the effect of drawing a desired population to the vicinity.[50]

The grand transformation of the waters of South Tyrol in the name of autarky and Italianization had correspondingly grand results. By the outbreak of the Second World War, South Tyrol's waterways generated one hundred times more electricity than in 1918. Their annual production of 2 billion kilowatt hours equaled 12 percent of Italy's total. Much of this electricity found its way to factories in northern Italy. But beginning with the establishment of a fertilizer industry in the late 1920s, the South Tyrolean branches of major Italian industries utilized this waterpower on the spot. South Tyrol produced 10 percent of the country's nitrogenous fertilizers. The operations of the Bolzano industrial zone eventually became an important part of Italian war production. The migration of thousands of workers to these workplaces, moreover, led to an abrupt jump in the city's population and sharply increased the number of Italians there. South Tyrol's new energy significance for the Italian state—and the substantial outlays required to create it—would remain an important argument for keeping South Tyrol Italian in the decades to come.

Conclusion

It was in the interwar period that the contours of the Alps as a mountain battery for West-Central Europe first became visible. The emergency of the First World War and its aftermath had created a desperate demand for energy. After the catastrophe of the war, Europeans were more willing to intervene drastically in the Alpine environment in the form of dams. They also countenanced building these dams ever closer to the peaks in the quest to capture as much potential energy as possible. Throughout the mid-1920s, politicians and engineers alike advocated electricity exchange as a means of promoting economic prosperity and peace. At the World Power Conferences, they argued that regions like the Alps, with abundant energy resources but comparatively small demand, should store and transport surplus power to Europe's centers of consumption. An early example of the possibilities came as Germany's Ruhr region forged a connection to the Austrian Alps. These ideas culminated in visions of a European electricity grid that, among other things, would exploit the energy characteristics of the Alps to make electrification in Europe more efficient. Just as the plans were announced, however, the world was entering the depths of the Depression.

Economic contraction and national self-interest dampened enthusiasm for European projects and encouraged an alternative point of view. This was the idea that the Alps held the key to economic autarky. Already in the 1920s, Fascist Italy had promoted Alpine hydroelectric development in its bid to become industrially self-sufficient. In the newly annexed region of South Tyrol, the Fascist regime simultaneously attempted to socially engineer a more Italianized population on the basis of hydraulic engineering. Alpine white coal supported the creation of a new Italian industrial zone in this borderland region. As the 1930s drew to a close, the Alps would become the focus of another bid for autarky—that of Nazi Germany. Though it is not often understood, the mountains would become a critical theater in Greater Germany's economic war effort.

SIX

The Alps and the Energetic Struggle for Existence

> Whoever understands how to read the border lines recognizes, that moreso than in the plains or in the hill country, attractions are here at work that do not allow a country to wait on one side of the slope if a pass leading up and over is open, that do not let the edge of a political territory rest at the foot of the chain, rather drive it upwards towards the sources of the rivers that stream from these mountains down to the land. With these forces innate to the mountains lay those from outside striving in strife, which attract the power of the great country encircling the Alps and anchor them in or project them above and beyond. This is a grappling that runs through the entire history of the Alpine peoples and states.
>
> —*Friedrich Ratzel (1896)*

FOCUSING ON THE WATERS OF the Alps during the Second World War reveals aspects of the conflict that have escaped widespread attention. As the consensus major alternative to coal for generating electricity in West-Central Europe, Alpine hydro automatically represented an important share of the region's material basis. But nowhere did it figure more heavily than in Adolf Hitler's Germany. Though it is not widely known, key people within the Nazi establishment viewed the incorporation of Austria's Alpine waterpower into the German electricity supply as one of the state's most pressing energy issues. German planners believed they needed to spare their coal resources in order to free up the substance to be used as a raw material in other processes. Chief among these were the creation of synthetic airplane fuel and rubber. They viewed substituting hydroelectricity as the best means, prior to the start of war, of conserving precious coal supplies.

For both geographic and historical reasons, the waterpower of the Austrian Alps presented itself as the most accessible option. But Nazis too believed that Austria's mountainous topography had to be utilized to store waterpower and make this fickle renewable behave more like a fossil fuel. German hydroelectric schemes centered around countering the perceived faults in renewable energy in order to help preserve finite fossil fuels. In the mind of Nazi planners, white coal was to help Germany bridge the gap until its conquest of additional *Lebensraum* would enable it to compete with its enemies who could draw on the advantages of continental resources and imperial networks. Whereas Germany early on seemed to set on a military strategy of "lightning war," in the energy sphere it opted for water. By prioritizing hydropower, however, it focused on a sector that required enormous upfront capital and long construction times—one that developed less at a lightning pace and more at a glacial one. At the very least, this seems to be a choice that was increasingly at odds with the country's strategic situation. That the Alps played both a role in Hitler's expansionist drive and in the failed German war effort only crystallizes the energy constraints that confronted the Third Reich.

This chapter looks at the longer history of Nazi efforts to incorporate the waterpower of Austria into the German electricity supply. By starting in the mid-1920s, it shows how white coal was bound up in the larger question of the relationship between Germany and the newly created state of Austria. The German word *Anschluss* roughly translates as "connection," and after the First World War it was used to express the wish of the majority of Austrians to be attached politically to the Weimar Republic. The peace treaties denied this desire, but Hitler would ultimately accomplish the *Anschluss* with his annexation of Austria in 1938. The same word in German, however, is also used refer to connections between technological objects such as electric grids. Reflected in the waters of the Austrian Alps, we can see an alternative history of the *Anschluss* via explicit efforts to accomplish an energy union in the interwar years via electricity.

Thereafter the focus shifts to a grand project to exploit the potential energy of one of the mightiest ranges in the eastern Alps, the Hohe Tauern of Austria's Salzburg province.[1] By looking at the most ambitious National Socialist dam scheme, it joins a growing body of research interested in the relationship between energy and the Second World War. Though historians have long recognized the importance of fossil fuels for the conflict, only recently have they begun to investigate other forms of energy like hydroelectricity.[2] Tracing Nazi German hydropower development in the Austrian Alps illuminates the problems that the state perceived in activating its war production and provides context for some strategic choices it made surrounding energy during the Second World War. In showing that the war's landscapes of militarized production reached all the way to the edge of the Continent's mightiest glaciers on the roof of Europe, it also represents a contribution to the environmental history of the Second World War.[3]

The history of the Third Reich's connection to white coal finally provides new insights into the National Socialist economy and the deranged calculus that underpinned Nazi aggression and war strategy. Since the beginning of the twenty-first century, a paradigm shift has occurred in historians' understanding of the Nazi German economy. Perhaps bewitched by the existence of modern, titanic corporations within the country, the assumption has long been that twentieth-century Germany was a global economic powerhouse. Research has uncovered, however, that many important economic indicators show a country that was by the interwar years run-of-the-mill by European standards. Most importantly, in comparison to powers like the British empire and the United States, Germany lagged far behind. Indeed, already in the 1920s, Adolf Hitler was preoccupied with the prospect of Germany (and the rest of the world) becoming economic vassals of the United States. The novelty of National Socialism lay in its insistence on violently challenging this perceived global shift. Based on resources alone, this was a conflict that Germany was all but guaranteed to lose. But it was precisely Hitler's warped racial worldview that convinced him that the only chance to prevent the inevitable economic subordination of Germany to the United States—which he understood as the extinction of the German race—was to desperately gamble on war before it was too late. The fact that Germany came as close as it did to prevailing comes down to its unprecedented success in mobilization after 1933. No capitalist economy has ever shifted a greater percentage of its national production to military purposes during peacetime.[4]

It is curious that with the renewed attention to economics and materials in Nazi historiography, energy as a theme is mostly absent. This chapter fills in the picture for hydroelectricity, which has received far less attention than coal or oil.[5] Nazi German deliberations about Austrian hydroelectricity confirm a state obsessed with mobilizing all available resources for war. They suggest, too, that historians' tendency to consider energy sources in isolation does not reflect the mentality of contemporaries. For Nazi energy planners, coal and hydro were inextricably linked. The goal of dams was to generate electricity that would free up coal for more critical uses—to make the rubber and synthetic oil that Germany was unable to requisition from continental holdings or overseas colonies as its opponents could. It is impossible to conclude precisely how Germany's energy policy influenced the war effort. The country faced so many economic and strategic limitations that it is doubtful whether even the most optimized electricity production plan could have altered the outcome. The decision to gouge gigantic dams into the high valleys of Austria, however, reveal a state whose energy policy concentrated on the long term when its military survival strategy required lightning-fast victories.

The First Austrian Republic and the Dream of *Anschluss*

The Republic of Austria came into existence in the fall of 1918, as the strains of the First World War finally began to dissolve the Habsburg monarchy into various new nation-states. Austria as a geographic concept actually dates back to the medieval period and refers to the "eastern reaches" of the Holy Roman Empire. This territory around the city of Vienna would eventually become the core holding of the Habsburg family. From then on Austria and the Habsburgs became inextricably linked. The Habsburgs gradually expanded their patrimony to include large swaths of the German-speaking eastern Alps. More significantly, luck and marital politics brought Bohemia and Hungary into the fold, making the Habsburg monarchy a great European empire. With the rise of nationalism in the nineteenth century, the Habsburgs had to contend with new separatist movements. Only as defeat in the Great War loomed did these centrifugal forces finally prevail.

One by one, the various nationalities of the monarchy declared their independence, ending what could plausibly be called Europe's longest-running major dynasty.[6] The Republic of Austria represented the majority of the leftover German-speaking Habsburg lands.[7]

At the proclamation of the republic, its founders had actually named the new state German-Austria (*Deutschösterreich*). This designation at once played up the Habsburg legacy and indicated the nationality of the future entity. It also related to the simultaneous declaration that the country would be a "component of the German Republic." At this point in time, politicians across the spectrum clamored for the *Anschluss*—the "connection" to a new German state. This idea of joining Austria and Germany has a long history and was the fervent wish of many of the revolutionaries of 1848. In 1918, most argued that it was necessary because German-Austria could not survive economically in its truncated form. Here energy concerns were at the forefront. As Otto Bauer, one of the founders of Austromarxism and foreign minister of the new state, declared in a July 1919 speech, "A land such as ours, that has no coal, and cannot produce food, and that additionally has no large export industry . . . cannot exist independently."[8]

Bauer's point about the coal situation, at least, was correct. In 1913 the territory of German-Austria burned some 12 million tons of coal annually, the bulk of it coming from the monarchy's richest mines in Bohemia. The creation of Czechoslovakia placed these seams under the control of a new state, creating a dismal fuel situation. In 1919, German-Austria produced 1.7 million tons domestically and managed to import some 2.5 million tons with great difficulty from Czechoslovakia and Poland.[9] The shortfall brought transport to a standstill and ensured that many industries remained shuttered almost a year after the war's end. At the peace negotiations in the Paris suburb of St. Germain, Austrian Chancellor Karl Renner warned the Allies that "German-Austria's desperate coal situation" could result in a catastrophic winter, particularly in Vienna. The victorious powers, however, thwarted Renner's preferred method of solving the problem. The Treaty of St. Germain ultimately prohibited the *Anschluss* and forbade the name German-Austria. Nevertheless, the situation began to improve in 1920. And Austria's general economic situation was perhaps not so hopeless as its leadership—keen to be treated kindly by the victorious powers—had made it out to be.

For while Austria did not have vast fossil fuel resources, it seemed predestined to be a country that would have to rely on its Alpine energy to survive. With just under a third of its area being composed by the eastern Alps, this was a state that could plausibly compete with Switzerland for the title of "Alpine republic." The watercourses coming out of the Austrian Alps drain into two of Europe's greatest rivers. Streams in the far western provinces make their way first into Lake Constance and from there on to the Rhine. The vast majority of the country's water, however, empties into the Danube. Austria's geomorphological situation ensures considerable hydroelectric potential.

Austrian leadership in Vienna saw substituting the country's waterpower for coal as the only means forward. This desire found institutional expression in the creation of a new government agency, the Water and Electricity Supply Office (*Wasser- und Elektrizitätswirtschaftsamt*, WEWA). Already the name of the office shows how inextricably linked water and electricity were in the young republic. The new bureau was the brainchild of Wilhelm Ellenbogen (1863–1951), a doctor and former Social Democratic member of the Austrian Reichsrat. In December 1918, barely a month after the collapse of the monarchy, Ellenbogen petitioned the State Council (*Staatsrat*) for the creation of the new authority. The councilor saw the necessity of promoting a unified, systematic approach to the development of postwar Austria's waterpower and viewed the creation of an overarching state authority as a precondition. Ellenbogen believed the WEWA should act as a broker between both the various sections of government interested in the question of the electricity supply and the various economic and social groups involved as well. On January 3, 1919, the provisional executive authority approved Ellenbogen's petition and placed the agency under its direct purview.[10] As Ellenbogen made clear in a January letter addressed to the governors of Austria's various provinces, one of the primary purposes of WEWA was to work against the particularist interests that stood in the way of the "rational" exploitation of the country's increasingly valuable white coal.[11] Here the doctor referred to the eternal struggle between the central state in Vienna and the federal provinces.

WEWA's main strategy for Austria's waterpower was to use it to run the country's most important transportation system. There were a number of reasons that using white coal for the federal railways appeared attractive.

Most importantly, it was an effective way to reduce coal consumption. The state railways devoured about a third of Austria's total coal consumption. Moreover, with Social Democrats holding some of the top posts in the new government, the issue of using state water to power state trains—a favorite of socialists throughout the Alps—finally gained traction. By 1919, even electrification's most dedicated enemy could no longer stand in the way. It is a measure of the new importance accorded to electric traction that the Austrian military administration begrudgingly agreed not to stand in its way.[12] A law passed in July 1920 mandated the electrification of federal railways. Before austerity governments terminated the program in 1928, it concentrated first on the routes going through the Alps. By 1930, almost all of the federal railways west of Salzburg had converted to electric traction.[13]

Another strategy involved exporting white coal. As we saw in the last chapter, efforts to integrate the waterpower of the Austrian Alps into Germany's energy system had started already in the mid-1920s. At that time the gigantic RWE utility gained access to the hydropower of the western Austrian Alps through the construction of Europe's most significant long-distance transmission line. Efforts such as these were often justified in the name of international cooperation. In interwar Austria, an additional rationalization surfaced. With regards to exporting power to Germany in particular, Austrians promoted an electrical *Anschluss*. In the German language, *Anschluss* can also refer to connection in the technological sense, as when two separate electricity grids are tied together. In a pamphlet entitled *"Anschluss* and Energy Supply," the head of Austria's WEWA argued that the fusion of Alpine waterpower with the German industrial regions in the north represented *Anschluss* by other means:

> The marriage of the sober power of the German coal region with the Romanticism of the Austrian Alpine waterpower is a symbol (*Sinnbild*) of that which we hope and long for from the *Anschluss* to the greater German fatherland.[14]

The tract concluded by admonishing Germans and Austrians to stop at nothing to complete the *"Anschluss* through the deed!"[15] It is important to understand such sentiment in its context before the National Socialist takeover of Germany. While *Anschluss* is often thought of as a right-wing goal, this was in fact a project that had backers of all political stripes. Socialists initially thought workers would be better served under the political union. Supporters of the Austrian republic, such as the personnel at WEWA, also kept the idea of *Anschluss* alive as a means of securing both Austria's and Germany's democratic futures.[16] Whatever the momentary prospects of political union between Austria and Germany in the interwar years might have been, support for an electrical *Anschluss* would continue to have powerful allies on both sides of the border.

There existed considerable economic interest to incorporate Austrian waterpower into the German electricity supply. After witnessing RWE's success in expanding its grid, Germany's other major utility—the original German General Electric Company (AEG)—moved to build its own transmission line into the Alps. Its target was the waterpower of the Hohe Tauern range, home to Austria's highest mountains and its highest peak, the *Grossglockner* (Figure 6.1). One enthusiast described Grossglockner as "the tremendous finger of God . . . the king of the Tauern." Locals rather noticed that the "closer to the Glockner, the thicker the skull."[17] In the shadow of the king, the AEG sought to outdo its rival in Rhineland-Westphalia by achieving a project of a sort that had not yet been attempted in the Alps. Rather than tap into the power of a single valley, AEG proposed to harness the energy of an entire mountain range. To do so would require building a series of interconnected, high-altitude reservoirs. All of their water would then be directed to the Kaprun Valley south of Salzburg. This steep, glaciated valley was well-suited for the construction of the centerpiece dams.

Most radical of all, the utility proposed to crisscross the mountain range with an extensive system of canals built into the slopes of the highest reaches. These so-called *Hangkanäle* would act like gutters on the roof of Austria's highest peaks and increase the energy bounty by impounding water as high as possible (Figure 6.2). Only such an extreme approach, AEG argued, would avoid a "wasteful depletion of precious natural resources" (*Raubbau*). All in all, the AEG plan called for some 1,200 kilometers of tunnels and canals to collect water from throughout the range. The utility calculated that the system would produce about half as much electricity as generated annually in all of Germany. This enormous amount of power would be enough to cover Austria's demands—especially for critical winter electricity—while leaving plenty of extra capacity for transport to

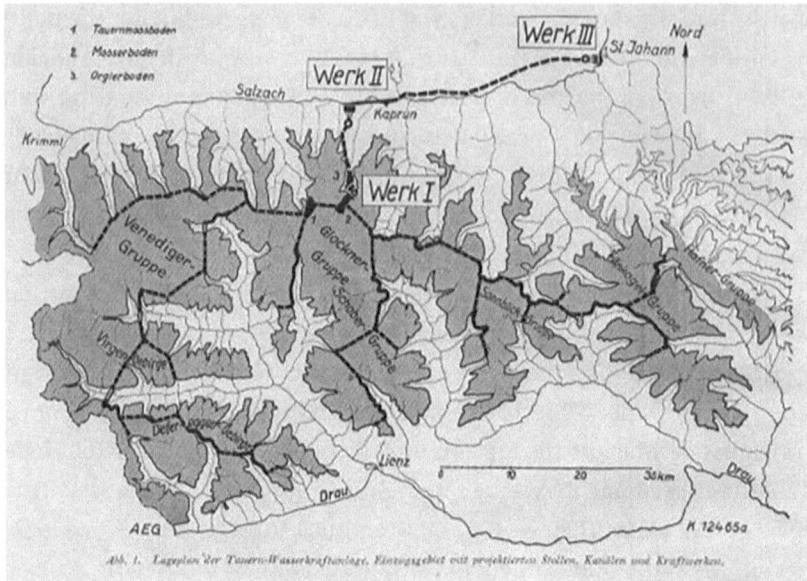

FIGURE 6.1 Layout of the Tauern waterpower plant, with the catchment area and projected tunnels, canals, and power plants. The dark areas represent the peaks of the Hohe Tauern range and its various mountain groups. The Salzach river drains the range to the north and the Drau to the south. Reservoirs 2 and 3 and plants I and II are located in the Kaprun Valley. Of particular importance are the dotted lines, which depict the "slope canals" intended to divert all high-altitude water throughout the range into the Kaprun Valley. Also missing from this German map is any recognition that this project involved diverting water from the provinces of Tyrol and Carinthia to Salzburg, a taboo for Austria.

Source: "Projekt einer Verwertung der Wasserkräfte im Bereich der Tauernkette," *Deutsche Wasserwirtschaft* 24, no. 2 (1929): 22.

the industrial region in eastern Germany.[18] AEG dubbed the facility the *Tauernwerk* (Tauern Works).

Not surprisingly, the plan drew a great deal of attention. It found a powerful backer in the governor of Salzburg province, where the main reservoirs and power plants would be located. He noted that far western Austria had already started exporting electricity to west Germany and applauded AEG for daring to propose such a colossal undertaking. Like his counterparts in western Austria, he recognized a golden opportunity to stimulate

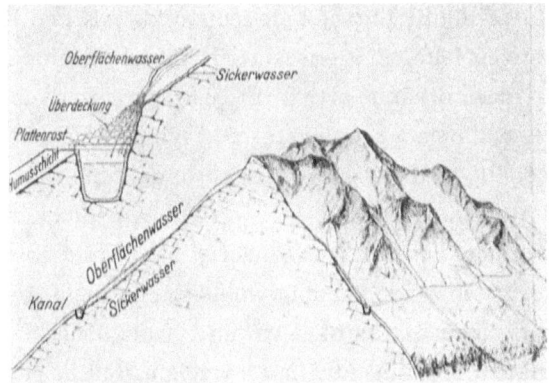

Abb. 4. Schematische Darstellung der Wirkungsweise der Hangkanäle

FIGURE 6.2 Schematic diagram of the functioning of the slope canals. On the right side of the diagram, the slope canal can be seen as a line traversing the mountainside. The idea indicates the degree to which some engineers sought to exhaust the energy potential of the Alps by capturing and collecting runoff at as high an altitude as possible.

Source: Wilhelm Münch, "Das Tauernwerk," *Deutsche Wasserwirtschaft* 26, no. 1 (1931): 3.

his province's economy.[19] Throughout the early 1930s, German and Austrian engineers discussed the scheme at length in the pages of trade journals. In the ensuing "battle of the projects," over a dozen different groups registered their opinion on the subject.[20] Attitudes varied by nationality, as well as by Austrian provincial identity. A consensus formed (among Austrian engineers at least) that the AEG project was outsized, even utopian in scale. Most engineers deemed the slope canals to be technically unfeasible, and the idea of diverting so much water over such distances to be too radical. Hermann Grengg, a confident hydraulic engineer and manager of Styria's electrical utility, emerged as one of the project's greatest opponents, attacking the AEG's centralization idea as a "propagandistic derailment."[21] Grengg opposed "centralizing" all of the power production in the Kaprun Valley, and argued that the Tauern power should be more evenly distributed throughout the range. His opinion would later attain crucial importance for the future of the Hohe Tauern.

Eventually international developments overtook the discussion. By

1930, the world was sliding into a severe economic crisis, and AEG's interest in the costly project faltered. The ascent of National Socialism in Germany also made German corporate activity in Austria an impossibility.[22] While the Nazi Party had plenty of adherents and sympathizers in Austria, the political leadership in the country sought to maintain the country's independence. After the establishment of a right-wing dictatorship in 1933, the new government banned the Nazi Party.[23] In its remaining few years, politics in the First Austrian Republic would be dominated by the effort to prevent *Anschluss* by maintaining Austrian independence.

Nazi efforts to undermine Austrian sovereignty reached a high point in the summer of 1934 when an abortive coup resulted in the assassination of Chancellor Engelbert Dollfuss. It is less-known that the widespread terrorist activity that preceded the murder also aimed at undermining Austria's energy infrastructure. A few days before the chancellor's death, a bomb was discovered at the Achenseewerk power plant, one of the largest lake-turned-reservoirs in Austria. Evidence indicated that the perpetrators had come from nearby Bavaria and that they had returned on foot over the border.[24] Ironically, had the action succeeded, the Nazis would have seriously undermined what would become one of their most important energy sources going forward.

Nazi Germany's Energy Policy and the Austrian Alps

After taking power, Hitler and the Nazis immediately began orienting the German economy for war. Initially this took place under the motto of strengthening the country's defenses (*Wehrhaftmachung*), but by 1936 rearmament and autarky to support expansion became the explicit aims. The launching of the Four Year Plan signaled this turn to becoming economically self-sufficient. As the focus shifted to preparing Germany to fight a war for European supremacy, a new interest in energy autarky also emerged. Planners recognized that while Germany was one of the world's leading coal producers, it was decidedly lacking in oil, especially as compared to powers such as Great Britain, the United States, and the Soviet Union. Under the circumstances, they concluded that Germany's best chance lay in using the waterpower of the Alps to free up coal for conversion into other raw materials.

We can see the nature of these views in the writings of Albrecht Czimatis. Czimatis, an otherwise obscure staff officer in the German War Ministry, would go on to attain significant influence over the German economy. He was among the first German officials to explicitly emphasize the connection between energy and the ability to conduct war. He expressed his thoughts on the matter in a 1936 tract entitled *Energy Supply as Foundation of the War Economy*. Czimatis saw the combat capability of any state as dependent in the first place on its energy supply. In writing the treatise, he hoped to persuade his audience that marshaling sufficient, reliable energy for war should be the goal of all economic policy. For Germany, this would mean using technology and political institutions to exploit "natural resources and natural forces to the outermost limits of their inherent energy and deliver them in the economically most perfect form for practical application." Czimatis christened this system a "high-quality energy supply (*Edelenergiewirtschaft*)." Germany's goal should not be simply to make the maximum quantity of energy available, but to take steps to ensure that all of its resources were being used in the most efficient manner possible.[25]

The need for such a system, Czimatis argued, grew out of Germany's nature and its position in the world. Surveying the globe, Czimatis noted that all countries based their energy supplies on one or another power source, depending on the geologic and geographic features of the land. Great Britain was rich with anthracite but had almost no lignite and very little waterpower. Switzerland possessed no coal, Italy almost none. Both instead developed their rich hydraulic resources. Moreover, if a country had access to a global empire—like Great Britain—then it had the option of obtaining key resources like coal and oil through trade. Germany's nature, Czimatis claimed, predestined the country to a more flexible energy strategy. Exchange and "supplementation" would be necessary to overcome some of the great difficulties rooted in "geographic disfavor and military-political limited space (*Raumenge*)."[26]

For Czimatis—as for many other Nazis—Germany's disadvantages also lay in the concept of space. Germany did not possess an empire or a continental territory that could ensure access to sufficient energy sources in the case of war. Above all, Germany's circumstances made procuring petroleum a real challenge. Still the country was not without hope in this regard. As Czimatis explained:

> Lack of oil can theoretically be compensated through transformation of coal; difficulties with raising capital and generating sufficiently cheap production will always place limits on this. For an oil poor country, the technological-chemical autarky in the arena of fuels is only achievable through the commitment of unusually great means.[27]

Among other things, converting coal into oil required massive quantities of cheap energy. For Germany, the key would be waterpower. Though waterpower plants were often located far from centers of consumption and had high upfront costs, building them could free up coal for other uses. Czimatis concluded:

> If coal must be used frugally or it should be reserved as raw material for the creation of more valuable chemical products, one can resort to drawing on the increased use of waterpower for electric production.[28]

This then, should be the goal of Nazi Germany's "high-quality" energy supply. Coal should be conserved as much as possible so that it could be employed to synthetically make up for Germany's restricted space. Hydroelectricity would be called upon to replace thermal electricity wherever possible. While the hydro resources of Scandinavia required a technically challenging overwater transmission, no such barriers stood in the way of accessing the hydropower of the Alps.

Throughout the 1930s, then, Germany remained interested in Austrian white coal. With the creation of the Four Year Plan and the ramping up of rearmament, German economic planners took a marked interest in the economic capacities of their southern neighbor. Albrecht Czimatis himself became the head of an agency at the heart of the Four Year Plan: the Office for German Raw Materials (*Amt für deutsche Roh- und Werkstoffe*). It was responsible for securing Germany's raw materials, including the all-important petroleum supply.[29]

Multiple German government agencies highlighted the importance of hydroelectricity in analyses of Austrian economic potential compiled after 1936. In 1937, the Reich Economic Ministry commissioned the construction of a high-voltage transmission line to stretch south from the industrial zone of eastern Germany toward the waterpower of the Hohe Tauern.[30] High-level actors within the Nazi state remained personally interested in the energy potential of the mountain chain. In a conversation just before the *Anschluss* for instance, Wilhelm Keppler, a German businessman and Nazi Party functionary, informed Foreign Minister Joachim von Ribbentrop about the significance of the Hohe Tauern range. The businessman had been instrumental in preparing the way for Hitler's legal assumption of power in Germany and had also served alongside Czimatis in a Four Year Plan agency. Keppler would go on to become Hitler's Reich Commissioner for Austria, charged with integrating the Austrian economy with the German after the *Anschluss*.[31]

That fusion finally took place in March 1938. Early that month, the Austrian chancellor made a last-ditch effort to preserve his state's independence from Nazi Germany by calling for a referendum. Austrians would be asked if they wished to have a "free, German, independent and social, Christian and united Austria." This was the last straw for Hitler, who ordered German forces to cross the border and invade Austria on March 12. Despite being plagued by missteps, the Austrian army followed orders and let the incursion unfold unopposed. In fact, the jubilation with which Austrians greeted the invasion surprised even Hitler. On the fly he made the decision to directly annex the country. Before a raucous crowd at the Heldenplatz in Vienna, Hitler triumphantly announced the *Anschluss*. Austria became a German province known as the *Ostmark* (Eastern March), and the Third Reich now controlled the water resources of the eastern Alps.[32]

If German interest in Austrian white coal had been strong before, National Socialist energy policymakers now found themselves in a kind of "water power delirium" in the words of one historian.[33] Energy experts gushed that Austria's water would finally put an end to the Reich's growing electricity supply problems and "place Germany in the position of a practically unrivaled energy-Great-Power."[34] RWE quickly dug up previously buried plans for the continued development of Austrian Alpine waterpower and were granted most of western Austria as their own sphere of influence. Perhaps because of National Socialist predilections for monumental works, the Tauern idea was swiftly prioritized by many, including one of the highest functionaries in the Nazi economy. The enthusiasm for Austrian waterpower and for the Tauern project in particular was infused with concern about Germany's long-term energy future. It would quickly become clear, however, that what the German war economy desperately needed was more electricity in the here and now. But by that point, the German economy had already invested valuable time and resources into arduous hydraulic projects that suddenly had little prospect of completion.[35]

A mere two weeks after annexation, a German government study on how to incorporate Austria into the German war economy identified harnessing the Tauern energy as a priority. The report emphasized the urgency of exploiting Austrian waterpower for use in the chemical industry, and it recommended reexamining a "mostly finished" plan for a large-scale power plant in the Hohe Tauern: the AEG plan. Based on the company's figures, the author of the report concluded that the Tauern hydroplant possessed "the greatest significance for the entire German electricity supply."[36] Three days later in a landmark speech in Vienna, Hermann Göring declared the immediate construction of a series of reservoirs in the Hohe Tauern and a hydroelectric dam on the Danube as priorities for the Austrian economy.[37]

Göring's speech was not mere rhetoric. A little over a month later, on the morning of Monday May 16, Göring traveled by special train to Kaprun to lead the groundbreaking ceremony for the first step in the construction of the Tauern Works. Standing at the head of the Kaprun Valley, surrounded by swastikas and villagers from the nearby town, Göring let the world hear what a fascist justification for large dams sounded like. "We want to be a tremendous (*gewaltiges Volk*) people," the field marshal proclaimed, "a mighty nation. We say to all, but especially clearly to those who are not happy to hear it: Germany above all!" In typical National Socialist style, Göring also emphasized the significance of action. "The opus is not accomplished and completed through speeches and celebrations; rather hard work alone leads to success." Finally, Göring declared that the time had come to put the natural resources of Austria to work. "We have enough mountains and water," declared the field marshal, perhaps anticipating criticism about the environmental effects of gigantic dam projects. "Now it is time to collect their powers. As the National Socialist movement once collected all the forces, all passionate currents, consolidated them, impounded them, and applied their concentrated powers, so the impounded forces of nature will create great value here, where they once, unrestrained, senselessly devastated the fields and annihilated the harvest."[38] After completing his speech, Göring reached for a shovel and broke ground at the site of the future powerhouse (in the wrong place, as it turned out, Figure 6.3). On the hillside above him a row of flags marked the future path of the penstocks.

In the meantime, the decisive circles in the German economy contemplated how precisely to utilize the hydropower bounty brought by the *Anschluss*. A memorandum produced by the Reichsstelle für Wirtschaftsausbau gives insight into the process. This agency was the successor to the raw materials section of the Four Year Plan. At its head, once again, stood Major Czimatis and his influence on its perspective is clear. The memorandum concluded that the core problem of utilizing Austria's waterpower "lay in the mastering of the balancing between electricity demand and the labile presence of energy." The author presented three potential solutions to the problem. The "natural" method would be to utilize the funnel-shaped Alpine valleys to store excess summer energy for the winter. Unusually for the time period, the memorandum also suggested that Germany might

FIGURE 6.3   Hermann Göring at the ceremonial groundbreaking for the Tauern Works in Kaprun.

Courtesy of the Austrian National Library, Vienna, S 108/67.

adjust its economy to fit the availability of waterpower. Certain industries, above all electro-metallurgy, might shift their peak production to the summer. The last resort would be to meet peak demands with electricity generated by burning lignite, either within Austria or via the grid from eastern Germany. However, following Czimatis, the memorandum asserted that "the protection (*Schonung*) of brown coal and its application for finishing processes (*Veredelungsverfahren*) is to be seen as the supreme rule (*oberstes Gesetz*) for future planning." Coal had to be preserved for its use in synthetic production industries. Priority was to be given to using the mountains to store energy for the winter. No mountain range possessed as much potential for storage as the Hohe Tauern.[39]

To finance and control the development of the Tauern power—as well as several other large hydroplants on Alpine rivers—the Reich created a state-owned corporation, the Alpen-Elektrowerke AG (AEW).[40] In the company's first annual report, it emphasized the necessity of developing Austria's unused Alpine hydropower as a means of substituting a renewable energy resource for coal wherever possible, just as Czimatis had advocated. The report also noted that Alpine waterpower would be key for the expansion of Austrian industry.[41]

As technical director in charge of the critical Tauern project, the AEW named Hermann Grengg, one of the most vocal critics of the earlier AEG plan. Among his other credentials for the job, Grengg (1891–1978) had joined the National Socialist Party in 1932, when membership in the Nazi Party was illegal in Austria. Despite the preference of the Salzburg provincial government and the Nazi bureaucracy for the AEG plan, Grengg managed to carry the day for a project that privileged Austria's regional electricity supply over a greater German one. To do so he had to engage in a battle that would result in delays for the project. Whether these were decisive or not is questionable. What the conflict reveals is that Germany had to make energy choices that its Allied opponents did not.

Over the second half of 1938, Grengg worked out his plan for the Hohe Tauern. While the rest of the world followed the drama of the Munich crisis, Grengg opted for a "decentralized" concept for the Tauern range. Instead of attempting to concentrate all of the power production in the Kaprun Valley, his plan aimed to maximize the amount of hydraulic energy that could be stored in the Hohe Tauern. He argued that rather than building cheaper and simpler dams that produced electricity constantly and could more quickly relieve Germany's critical coal supplies, AEW should focus on more expensive storage dams for the long term. Only in this manner could the Tauern fulfill its true destiny to serve as auxiliary power for a growing and modern Austrian grid.

Several reservoirs and accompanying hydroplants would be built in a number of valleys. These would be built in stages, in accordance with the demand for electricity. Since the Kaprun Valley offered the greatest energy potential, construction would begin there. It would consist of a main stage and an upper stage. Incorporating a new technique, a pump facility would make it possible to use excess electricity to lift water from the lower into the higher reservoir, where it would be more valuable. Grengg did not reject all aspects of the earlier AEG scheme. To help fill the reservoirs in the Kaprun Valley, he adopted the idea of incorporating the runoff of the Pasterze, Austria's largest glacier, into the energy bounty. The catch was that the Pasterze lay directly below the Grossglockner on the south side of the main ridge of the Alps, while Kaprun lay on the north. A tunnel through the ridge solved this problem, and the Pasterze meltwater that previously drained off the south slope would now flow north (Figure 6.4).

According to the AEW, their plan was also more reasonable from a water management perspective. Grengg professed to take seriously the idea that the German nation needed to balance competing claims and interests for the use of water throughout the Reich, including in the Hohe Tauern. The AEW plan, he argued, left more irrigation water for the farmers that utilized the higher valleys for pasturage—a significant aspect of Alpine agriculture. In fact, he recognized only one true, irreconcilable conflict produced by his works: that of "turbine and the waterfall." Unfortunately, there was no way around destroying waterfalls, the "crown jewel of an important German recreational landscape," which would become ever more significant as German space became more crowded and altered by technology. The engineer could only make vague promises that all losses in this area would have to be compensated. Grengg even presented the rerouting of water from the southern to the northern side of the Hohe Tauern through a *völkisch* lens. This measure would have the added benefit

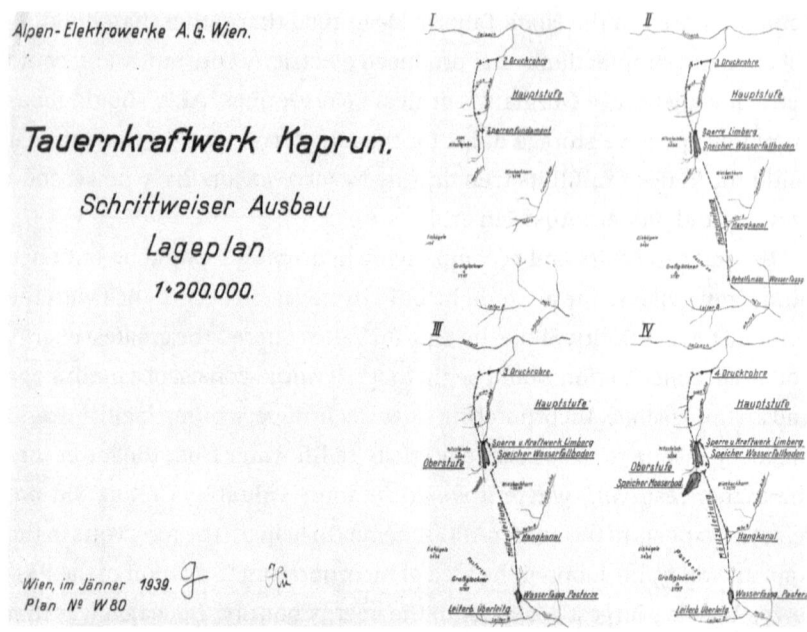

FIGURE 6.4 Step-by-step plan for Tauern Works Kaprun, January 1939. These are Grengg's proposals (along with his "G" signature) for incremental development of the storage in the Kaprun Valley. The main reservoir and the reservoir to capture the Pasterze glacier runoff are discernible (under III). The final step in the plan entailed creating an "upper stage" reservoir, into which water from the lower one could be pumped in times of excess electricity production.

Courtesy of Salzburger Landesarchiv, RStH V/3 225.

of diverting water that would soon flow into Yugoslavia back onto German soil, where it would improve the German stretch of the Danube's function for shipping.[42]

Developing hydropower in a high-Alpine setting brings with it many environmental peculiarities. One prominent wrinkle is the presence and influence of glaciers, and these were very much on Grengg's mind as he drew up his plans. As projects like this zeroed in on Alpine glacial zones themselves, Europeans were forced to reflect more closely on how these ice sheets fit into hydropower development. Grengg found that they offered both huge advantages and raised great uncertainty. From an energy perspective, glaciers are not unlike fossil fuel stores. They are reservoirs of potential energy created long ago. While the water molecules in a given stretch of a river in a non-glaciated region may have been part of the ocean a mere week before, the water that flowed off the Pasterze glacier on the Grossglockner had likely been there for centuries or even millennia. Grengg likened the glaciers of the Alps to "immense reservoirs" that stood between precipitation and drainage. They stored a vast multiple of the amount of water that even the most intense dam building could achieve. The Tauern dams he envisioned, for instance, mustered only a total of 13 square kilometers of artificial storage area compared to the region's 183 square kilometers of glaciers. Glaciers ensured that the differences between annual runoff flows in the Alps were comparatively small. In glaciated catchment basins, waterways ran high even in warm and dry summers thanks to glacial melt. But, and here lay the uncertainty, the glacial function as reservoirs depended heavily on much longer cycles of ice expansion and retreat. Grengg noted that the Alps had been in a period of "extraordinarily strong" glacial retreat for quite some time (he referred to this as a "fair weather" era). This glacial melt meant that many rivers had higher annual flows than the precipitation they received. Engineers prized precisely this additional runoff, as it could offset the dry periods of the rivers of the plains. But what if glaciers began to grow again? Could they possibly reach down to and destroy dams? Certainly less water would flow down from the heights. Reservoirs—the most expensive elements of the project—would go unfilled. These questions made it difficult to calculate how large the dams should be. Grengg concluded that to keep the project economical, they should not be larger than necessary to even out the flow over the course of the year.[43]

AEW intended their projects in the Hohe Tauern to provide winter energy and stability for an electricity supply that would be based primarily on the hydropower of Austria's non-glaciated, lower lying rivers. Grengg reckoned the Kaprun main stage could begin to produce power by 1941 and be completed by 1943. On this basis, construction proceeded. In the meantime, however, questions about the AEW project swirled at the highest levels. By 1939 the electricity situation in Nazi Germany was becoming acute. Thanks to the industrial expansion accompanying the Four Year Plan, the Reich no longer retained any reserve capacity.

In response, Hermann Göring, point person for the German economy, created a special position designated to take all measures to increase production and streamline distribution. As Plenipotentiary for the German Energy Supply (*Generalbevollmächtigte für die Energiewirtschaft*), the field marshal tapped Just Dillgardt.[44] Dillgardt—the mayor of Essen, home of RWE—joined other recently anointed plenipotentiaries for matters such as motorized transport, construction, and machine production. After his appointment in January 1939, Dillgardt set about studying Germany's electricity situation. Three months later in a letter to Josef Bürckel, the man Hitler had tasked with carrying out the reunification of Austria to the Old Reich, Dillgardt explained his conclusions: "I am convinced that the exploitation of Alpine waterpower and its activation in the energy supply of the Ostmark as well as the whole Reich must be the most important matter for the German energy supply for the near future." Since he had also been contacted by a number of people about energy issues in Austria, Dillgardt requested a meeting with Bürckel to help him make the final decisions.[45]

As it happened, white coal was also at the front of Bürckel's mind. The very day that Dillgardt sent his letter, Bürckel held two meetings at his office in Vienna to hear representatives present their plans on the two main variants for the Hohe Tauern. He had also tasked two German engineers—Thürnau and Schmidt—to appraise various potential hydroelectric projects in Austria. First on this list was the AEW project.[46] In their written evaluation, Thürnau and Schmidt came out in favor of AEG's centralized project. This was not entirely surprising. During the "battle of the plans" earlier in the decade, Thürnau—a professor at Leipzig—had written an article in favor of the AEG project.[47] Thürnau and Schmidt concurred that AEG's designs deserved support because they came closest to what they called the "optimal final development (*Endausbau*)." For them, it was irrelevant which project promised an easier start, or which design was capable of delivering electricity into the grid at the earliest juncture. Rather, the best plan was the one that harnessed the maximum amount of raw waterpower that was technologically and economically possible. This meant concentrating and storing as much Tauern water as possible in the steep Kaprun Valley. All calculations about how big to make the Kaprun reservoirs were superfluous. The motto needed to be "as big as conceivably possible." Restricting the dams to this one valley would make the development of the Tauern power more cost-effective by reducing the amount of materials and the number of construction sites. To Grengg's assertion that the geology of the Kaprun Valley did not permit the construction of dams tall enough to hold all the Tauern water, Thürnau and Schmidt countered that engineers should "master" geological difficulties through technical means.[48]

All of this effort, according to Thürnau and Schmidt, was necessary and justified to help waterpower fulfill its critical purpose in the German war economy: to substitute for coal in electricity generation. Here again the argument formulated by Czimatis surfaced. Coal was urgently needed for specific industrial processes critical to the war economy, especially in the production of synthetic fuels for military aircraft and vehicles. Energy experts, according to Thürnau and Schmidt, knew that coal was far too precious a commodity to be wasted firing boilers. Waterpower, on the other hand, was "inexhaustible" and had much lower operating costs. In the near future, comparative costs of waterpower and coal would not even matter, as the latter fuel would become invaluable as an input for the chemical industry. Both engineers admitted that it would probably never be possible to satisfy all of Germany's electricity demand with waterpower. But they argued it was "a national duty" to cover as much of this demand as possible with waterpower. This purpose could be satisfied by constructing the Tauern Works to produce the greatest number of kilowatt hours possible.[49]

Grengg now had to defend his choices. Most of his case rested on the devastating environmental impacts of the rival scheme: "If the centralized system of slope canals fulfills even half of its intended purpose, then it will deprive numerous valleys in the provinces of Carinthia and Salzburg of their water." Grengg argued that such a system would contradict the precepts of the "new German water management." In this case, he claimed, agriculture would suffer mightily, a prospect that would not please the Reich Food Estate, responsible for organizing Germany's food supply. Grengg concluded by noting that alterations to the project could now only be achieved by casting aside all of the progress that had been made. It would require compiling a detailed centralized project and renewed negotiations with concerned parties. "The Tauern waterpower," he argued, "would thereby bow out of the energetic struggle for existence on the part of German water management." This mountain energy would not be able to aid Germany in its struggles at any time in the near future. Sacrificing such an important source of power would only be justified if the AEG plan had better future prospects. Grengg was convinced that this was not the case.[50]

In the end, Dillgardt decided against the centralized solution and ordered the continuation of the AEW project. His ultimate decision had little to do with the kinds of technical and environmental objections that Grengg had marshaled in defense of his program. In a meeting in Bürckel's office in late May 1939, Dillgardt explained that, above all, "military reasons" spoke against a centralized exploitation of the Hohe Tauern waterpower in a singular plant. He presumably feared how focusing all energy production in the Kaprun Valley would make it easier for enemy action to disrupt the plant. As a secondary consideration, Dillgardt also mentioned agricultural concerns. Here he echoed Grengg's arguments about the effects of diverting so much water over such long distances via the slope canals.[51] Regardless of the reasoning, the energy czar's pronouncement at the beginning of the summer of 1939 set in stone a considerable part of Nazi Germany's energy future (Figure 6.5). The expanses of the Kaprun Valley would be called upon to store the energy necessary to integrate Greater Germany's "southern waterpower" into its rapidly escalating war economy.

The Second World War

These plans were dashed by the coming of the Second World War. Although the Plenipotentiary for the German Energy Supply had declared the construction project as "politically significant" and "urgent," it almost immediately ran into war-induced difficulties. Already since 1937, the German state had been rationing iron to war industries. From the perspective of the electricity branch, their sector had been denied sufficient materials for expansion.[52] In a meeting at the Reich Ministry for the Economy in Berlin on September 19, 1939, the AEW was informed that only small amounts of iron could be made available for its most important waterpower projects. The Tauern Works and a dam project on the Drau were included in this privileged group. But it is a testament to the already difficult situation that construction on a hydropower dam on the Danube had to be halted completely. Kaprun, moreover, would not receive the iron it needed for its

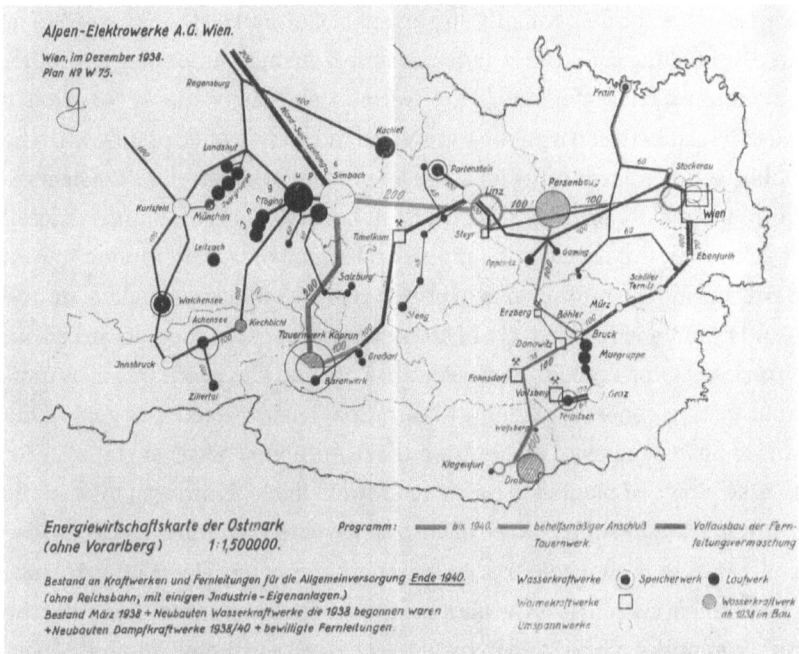

FIGURE 6.5 Energy supply map of the Ostmark, December 1938. An energy map of Austria without the far western province of Vorarlberg. It would remain the basis for the AEW plans in Austria. Vienna (Wien) is in the east and Innsbruck in the west. The map shows the general concentration of hydropower centered on the German and Austrian Alps (hydroelectric plants are represented by circles), the superlative size of the Tauernkraftwerk Kaprun, as well as the effort to build new hydro dams in Austria (circles with hash marks). The thick lines with large numbers depict transmission lines and their projected voltages. They were to be built in order to transport white coal north to the eastern German industrial zone and to the industrial centers in eastern Austria.

Courtesy of Salzburger Landesarchiv, RStH V/3 227.

electrical equipment or for its penstocks. It was estimated that these disruptions would delay the plant's initial launch by at least a year and probably more.[53] The situation worsened considerably when the project lost its urgent wartime status in 1943.

At the outset of the war, the Third Reich's electricity supply was predominantly reliant upon fossil fuel. Though this would fluctuate over the course of the conflict, roughly 80 percent of German electricity was generated by burning coal, and 20 percent came from hydraulic energy. Another outstanding characteristic of the German electricity supply was that it was divided between a public supply and industry-owned plants, with the public sources accounting for some 60 percent of generation. The heat required to make Germany's electricity necessitated large amounts of fossil fuel. In 1942, the public grid's plants consumed at least 13 million tons of hard coal and 46 million tons of brown coal. These sums equalled around 6 and 17 percent, respectively, of domestic coal consumption. It worked out to an average of 0.76 kilograms of anthracite and 2.38 kilograms of bituminous coal to generate a kilowatt hour. Coal use for electricity generation would have been even higher than these numbers, however statistics for industry-owned plants are not as readily available. From 1939 to 1945, the Third Reich intensified its use of anthracite, with the ratio of hard coal use compared to brown coal increasing about 10 percent. By far the greatest consumer of current was German industry, which increased its share of the public supply from 70 to 80 percent in the course of mobilization and war. Industrial concerns also used the entirety of their proprietary generation.[54]

The wartime experience of Germany's grid reveals the difficulties that the Third Reich grappled with thanks to an electricity supply based on coal and an increasing contingent of Alpine waterpower. The managers of the grid had to contend not only with scarce materials for plant maintenance and aerial bombings, which occasionally managed to shut down industrial production in western Germany, but they also had to struggle with the terrestrial flows of power on which the network depended. Reports from government regulatory agencies are instructive. One from September 1940 highlights the daily balancing act necessitated by fluctuating flows of water. In the first winter of the war, 1939–1940, Germany experienced what the report called a "monstrous cold snap."[55] The cold weather both lowered the capacity of the hydropower plants in the south and hardened the brown coal that provided much of Germany's thermal electricity, making it more difficult to mine. The report proudly confirmed that dispatchers overcame these shortfalls by wheeling current around the country. In this manner, the particularly important deliveries to the aluminium and chemical industries could be maintained. But it came at the cost of burning more fossil fuel.

The inverse situation caused a temporal miracle of sorts over the Pentecost holidays of 1943. In those June days, hefty rains over the Austrian Alps caused an unexpected surplus of hydraulic energy. To absorb this windfall, German load dispatchers switched a synchronous grid, reaching from Austria all the way to the North and Baltic Seas, into existence for the first time. The presence of Alpine waterpower in the German network allowed thermal plants in northern Germany to ramp down, saving coal for the future. From that day forward, dispatchers recreated this circuit as necessary. War had finally offered a preview of how the chain could serve as Central Europe's mountain battery.[56]

In the fall of 1940, Nazi aspirations to yoke the power of the Alps reached a high point. Germany's conquests in the first half of that year kicked off visions of establishing a continental energy supply to rival its imperial enemies. Particularly, the capture of Norway inspired further deliberation about the necessity of feeding more hydraulic energy into the German grid. When in the fall of 1940 authorities began putting together a Second Four Year Plan, increased development of waterpower represented something like its basic principle. While the first edition had of necessity focused on Germany, the second would take advantage of the new possibilities offered by the *Anschluss* with Austria and the new European conquests. To overcome the bottlenecks of coal and electricity production that were stifling synthetic fuel production, the plan authorities concluded that "the second Four Year Plan can only be executed . . . if the current ratio between thermal and waterpower is changed fundamentally in favor of waterpower as we continue to develop energy production."[57]

This energy would come primarily from the southern German Alpine area and, in the future, Norway. At the same time, Germany's energy czar commissioned his own report on how the country should continue to develop its waterpower to conserve coal. Ecstatic about the successful war effort, it began to reckon how Germany would grow its electricity supply over the next fifteen years, into the mid-1950s. For the first phase, Germany's Alpine resources would be developed as they represented lower hanging fruit in comparison to Scandinavia. The Reich's electricity experts also began to consider other worthy projects in Europe that could be developed after a "successful denouement" for the war. First on the list were waterpower sites of the Swiss Alps, especially those that could provide winter peak energy. The report left unclear whether this energy would be accessed by Switzerland's continued cooperation in delivering electricity to Germany—or by conquest.[58]

Hydropower development became so important for the Nazi leadership at the beginning of 1941 that even the Führer himself had to contemplate questions of energy policy and the role of the Alps. From what little evidence exists, it seems that Adolf Hitler had his own opinions on the merits of water as an energy source. Perhaps not surprisingly, he harbored Romantic notions of the role that waterpower had played in Germany before the advent of electricity. He also found the idea of using a renewable energy source like water appealing. During one of his many conversations in the Führer headquarters, Hitler held forth on these views. While he understood that the chemical industry had a need for large dams,

> For the rest, we will have to incentivize the production of every single horsepower in the style of our earlier mill power usage. Water flows, one need only create a fall and one has what one needs. While coal will one day run out, water always returns.[59]

Hitler concluded these thoughts by opining that waterpower usage did not have to be oversized, that it could return to its small-scale communal roots.

Interesting in themselves, these Romantic views are significant for another reason. They led Hitler to work at cross purposes with the majority of his advisors at a crucial moment. Early in 1941, the head of the province in Austria where Hitler was born—and one of Hitler's closest confidants—complained to his friend about an energy problem in Linz. The "city of the Führer's youth" was unable to break ground on a badly needed new hydropower dam on the Enns River. Hitler's crony chalked up the delays to the "capitalistic" behavior of the Alpen Elektrowerke, which concentrated on securing resources for its grander projects.[60] Disgusted by the story, Hitler created yet another new governmental position and appointed Fritz Todt to finally remedy the inability to bring new power plants online.[61] Todt, the most prominent engineer in the Reich, spent the spring drawing up decrees that Hitler would deliver on the future of energy policy. Hitler, for his part, could not be swayed from his position that the AEW were acting in a capitalist and completely "un-National Socialistic" manner despite the corporation's protests that they were 100 percent government owned (and holders of a stake in the Enns River utility). That, Hitler argued, only made him judge the behavior more harshly.[62] The episode showed Hitler espousing ideas that contradicted his own state's efforts to maximize energy production during an escalating war. His decision to make Todt the head of a new energy agency set off months of bureaucratic fighting as government agencies struggled to maintain their spheres of influence. In July 1941, Hitler signed the decree creating the new authority. Showing how closely intertwined the Nazis viewed the two offices, Todt was to be General Inspector for Water and Energy (*Generalinspektor für Wasser und Energie*). He would lead the inspectorate until his death in a suspicious plane crash in 1942. Thereafter Albert Speer, known primarily for his work as munitions minister, took over.

By this time, Germany's efforts not just to efficiently distribute electricity but expand capacity ran up against the dictates of total war. Everywhere, the lack of construction material, equipment, and labor made progress difficult. After the war a German energy expert framed this, perhaps *the* economic dilemma of Nazi Germany, in these terms: "One cannot simultaneously stamp armies, weapons, and power plants from the soil." Airplane and tank production received priority status before new electrical generation.[63] A major dam on the Danube—one that Hitler himself waxed philosophic about—got almost nowhere.[64] In the province of Tyrol, a project to develop the considerable waterpower of the Ötz Valley never advanced past the initial phase. Planners, however, found a creative use for some of the excavations and existing hydraulic power. They used the tunnels drilled for the first stage to begin constructing a water-powered wind tunnel to test the super weapons—jet aircraft and rockets—supposed to turn the war effort for the Nazis. These, as we know, never achieved their desired effects. Much more successful, by contrast, were the efforts to harness the power of the far-western Ill Valley on the border to Switzerland.[65] Given the lack of men and materials, however, there seemed to be little hope in hydropower dams being able to satisfy the hunger for electricity.

Finally in the summer of 1942, the perspectives of Nazi energy experts began to shift. A government report made clear that earlier hopes placed upon Alpine hydropower could not be expected to meet Greater Germany's future electricity demands. The ravenous electricity hunger of the Reich's new raw materials industries, chemicals above all, had practically devoured the country's remaining reserves. The report recommended the immediate construction of ten standardized coal-powered plants, as these could be most rapidly thrown into the breach. In September 1942 Albert Speer approved the Thermal Power Crash Program (*Wärmekraft-Sofortprogramm*) that would remain the primary directive until the end of the war. While this plan is often mentioned in economic histories of Nazi Germany, the environmental implications of the break are scarcely appreciated. Heat from Germany's coal, not water from its mountains, was now seen as the only way to stave off a disaster for the electric grid. But all fossil fuel sacrificed for electric current could no longer be used to make up the Reich's raw material deficits.[66]

Beginning in 1943, intensified aerial bombardment of Hamburg and Rhineland-Westphalia led to the intermittent failure of hundreds of megawatts of grid capacity. By September 1944, an increasing discrepancy between available capacity and demand became irreversible. Bombing, territorial losses, and the inability of the *Reichsbahn* to deliver anthracite to the boilers all took their toll. Mandatory curtailments became necessary and reached one-third of peak demand in January 1945. In March of 1945, Switzerland finally suspended its contractually obligated provisions of hydroelectricity from its dams on the Aare River—a step that still seemed to wound German engineers even decades after the war.[67] That spring Allied strafing of high-tension wires also severed RWE's vanguard connection between the Austrian Alps and western Germany's industrial heartland. The orderly supply of electricity finally broke down in many parts of the Reich. A week after Adolf Hitler took his life amid the Battle for Berlin, the capital's electricity supply completely collapsed.[68]

Wartime events also made their impact felt on the Kaprun project. When, on May 16–17, 1943, British bombers attacked several dams in western Germany and succeeded in seriously damaging the reservoirs of the Möhne and Eder valleys, questions arose about the security of the dam designs in Kaprun. The German military eventually vetoed the buttress dams planned for the Hohe Tauern. The primary virtue of these slimmer structures had been their cost-effectiveness. Instead Grengg opted for sturdier arch designs, further delaying construction.[69] In late 1944, the AEW finally managed to install a temporary barrage that allowed Kaprun to begin feeding small amounts of base load into the eastern Austrian grid. But this was nowhere near the hoped-for bounty. As Germany's military situation deteriorated in 1945, all further progress came to a standstill.

Only the use of thousands of foreign and forced laborers in the Hohe Tauern had made the partial activation possible. Since the early stages of Nazi development of Austrian hydropower, labor shortages encouraged leadership to consider the use of foreign labor—despite supposed ideological reservations. In Kaprun, management hoped to attract Italian laborers but also settled for those from the "protectorate" of Bohemia and Moravia. With the start of the war, however, much of this work ceased to be voluntary. Foreign workers were deported from their countries or forced to work as POWs. At this point, forced laborers and prisoners became critical for several hydropower projects in Austria.[70] Of all energy projects in Austria, Kaprun received the greatest number of foreign workers by far. They quickly made up close to 90 percent of the total workforce in the Hohe Tauern. All in all, some 6,300 foreign workers have been identified for Kaprun. Initially, the POWs came primarily from Poland and France; after 1943 they were joined by workers from the Soviet Union—so-called *Ostarbeiter*. The latter—perceived to be "racially inferior"—received the harshest of treatment and lived in truly miserable isolated conditions at high altitude. Over twenty different nationalities have been accounted for, including thirty (presumably Austrian) Jews. The use of forced labor reached a peak in 1943 and declined precipitously thereafter. A historical commission confirmed fifty-six deaths among foreign workers at Kaprun though the actual number was certainly much higher. A postwar Soviet memorial to its citizens who died at the job site commemorated eighty-seven souls.[71]

Kaprun had another connection to Nazi atrocities. During the war, AEW began building a high-voltage transmission line to link Austria directly to the industrial center in Upper Silesia. Before the end of the war, the northeastern section of this linkage came online, primarily to feed electricity to the gigantic IG Farben synthetic fuel plant at Auschwitz. Just such an interconnection had been advocated by those who sought to build a European grid to encourage international economic cooperation in the interwar period. During the war, however, the linkage aimed to fulfill the visions of Nazi energy planners. Austria's mountain water could now help Germany overcome its raw material deficits by aiding in the process of converting coal to oil. Kaprun and Auschwitz, separated by some 500 kilometers, were connected by an electrical *Anschluss*.[72]

Conclusion

Focusing on the energy of the Austrian Alps in the interwar years reveals ways in which, thwarted in the political arena, Austrians and Germans sought to accomplish an *Anschluss* via electric power. After 1933, white coal also began to figure heavily in German plans for economic mobilization and autarky. When Hitler finally made political union a reality in 1938, ideas to utilize Austrian energy became concrete in the form of the Kaprun project. While Kaprun contributed only a small fraction of its expected potential, Nazi Germany did manage to wrest a fair amount of energy for its war effort from the water tumbling off the Austrian Alps. In the seven years that Austria was part of Hitler's Greater Germany, over ten new dams were placed astride the country's waterways. Compared to the grand visions of using Austria to transition the Reich to a land that used renewables to conserve coal as much as possible, however, this achievement fell far short.

While it is impossible to conclude precisely what role Nazi Germany's concentration on Austrian white coal ultimately played in its eventual defeat, the episode provides a window onto the energy and resource concerns that underlay all combatants' participation in the global conflict. It also underlines the resource constraints that confronted Germany when compared to its main rivals, who could draw upon continental and imperial bases to wage war. Still, the ramifications of German energy policy in Austria would not expire with the *Götterdämmerung* brought on by Hitler's deranged visions. The new hydroelectric projects developed by the Nazis mostly survived the war unscathed and represented an extra two-thirds of Austria's electrical capacity prior to the *Anschluss*.[73] In addition, about one-third of the work for constructing the Kaprun dams had been completed. The next iteration of the Austrian state would benefit greatly from this investment, as it and a shattered continent looked once again to white coal to provide energy for Europe.[74]

SEVEN

Completing Europe's Battery

IN THE YEARS BEFORE THE Second World War, a consensus in Europe formed that the continent's primary mountain chain should be utilized above all to store hydraulic energy for a burgeoning electric grid. The outlines of this newly created, natural-technological hybrid energy landscape could already be detected prior to 1945. In the 1920s, Germany's western industrial hub established a long-distance link between the power plants of its coalfields and the Alpine reservoirs of far western Austria. Nazi Germany was in the process of constructing a similar connection leading from newly incorporated Austria to the eastern German industrial zone when its plans were interrupted by defeat at the hands of the Allied powers. The maturation of this landscape truly took place, however, in the years and decades after 1945. Not only did the postwar period witness a dramatic increase in the sheer number of Alpine dams being constructed, but the explicit purpose of these dams was not for immediate energy production but for storage. In search of this scarce energy commodity, European states would venture as never before into the highest reaches of the chain. The calculation was in part economic, as a rational electricity supply required reliable inputs of waterpower. But the desire for greater control over nature played a role as well.

In many respects, after 1945 the interwar liberal visions of a Europe bolstered by international economic integration came to pass. In part this was because the defeat of Europe's fascist powers removed a primary obstacle to these visions. But the destruction of the Second World War also created a continent hungry to rebuild its economy. Decisions made after 1945 ensured that reconstruction would aim at establishing continuous economic growth and would be powered by abundant, cheap energy. Completing Europe's mountain battery became a part of the burgeoning Cold War.[1] As other historians have shown, programs like the Marshall Plan were instrumental in "Americanizing" Europe and paving the way for the development of consumer society.[2] The view from the Alps reveals that Americanization impacted the environment as well. It included support in financing the transition to the high-energy society that broke through in Europe in the decades after the Second World War. In this chapter I will trace some of the conditions that encouraged this postwar boom in Alpine dam building, using examples from Austria and Italy to illuminate the most relevant trends, as well as the dangers that attended extracting power from the labile high-altitude environment. The creation of this landscape of energy storage represents one of the most important environmental consequences of postwar European energy development. As such it was part of the foundation for an era of unprecedented, human-driven environmental change many have begun to call the Anthropocene.[3]

White Coal, European Reconstruction, and the Cold War

After the Second World War, the electricity situation in Europe, like the wider economy, was in shambles. In many places the war had actually destroyed power plants and electrical infrastructure. This was particularly the case in industrial western Germany. On the whole though, the Allies had neglected to target the Nazi electricity supply in their strategic bombing campaigns. This fact perplexed Nazi leaders such as Albert Speer, who openly wondered to Allied interviewers why the electricity grid had not been the focus of attacks.[4] In France on the other hand, Resistance sabotage frequently targeted electric infrastructure. By 1945, electric demand in Europe was 50 percent higher than it had been in 1937, whereas the available capacity had actually decreased somewhat. The war economy had also demanded that utilities forgo regular maintenance, meaning that existing electrical equipment was severely degraded. Furthermore, the capacity of the electrical industry to produce the necessary equipment was probably less than it had been before the war, largely because of the loss of Germany's manufacturing resources.[5] While most Alpine dams had been spared destruction, their integration into damaged regional supply systems meant that they too were impacted by the war.

Under these circumstances, white coal would be called upon to play a major role in European economic recovery. In histories of European economic recovery, the energy source coal, and its product steel, have rightfully received a great deal of attention. Coal was the dominant industrial fuel and accounted for a large portion of electric generation. Institutions built to manage international access to this heavy industrial staple would become forerunners to the European Union.[6] It has been less recognized that the prospects of fossil fuel and water were inextricably bound in this setting. Hydroelectricity remained the most important alternative to coal in the European energy system, and substituting water was one certain way to conserve the fossil fuel. Immediately after the war, European states turned to the Alps to make up for shortfalls elsewhere. The state of affairs manifested itself in the Alpine landscape in the form of drained lakes and reservoirs. The situation continued to be so acute in the Alpine area that almost three years after the end of the war, reservoirs were being emptied in the winter to provide emergency energy. In Bavaria, for example, a dry fall in 1948 caused, in the words of the American military governor, a "catastrophic power shortage" that winter. Bavaria passed a law permitting the lowering of its most important reservoir, the Walchensee, to the maximum technically possible. This was 6.6 meters, 2 meters below the previously permissible level. Residents complained about the damages this caused, but American officials could only suggest that the community should appreciate that its sacrifice had enabled thousands of German laborers to work instead of being idle.[7]

No place in the Alps would be as impacted by the postwar push for white coal as Austria. For the newly reformed Second Republic of Austria, hydropower development after World War II held a similar significance as the Tennessee Valley Authority for the United States during the New Deal.[8] That the country existed at all can be chalked up to strategic considerations. Already during the war, the Allies decided to declare Austria the first victim of "Hitlerite aggression." This, it was believed, might encourage

Austrians to fight for the liberation of the country. Though decisive resistance never materialized, the choice set the stage for the state's revival at the end of the war. The Soviet Union took the first step. In April 1945, the Red Army invaded eastern Austria, liberating and occupying Vienna. Late that month the Soviet Union recognized a provisional government of the Austrian Republic. Austria, like Germany, would become subject to Allied occupation. The British, French, and American zones were located in the mountainous southwestern portion of the country, with the US commanding the land's highest peaks in Salzburg province. The Soviet Union controlled Austria's eastern reaches. As in Germany, the capital of Vienna was subdivided between the powers and lay completely within the Soviet zone. This situation would continue until 1955, when the Soviet Union acceded to Austrian independence in exchange for its "eternal neutrality."[9]

To Austria's new leadership it was clear that the country's economic future depended on activating white coal. Hydropower was the only abundantly available source of potential domestic energy. But the small size of the country dictated that it could only be developed with an eye toward exporting some of this energy internationally. The situation in Austria, moreover, was dismal. Available energy equaled just one-quarter of the demand.[10] The Second Republic not only dealt with problems arising from the destruction of electric infrastructure during the war, but the Soviet Union immediately began to dismantle and remove large amounts of industrial equipment in its occupied territory as a form of reparations.

One indication of the importance of developing hydropower for the Austrian state was the creation of the new Ministry for Energy Supply and Electrification by its first democratically elected government. The result of national elections in November 1945, this government, with the conservative Leopold Figl serving as chancellor, was nearly evenly split between the conservatives and socialists. The Austrian Communist Party, which had anticipated a much better result, barely garnered the 5 percent of the vote necessary to enter parliament. Nevertheless, Chancellor Figl awarded the communists the newly created energy post. Figl charged Karl Altmann with improving the electricity situation in Austria, particularly in the critical area around Vienna. In a conference he called immediately after his appointment in early 1946, Altmann expressed the official viewpoint. Quite apart from its quotidian importance, the electricity supply in Austria was simply one of the most important economic assets the country possessed. It was, he explained, "the prerequisite for Austrian economic life at all."[11] As a communist, he also stressed the value of electrification in terms of job creation.[12] With its own means—and support from abroad—he argued that Austria would be able to capitalize on the natural bounty of its hydropower. But the country would have to act. Addressing the widespread skepticism that had accompanied the creation of the First Austrian Republic, Altmann stated, "If Austria has a will to live, then it is viable."[13]

Throughout the Alps, the pressing question reemerged about who would control the bounty of hydraulic energy—the state or private enterprise.[14] As with most Alpine states after the Second World War, the first elected Austrian government immediately opted for public ownership of these resources.[15] In many ways, this was not a departure at all. Throughout the existence of the First Austrian Republic, most of the electricity supply had been under state control anyway. The various Austrian provinces had each created their own utility to direct electrification, and large municipalities like Vienna also possessed their own power plants. The nationalized rail authority also operated its own facilities to provide electric traction on main routes.

But postwar Austria had a unique reason to favor nationalization. As part of Hitler's Greater German Reich, Austria had benefited massively from economic investment. As we have seen, this was also the case in the energy sector, where the Reich-owned Alpen-Elektrowerke in particular funneled enormous sums into the Austrian electricity supply. It had also completed several dams, including part of the gigantic Kaprun project. Now, after the collapse of Nazi Germany, the question of what to do with these so-called German assets arose. In the quest to hold onto the German assets, all of Austria's main political camps recognized that nationalizing them was the surest strategy. The new energy ministry concluded that while the pre-*Anschluss* arrangements would be left alone, a national company of some sort would have to be created to absorb the German assets and to supervise the construction of a national grid.[16] In March 1946 and March 1947, the state followed Altmann's wishes, passing two laws that paved the way for the nationalization of the electricity supply. These measures left electric production and generation overwhelmingly in public control. In the form of the Österreichichische Elektrizitätswirtschafts-AG (commonly known as the *Verbund* or "Grid"), the laws also created an overarching national company to build and operate large power plants, as well as any infrastructure that involved more than one Austrian province.

Austria also knew that it could not achieve its electrification plans without help from abroad. Raising the capital needed to build large dams would be a serious challenge. Additionally, Austria's supply of hydropower far outstripped the small country's demand. It would be reliant on an export market to sell its electricity. Put another way, Austria could not easily take advantage of one of its most important economic assets without coordination with its neighbors. After the destruction of the Second World War, much of Europe found itself in the same boat. This intractable problem was one reason for the sluggish economic growth experienced in the first few years of peace. It is one reason why, when United States Secretary of State George C. Marshall floated his plan for an economic aid program for Europe in the spring of 1947, he insisted that it be cooperative. Immediately, Austria and the rest of Europe spied an opportunity. The foreign ministers of Great Britain and France invited twenty-two European nations to Paris for a meeting in July to formulate a joint European response. Austria eventually accepted the invitation to attend, though not without some trepidation. For the Soviet Union refused to participate in the "imperialistic" program and ultimately forbade its satellite states from being involved as well. Apparently, Energy Minister Altmann did not receive clear messages from Moscow. The Austrian delegation, however, knew it had to proceed cautiously.[17]

On July 12, sixteen governments of western European nations (and the western occupation zones of Germany) met in Paris to negotiate a response to the American offer. In the mind of one Austrian principal, the talks in the summer of 1947 did nothing less than resuscitate the idea of a European grid that had been abandoned with the Great Depression. The Paris talks revealed a substantial energy demand in northwestern Europe. From the Austrian delegation's perspective, nothing could be more obvious than to assign to the Alps in the western part of their country a "central and balancing role in the European electricity supply." In fact, this role would materialize almost automatically, if the longstanding plans to enable an energy exchange from northern Italy to the area of the lower Rhine were realized.[18] For the assembled, Austria had a special mission to fulfill as Europe's grid expanded to meet its burgeoning electricity consumption. The Austrian contingent concluded:

> Austria believes that it will benefit from the economic integration of Europe and its reconstruction with the help of the expected creation of new potential in some sectors of its economy by the Marshall Plan. Austria is aware that one should not just take, but must also give, and it is ready to contribute to the best of its power to the buildup of Europe. This will be possible by harnessing the available waterpower."[19]

When the participating countries reported the results of this meeting to Marshall in the fall of 1947, they declared building dams in the Alps a priority, as increased hydropower capacity would allow for the conservation of coal.[20] After the Marshall Plan passed Congress in the spring of 1948, the United States set about administering the aid program in the participating countries.

At almost precisely the same time as the Marshall meeting was taking place in Paris, in Geneva the United Nations was also zeroing in on the Alps as a solution to Europe's electricity woes. On July 12, the Economic Commission for Europe (ECE) created an Electric Power Commission to study the measures necessary to secure the future supply in Europe. The Electric Power Commission was first among a handful of subsidiary groups created by the ECE.[21] According to its mandate, it should explore "the means, that best allow for the coordinated development of electrical energy in Europe."[22] The electric committee singled out mountain hydro in its early study of the electrical problems in Europe. "Among all European sources of electrical energy white coal is one of the most significant."[23] The group estimated that about half of Europe's electricity came from hydropower and anticipated that its share would grow significantly up until 1955—this for the simple reason that the renewable source was the most promising avenue to satisfy growing demand (Figure 7.1).

The committee identified three main groups of countries producing hydroelectricity. First on this list were the "countries of the Alpine massif," which were responsible for about 55 percent of waterpower in Europe. The Alps were followed by Scandinavia and then the Danube riparian states. In recognition of the mountains' importance, the committee formed an

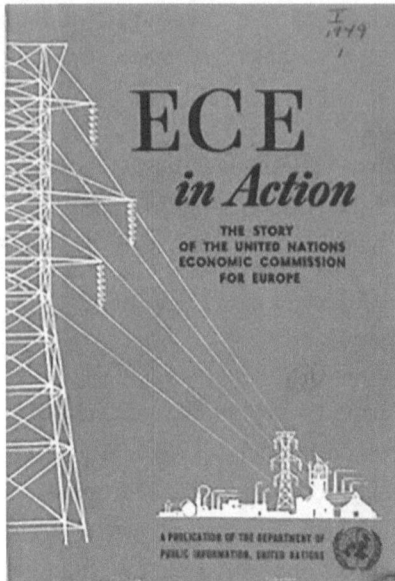

FIGURE 7.1 Economic Commission for Europe. Cover and frontispiece of a 1949 informational booklet on the history of the Economic Commission for Europe, published by the United Nations. After a foreword by Executive Secretary Gunnar Myrdal, an image of the Grimsel Dam in the Bernese Oberland led off the volume. Though the Grimsel was completed already as one of the first high-Alpine dams in the 1930s, the image suggested the type of bold action the commission wanted to take in places like Austria. Increasing the electricity supply clearly stood at the forefront of the ECE's public relations.

Source: United Nations, Department of Public Information, *ECE in Action: The Story of the United Nations Economic Commission for Europe* (Geneva, 1949).

Alps regional study group that met in Lake Como in April of 1948.[24] Though its recommendations were not legally binding, the committee served as a center to promote international electricity exchanges. In its first few years of operation, it facilitated the signing of short-term contracts, most of which involved transmitting excess Alpine summer waterpower to industrial centers in western Germany, Poland, and Czechoslovakia. It also put together a report on which energy sources in the Alps should be developed to meet Europe's future needs.[25]

Within this framework of international cooperation and financing, the

Austrian republic moved to transform its mountain landscapes to produce electricity. The first order of business was to complete those promising projects that had begun during the National Socialist regime, what one engineer characterized as "picking out the raisins."[26] The choicest fruit were the storage dams begun by the Nazis in the Kaprun Valley. Kaprun continued to capture the imagination of Austria's leading energy authorities as both critical for the country's expanding grid and for the symbolic importance that a mighty dam could have for the recovering nation. However, since the Americans had liberated the region in May 1945, the uncertain political situation precluded much progress. It was not until the summer of 1946 that the Austrian government even gained legal control over the construction site, when the United States transferred it and the shares of the Alpen-Elektrowerke that it had seized from the Reich, in trust. The next year, after the nationalization of the electricity supply, a federally owned company called the Tauernkraftwerke AG assumed control of Kaprun as well as responsibility for its completion. The question was how to pay for it.

Here the Marshall Plan made its impact. In Austria, as elsewhere in Europe (particularly West Germany), the overwhelming majority of Marshall Plan money flowed into the electricity supply. In all, Marshall Plan loans would cover 70 percent of investments the country made in its electric infrastructure until 1955.[27] From 1948 until its completion in 1955, Kaprun received almost half of all Marshall Plan funds invested in the Austrian electrical sector, making it by far the largest single project financed in all sectors of the Austrian economy. Financing for the completion of Kaprun's main stage dam came rather easily. The Americans hoped to achieve quick results with their investments and recognized that in order to take full advantage of the capital and material investments made during the Nazi period—it is estimated that the AEW completed about one-third of the entire project—the reservoir had to be finished. In 1952, this portion of the complex went into full operation.

The Austrians later convinced the Americans of the necessity of completing the upper stage dams as well, and their insistence shows how ingrained the idea of utilizing the Alps for storage had become. While the Yankees would have preferred to fund a less expensive run-of-river dam in the lowlands, Austrian leadership insisted on the necessity of storage works to combat the so-called *Winterlücke* ("winter gap") that resulted from reduced streamflows in the wintertime (Figure 7.2). According to one American official, the pricey upper stage only received funding approval because he was in the hospital at the time of the decisive meeting.[28] These additional dams were completed in 1955, and the entire complex received the name Tauernkraftwerke Glockner-Kaprun to emphasize its function in processing the power of Austria's mightiest range and mountain. Since then, many have argued about the significance of the Marshall Plan in Europe's postwar economic expansion. In the case of Kaprun, one expert is certain. Without Marshall aid the dams might never have been raised, and they definitely would have been delayed by at least a decade.[29]

Kaprun would go on to play a crucial role in the history of the Second

FIGURE 7.2   Why save electricity? This poster from a socialist Austrian printing house in 1948, explains that electricity production sinks by two-thirds in winter. Only by completing the storage works would Austrians have the necessary electricity for winter. The poster shows the connection between winter electric deficits and storage dams. In this manner Austrians were encouraged to link dams with future energy abundance.

Courtesy of Salzburger Landesarchiv, Plakats. 601.

Austrian Republic. It was the single most important infrastructure project of the state's early years, supplying the rebuilding society with electricity. The new state and its elites instrumentalized Kaprun as the outstanding symbol of Austrian reconstruction. Particularly in the 1950s and 1960s, school textbooks, newsreels, feature films, cigarette advertisements, and numerous postage stamps commemorated the achievement.[30] These media representations portrayed the dam's completion as a result of Austrian hard work and ingenuity. That the idea for the project had emerged in the time of the First Austrian Republic, that the project had actually commenced when Austria was a part of Nazi Germany, and that construction had depended upon the heavy use of forced laborers, prisoners of war, and American aid, usually went unaddressed. In this manner, Kaprun attained mythical status for the new republic.[31]

The Kaprun dams are concrete proof of the appeal of these structures to all manner of political regimes—and a testament to the significance of electricity as a panacea for a myriad of economic problems.[32] They also demonstrate that history has no shortage of irony. In the postwar period, Austria—with massive assistance from the United States—completed a project begun by National Socialists, and for many of the same reasons. The National Socialists had hoped to enlist the waterpower of the Hohe Tauern to free up critical coal resources for their expansionist struggles. These included, among other things, conquest of *Lebensraum* from the communist Soviet Union in the east. In its bid to vanquish the Soviet Union, Nazi Germany at once ensured the unavailability of the resources necessary to complete the dams and gained access to much of the slave labor that permitted what meager progress they actually achieved with these projects. After 1945, the Americans provided Austria with that which Barbarossa took from the Nazis: materials and time. In Kaprun, Austrians saw a means of stabilizing their electricity supply, improving market prospects for its Nazi-built aluminum industry, and legitimizing the newest iteration of Austrian democracy. For the United States, Kaprun's hydroelectricity was a means to fight communism by bolstering standards of living in Austria and Germany and integrating western European capitalist economies. One final irony was that the Marshall money that poured into Austria basically equaled the value of economic assets that the Soviet Union removed from their occupation zone.[33]

To the south of Austria, Alpine energy was becoming entangled in the Cold War as well. With the defeat of Fascist Italy during the war, the question arose once again about what to do with the Alpine territory of South Tyrol (Südtirol/Alto Adige). This predominantly German-speaking region had formed the southern portion of the Habsburg crownland of Tyrol until it was annexed by Italy after World War I. After 1945 Austria hoped to regain the province. But its leaders recognized that the intense energy development that Italy had launched there during the interwar years made the region even more valuable. Austrian Chancellor Figl proclaimed his country did not want to confiscate the dams that Italians had built, but was more interested in the land's abundant fruits and vegetables—of which he argued the Italians had enough.[34] Italy protested to the Allies that South Tyrol was its most important electricity-producing region after Piedmont and that the country might turn communist with the loss of such economically and politically important real estate. For related reasons, the British Foreign Minister Bevin dashed any hopes of a border change in March 1946. Bevin worried that returning South Tyrol to Austria could become problematic if the latter fell to a Soviet invasion. In this case the Soviets controlling territory south of the Brenner Pass would gain an effective energy lever with which to influence Italian politics. To the Allies, guaranteeing Italy access to its mountain power seemed necessary to stave off communism (Figure 7.3).[35]

Italy's postwar energy hunger resulted in one of the most controversial dam projects ever to appear in the Alps. In 1950, in the upper reaches of the Etsch/Adige River, the Italian chemical giant Montecatini completed its damming of the natural lake Reschensee. The idea of turning this lake into a reservoir stretched all the way back to 1911, but it remained only a vision until the Italian annexation of South Tyrol and the rise of the Fascist state. For damming the Reschensee would result in the flooding of valuable agricultural land and three villages as well. Nevertheless, Benito Mussolini took a personal interest in the scheme, and construction began during the Second World War.

Progress ceased with Italy's exit from the war and the German occupation of northern Italy. In the postwar period, however, Montecatini revived the project with the help of a sizable loan from the Swiss Eletkro-Watt Bank.

FIGURE 7.3 Electric infrastructure of the Montecatini chemical concern. The South Tyrolean power plants are located in the upper righthand corner, in the northern vicinity of the city of Bolzano. They extend to the Reschen (Resia) reservoir and up to the Brenner Pass (Brennero). This cluster and the region's future potential was a main reason that the Allies ensured it remained part of Italy.

Courtesy of Südtiroler Landesarchiv, Abt. 27a (Raumordnung), Nr. 680.

In exchange for financing a quarter of the costs, the Swiss would receive cheap winter energy for a period of ten years. The energy was particularly important, as Swiss citizens had just rejected a similar dam proposal that would have resulted in resettlements. The inhabitants of the doomed South Tyrolean villages were not so lucky. They formed a committee headed by the parish priest to stop the dam. Despite an audience with the pope, however, the reservoir proceeded unabated. The Italian state expropriated the land that would be drowned by the rising lake, compensating some property owners with houses higher up on the hillside in "new" villages. When in 1948 the company began a test to fill the reservoir without warning, local newspapers reported dramatic scenes of inhabitants being surprised by

FIGURE 7.4 South Tyrolean village of Reschen. The first houses are being submerged by the filling of the reservoir. Eventually, a church tower in the neighboring village of Graun would become the only visible evidence of the submerged villages' existence. Throughout the mountain range, dams necessitated the relocation of villages and destroyed valuable agricultural land.

Courtesy of Südtiroler Landesarchiv, Sammlung Ansichtskarten, Nr. 1324.

water in their homes. One village was completely inundated by the new, 22-meter-deep body of water. To this day, the tip of its church tower sticks out as a reminder (Figure 7.4).[36]

Risky Energy

Some of the most dramatic consequences of activating Alpine energy, from a human standpoint, have been the failure of the dams that concentrated mountain waterpower. From the beginning of the dam-building era until the present, the Alpine zone has experienced three catastrophic dam disasters, two of which resulted in over ten fatalities. In 1888 the failure of the small Sonzier reservoir in Switzerland caused several casualties. Then in the 1920s, the Gleno Dam in the Lombard Alps burst after the reservoir's first filling, claiming 600 lives. As the postwar period witnessed the greatest expansion of Alpine hydropower, it is not coincidental that the most severe dam accident in the Alps concerned a structure that had been built in this time period. In 1963, a landslide into the reservoir of the Vajont Dam in the Alps north of Venice triggered the overtopping of that reservoir.[37]

The Vajont disaster stands as one of the worst dam catastrophes to date. The 260-meter-high Vajont Dam belongs to the world's highest and closes off a limestone canyon at the foot of Mont Toce in Italy's Friuli region. Construction on Vajont had begun in 1957 by the Adriatic Electric Company (SADE). SADE had built off of earlier plans for a smaller reservoir to create "il Grande Vajont," a project that aimed to collect the runoff of the Piave and its tributaries to even out dry periods and continually feed two power plants. In the fall of 1960, as the reservoir began to fill for the first time, seismic shocks were recorded, and portions of the gorge began to slide towards the reservoir. Operators partially drained the lake, and the seismic activity almost completely ceased. The reservoir was filled again in the spring of 1962, and the tremors returned. Engineers, however, believed that the rockslide the seismicity set in motion would fall harmlessly into the reservoir.

Heavy rains in the summer of 1963 filled the reservoir to unprecedented heights, and by early September the seismic shocks had increased dramatically. Within the span of a few hours on October 9-10, 1963, the slope's disintegration began to accelerate until a section suddenly broke loose. A 2-kilometer-long piece of the hillside composing some 300 million cubic meters of material slid at high speed into the reservoir. The impact created a wave that crashed first into the opposite side of the valley at a height of 200 meters, laying waste to the banks of the reservoir. Finally, a flood wave overtopped the dam by nearly a hundred meters—the height of a twenty-story building—sending approximately 25 million cubic meters of water into the Piave Valley below. About two minutes later the water reached the nearest village of Longarone, devastating the community and killing nearly all of its inhabitants. Several other villages also washed away. All told, the catastrophe caused over 2,000 deaths. With the exception of some damage at its crown, the dam held, a fact that many engineers at the time interpreted as evidence of the safety of well-founded arch dams.

The accident prompted the evaluation of the integrity of the landscapes around reservoirs throughout the Alps and promoted the study of rock mechanics to better understand the dynamics of landslides. Many suspect that the Vajont episode represents a case of reservoir-induced seismicity (RIS), meaning the existence of the new lake helped trigger earthquakes that led to the landslide. While the science behind RIS remains unclear, most believe that the increased seismic activity near some reservoirs stems from the extra water pressure in subterranean microfissures caused by the added weight of the water. To this day, the Vajont reservoir remains unused.

Italian environmental historian Marco Armiero considers Vajont to be one of the worst catastrophes of twentieth-century Italian history and a clear case of environmental injustice. As he shows, several experts and scores of valley inhabitants warned of possible dangers arising from the locally unstable geology. The most important of these figures was the journalist Tina Merlin. Years before the tragedy, Merlin repeatedly predicted the possibility of landslides. A member of the Italian Communist Party and former partisan, Merlin reported on the efforts of locals who feared the instability of the mountainside surrounding the reservoir. Four years before the disaster she noted in print that the dam stood in an unstable area. In 1961 she declared that the slope was in the process of collapsing, with consequences that could be disastrous. SADE sued the journalist for spreading "false and misleading information," though a court in Milan eventually acquitted her. In 1971, the Italian High Court of Appeals convicted one government and one company engineer, the only two people to be sentenced for wrongdoing in connection with the disaster.[38]

The safety record in the Alps is a sobering one. Outside of China, some 12,000 humans have perished due to dam failures in the twentieth century.[39] The Gleno and Vajont Dam accidents by themselves account for about 20 percent of this figure. In terms of numbers of disasters, there have been about 200 large dam failures since 1900. The International Committee on Large Dams estimates that 2 percent of dams built before 1950 have failed and less than 1 percent of those constructed since then. The Alps show that dam failure is not simply a problem in developing regions.

Conclusion

Kaprun and the Reschensee project are examples of two dams completed in the immediate postwar period. Similar projects could be found in all corners of the Alps and would continue to emerge through the decades of astonishing economic growth that characterized postwar Europe. Early on, projects like those at Kaprun and the Reschensee were justified as necessary for basic economic reconstruction. In many cases, their construction was inextricable from the expanding Cold War, as they benefited from significant American financial support in the form of Marshall and other aid. By the mid-1950s, proponents justified new dams to meet the continually growing hunger for electric power. All the while, the focus shifted ever more towards exploiting mountain topography to impound water. In the period from 1955 until 1970, the number of reservoirs increased by 50 percent while the usable storage capacity of the reservoirs more than doubled.[40]

In this time period the greatest of all high-Alpine reservoirs, the Grand Dixence in the uppermost reaches of the Rhone catchment, emerged. Conceived by western Switzerland's public electric utility, the reservoir was the centerpiece of a complex aimed at storing enormous amounts of high-altitude water for use in the winter. Construction started in 1950 and was completed in 1961. With a storage volume of 400 million cubic meters and a capacity of 2,000 megawatts, the Grand Dixence is by far the largest hydroplant in Switzerland. It also boasts the largest fall in the entire chain. At 285 meters in height, Grand Dixence Dam stood as the world's tallest barrage until displaced by the Nurek in Tajikistan in 1980. It remains the world's largest gravity dam.[41] It and a handful of other less superlative dams of the postwar period represented the finishing touches on the landscape of energy storage Europeans had been theorizing since Aristide Bergès began searching for a high-altitude reservoir for his paper mill at the turn of the twentieth century.

The early 1970s seem to have marked the high tide in Alpine hydropower development, at least for the time being. At this time, the pace of construction of major dams in the mountains slowed considerably. The reasons for this development are complex. From an economic point of view, hydropower became relatively more expensive. The labor and material costs of constructing dams grew along with the general economic upturn in postwar Europe. Some additional outlays came with the rise of environmental resistance to large-scale dam projects. More importantly, burgeoning environmental movements throughout Europe dampened societal enthusiasm for major projects.[42]

Then, just as hydropower was losing some of its competitive advantages in the early 1960s, potential new sources of energy appeared on the horizon. None was more alluring than nuclear power, which had some believing that it would produce electricity so cheap that it would not require metering. The 1961 annual financial statement of a major Swiss engineering firm expressed what many utilities in the Alps believed about the future of white coal: "Due to their less favorable locations, price and wage increases, and additional burdens, the costs of the yet-to-be-built hydroelectric plants have increased in such a way, that the energy from these facilities is scarcely cheaper than that from thermal power plants."[43] In the event, Switzerland did switch to thermal energy, above all nuclear power. The country's first commercial reactor went online in 1969. Germany and especially France joined in this shift to nuclear power.[44] The calculus that had made white coal the most important alternative to coal in electricity production since the turn of the twentieth century had changed. Interestingly, this occurred more or less at the moment that large dam building took off in the developing world. But the era of large dams in the Alps might not yet be a closed chapter of European history.

Conclusion

> As I said this I suddenly beheld the figure of a man, at some distance, advancing towards me with superhuman speed. He bounded over the crevices in the ice, among which I had walked with caution; his stature, also, as he approached, seemed to exceed that of man. I was troubled; a mist came over my eyes, and I felt a faintness seize me, but I was quickly restored by the cold gale of the mountains. I perceived, as the shape came nearer (sight tremendous and abhorred!) that it was the wretch whom I had created.
>
> —*Mary Shelley,* Frankenstein, *vol. II, ch. 2 (1818)*

TO UNDERSTAND THE SIGNIFICANCE of the making of the new Alpine energy landscape, it is helpful to envision how the uplands have influenced the hydrological cycle in Europe at the microscopic level. Since their creation some 100 million years ago, the mountains have directed innumerable single water molecules on epic terrestrial passages. The Alps themselves helped to wring water out of the sky, their peaks forcing the moist air masses passing over Europe to rise and shed some of their liquid freight. A water droplet that fell on the main ridge during winter might take its place beside countless others frozen on a vast Alpine glacier and wait. There it might rest for hundreds of years until a hot summer day, when the water particle might finally leave the snowfields in liquid form. Now, as part of a mountain torrent, the molecule would crash down from the heights, performing imperceptible work all the way. Its voyage might then be interrupted by the cascade's confluence with the slack waters of a mountain lake. Eventually, this bit of water might be escorted out of the uplands by one the region's larger rivers: the Rhône, Rhine, Po, or Danube. It is conceivable that the same single particle may have made this journey several times over the eons. Over the course of the previous millennium, this droplet may have also done its small part to turn one of the many waterwheels then proliferating in the Alps.

Since the beginning of the twentieth century the waypoints along this hypothetical trip have changed drastically, and with them the consequences. While it may have still included a glacial sojourn, a shooting of rapids, and a lake visit or two, water flowing to the sea now encountered more and different obstructions, resulting in different types of work. Water molecules are now far more likely to be waylaid behind a gigantic concrete wall in an artificial reservoir, or to rest a while in a lake modified for storage. Their journey might now include a trip through a tunnel connecting once separate watersheds and a rapid descent in an iron pipe onto the blades of a turbine. Any water molecule that happened to cross paths with a turbine coupled to an electric dynamo kicked off the voyage of another caravan of subatomic particles. For this action set in motion a flow of electrons that might, in the instant after their generation, be transported overland to perform any number of new tasks. These electrons were set loose inside an entirely different network reaching through all corners of Europe. Indeed, to some degree this new grid was created precisely to pool Alpine energy. In this book I have tried to tell the story of the making of this new landscape, how the Herculean efforts expended in making it reflect its centrality in European history. What follows is a consideration of this new landscape's place in the world and the ways in which diverting the hydraulic energy streaming down the mountains have altered the European biome.

The Alpine Damscape

Though their relative importance has perhaps diminished since the early postwar decades, the Alps still play a crucial role in the European electricity supply. For perspective, Europe presently derives 15 percent of its electricity from water. Today Alpine generators make up 40 percent of installed hydropower capacity in the entire European Union, producing around 166 terawatt hours per year. This represents a savings of more than 180 million tons of carbon dioxide annually.[1] To produce this amount of energy would require the most modern of coal-fired plants—which manage to convert only about half of the fuel's thermal energy into electricity—burning 23 million tons of anthracite each year.

But it is not the quantity of white coal that matters so much as its quality. Europeans modified the Alps expressly to store energy, to create a renewable energy source that could be tapped on demand. The figure for the installed capacity of all Alpine hydropower plants is perhaps most important. Mountain water supports the potential for 40,000 megawatts, much of which exists to provide the critical peak power and load balancing that allow Europe's grids to function efficiently. Over twenty different long-distance transmission lines now crisscross the Alps to spread this capacity over the European grid. The Alpine contribution to the European electricity supply is comparable then to other major energy landscapes such as French nuclear power, German coal, and North Sea natural gas.[2]

European ideas about how to use Alpine energy made the mountains one of the premier dam landscapes—or damscapes—on the globe. Until the mid-1990s, three of the world's ten highest dams could be found in the Alps, two of them in Switzerland and one in Italy. For almost two decades, from 1961 until 1980, the Grand Dixence Dam in Switzerland stood as the tallest in the world. It remains the largest gravity dam—holding back water using the weight of its materials. In terms of reservoir volumes and hydroplant capacities, the Alpine dams cannot compare to those that impound the planet's great rivers. By way of comparison, one of the world's largest hydroelectric plants, Itaipú on the Paraná river between Paraguay and Brazil, has an installed capacity of 12,600 megawatts, six times that of Grand Dixence and greater than all of the hydroplants in Switzerland combined.[3] But a main point of this book has been to try to explain that in some ways, comparing Alpine dams to projects like Itaipú or Three Gorges in China is like comparing apples and oranges. In the Alps, Europeans focused not on fashioning the mightiest dams astride its greatest rivers, but on seeing mountain waterways as part of a broader natural system that could be exploited to rationalize energy use and enable greater human control over the hydrological cycle.

To illustrate this contrast, the largest Alpine dams lead the planet in a different metric: reservoir power density. Measured in watts per square meter, this figure indicates how concentrated a power source is in terms of land area. Thanks to their great relief, many Alpine reservoirs surpass 100 watts per square meter.[4] The dense constellation of reservoirs reaching up to the mountain summits is unique worldwide. In New Zealand, for example, dams have been located primarily in the country's pre-Alps. These reservoirs take up an enormous amount of space, but they submerged land that was mostly uncultivated and free of permanent settlements at the time.[5] A quick glance at the North American Rockies reveals another mountain region whose waters have been harnessed in a much different manner than the Alps. Here sparse population, aridity, and a different national energy picture resulted in a completely different hydroscape. In the Alps, a high mountain range that had been settled by agricultural societies for millennia, and one that was located in close proximity to the densest urban and industrial clusters in the world, different circumstances obtained. Unlike many other damscapes, the Alpine one was almost exclusively about power production.[6]

The creation of this Alpine damscape also has a global legacy. Making the Alpine energy landscape created national industries and expertise that states have sought to support by promoting hydro development abroad. One of the most important avenues for doing so is in the provision of "aid" money to finance projects around the globe that otherwise might go unbuilt. Countries like Austria—home to some of world's foremost dam construction firms and equipment suppliers—can lend money for hydro development abroad secure in the knowledge that much of those funds will return to Austria to purchase the necessary goods and services. In the desire to support the domestic firms that cut their teeth building and equipping the dams of the Alps, European governments have subsidized the export of their practices abroad.[7]

Environmental Consequences

Creating this mountain battery entailed massive environmental changes. As Europeans increasingly began to think first in environmental and then ecological terms over the course of the postwar period, they began to identify the changes wrought by hydropower development in the Alps as problematic. An early sign of this concern was the creation in 1952 of the Commission Internationale pour la Protection des Régions Alpines (CIPRA), the first international NGO devoted to protecting Alpine nature. The organization's outlook remained fundamentally similar to nineteenth-century forerunners in that the nature it sought to protect was largely aesthetic.

From the outset, CIPRA possessed a strong organizational network. It had been spun off of the most important worldwide nature conservation organization, the International Union for Conservation of Nature (IUCN), and most of its members were academics or professionals. Nevertheless, as a primarily consultative body in an era when European states strove above all for economic growth, CIPRA's influence was limited. With the rise of a global environmental movement starting in the 1960s, CIPRA shifted to a more ecological approach. By the mid-1980s, CIPRA made its goal the achievement of a circum-Alpine environmental policy with teeth. In 1991 the organization celebrated an important victory when the environmental ministers of the Alpine states signed a framework agreement for the protection of the Alps, broadly known as the Alpine Convention. The Convention enshrined the mountains as one of Europe's largest contiguous natural spaces with a unique culture and history. It also identified the Alps as being threatened ecologically by economic activity and proclaimed the need for harmonization. In the intervening time, implementation of the Convention's measures has been lackluster. But CIPRA and the Convention have lent impetus to diagnosing the environmental problems of the Alps (Figure C.1).[8]

The environmental impacts of the modification of the Alps are a function of the hydrological changes required to store water and generate electricity. The results can generally be distinguished between those impacts generated by the creation of so-called run-of-river schemes, and those caused by storage works.[9] From the perspective of environmental scientists, the consequence of run-of-river dams has been the widespread alteration of floodplain habitats caused by modification of watercourses. Run-of-river works are built directly into the main bed of a river, which then directly drives the turbines. Modifications necessary for run-of-river plants take different forms and have different ecological impacts. One

216   Conclusion

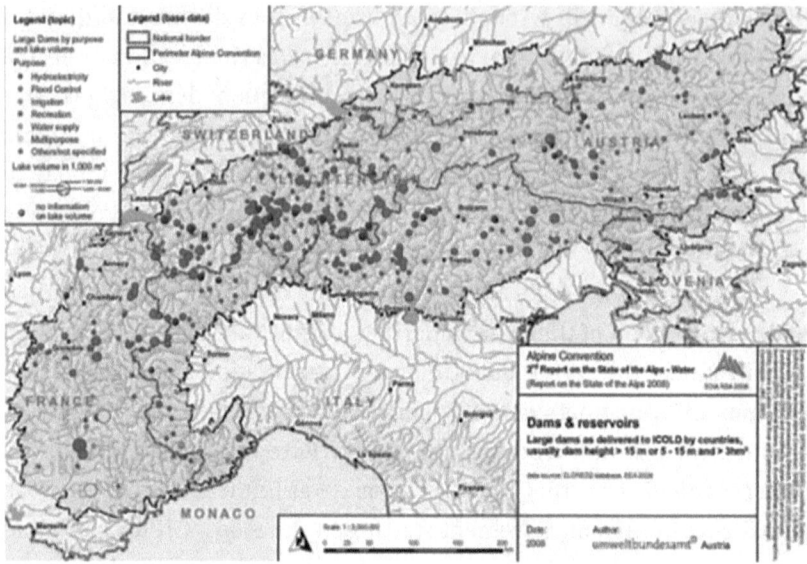

FIGURE C.1   Map of large dams and reservoirs of the Alps. The circles indicate large dams, those over fifteen meters in height or lower ones with significant volume. The map shows how the mountain landscape has been exploited primarily for producing energy; multipurpose dams and other types are comparatively few. This map was produced by the Alpine Convention for its "2nd Report on the State of the Alps—Water (2008)," a publication devoted especially to water issues.

Map produced by Austrian Environmental Ministry and printed by Alpine Convention © Umweltbundesamt/Austria.

broad category is channelization, the forcing of waterways into a straightened streambed to create steeper falls. The river, which may once have had several branches, comes to resemble a single canal. A recent study calculated that approximately 500 kilometers of the largest Alpine rivers (or 4 percent of their previous total) have been lost due to channel straightening. As a result, only 15 percent of Alpine rivers retain their earlier multiple channels and therefore their habitat heterogeneity. Energy development was one of the primary drivers of these changes.[10]

Building a weir or dam into the path of the river of course disrupts its previous flow. The most manifest result is the fragmentation of the waterway, impacting migrating organisms. Today Swiss rivers are blocked by more than 100,000 impediments. In Austria the number is 46,000, and only half of these can be skirted by fish. These obstructions divide the country's watercourses into sections averaging less than a kilometer in length.[11] The interruption of sediment flows by weirs and barrages can also cause rivers downstream of the obstruction to scour into their beds in a process known as incision. As the river deposits some of its sediment after being slowed by a dam, it passes the barrage with increased energy that causes cutting into the channel. In the worst cases, rivers can scour down several meters into their bedrock. Scientists studying the phenomenon in the French Alps note that it represents a reversal of the traditional trend of aggradation: the slow raising of bed level by the deposition of sediment. Particularly by lowering the surrounding water table, incision has considerable impacts on riparian ecosystems, lowering their biodiversity. Interestingly, the greatest impacts are felt in those habitats farthest from direct contact with the running river.[12] Both channelization and damming accelerate flood waves. On the Inn River, for example, the time it takes for a flood to pass has dropped by two-thirds. Retention ponds constructed to compensate for accelerated flooding are prone to sedimentation, thus quickly losing their effectiveness.

Storage works can cause similar disruptions but also come with their own peculiar impacts. By definition they submerge land, thereby altering the distribution of species, nutrient cycles, water temperature, sedimentation, and groundwater levels. Reservoirs change microclimates and can sometimes impact regional climates as well. Researchers have determined that some 12 percent of the courses of larger Alpine rivers are obstructed. This figure must be seen as an absolute minimum estimate, however, as it is difficult to measure how far upstream of an impediment the interference persists. It is also somewhat misleading, for as this study has shown, Europeans purposefully sought to store water at high altitude. Thus reservoir building has focused on the smaller rivers and torrents of the steeper uplands.[13]

Both run-of-river and storage plants rely on water diversions that leave some unfortunate valleys bereft of water—and others with a new surplus. Stretches downstream of the diversion often exhibit extremely low (to nonexistent) discharge rates, which lead to fundamental changes in typical habitats. The long-term result is a change in species composition and ultimately lower biodiversity.[14] One example described in this book is the diversion of the upper Isar River in Germany to fill up the Walchensee reservoir. From the moment the power plant went online in 1924, the riverbed in the "Isar Bend" ran dry for most of the year. In part to make up for the diversion of Isar water from the region, the Bavarian state placed a reservoir in the bend in 1959. The resulting habitat destruction reached all the way up the food chain to affect humans, as the hundred or so inhabitants of the village of Fall were forced to resettle several hundred meters south. Now only during long drought periods or purposeful lowering of the reservoir can the foundations of the old houses be seen.

Taken together, energy development transformed the upper Isar from a typical torrential stream with regular seasonal floods that remade its banks to a drier, less industrious channel. Gone (or severely reduced) are the hardier organisms like the German tamarisk (*Myricaria germanica*) adapted to the previous dynamic. In their place are the stubby grasses and heath shrubs that can thrive in the new, impoverished floodplain. As the years go by, woodlands will claim these locations, reducing biodiversity in the process. Not surprisingly, these interventions have led to the almost complete disappearance of migrating fish and other organisms in the area. A locality that once supported a thriving fishery now primarily furnishes electricity. In an attempt to restore some of the river's previous wild character, in 1990 the Bavarian state allocated a small, token flow be taken away from the Walchenseewerk as a sop for the earlier ecosystem. Hundreds of other catchment basins could testify to similar changes.[15]

The phenomenon of *hydropeaking* also causes important environmental changes. Hydropeaking occurs when a reservoir power plant periodically releases water to cover energy demands. For a period of time, a watercourse's ecosystem experiences a flood that then ceases again. In terms of environmental impacts, it is the frequency and speed ("ramping rate") of peaking that has been found to have the most significant effects. Research has found that organisms are often unable to cope with the initial velocity of the peaking flow and therefore get swept away downstream. When the flood recedes, other species can be left stranded in pools disconnected from the main channel.

Unlike natural floods, hydropeaking is not accompanied by signals—like a rise in groundwater level—that might allow organisms to make behavioral responses. Studies of hydropeaking on the benthic (bottom-dwelling) invertebrate communities of Austrian rivers have found a range of outcomes. On the Drau, hydropeaking caused the complete extinction of all species except those that lived at the deepest points. On the Salzach, total biomass has remained unchanged, but there has been a significant change in species composition. Both of these rivers lay in the catchment of the storage plants (like Kaprun) built in the Hohe Tauern range. In the Bregenzerach feeding into Lake Constance, one finds a reduction of biomass without a change in the taxonomy of species. Fish have been similarly impacted. In the upper reaches of natural rivers, one could expect to find about 170 kilograms of fish per hectare. In the sections of the Bregenzerach affected by hydropeaking, scientists found only 12.4 kilograms.[16]

Hydropeaking also occurs regularly as a cheap means to flush out some of the sediments that build up in reservoirs. An extreme example of what can result came in 2013, when operators at a dam located within the Swiss National Park inadvertently flushed massive amounts of mud from the reservoir into the water downstream. The result was the near total annihilation of the trout population directly below the dam.[17] Alpine reservoirs also cause hydropeaking on much larger time scales. True to many reservoirs' purposes, the dams store summer floodwaters to be released during the winter months. Throughout the Alps then, reservoirs have caused a shift in the seasonal distribution of water. In the Bernese Oberland of Switzerland, the reservoirs of the Grimsel can shift up to 22 percent of the summer runoff to the winter. Biota that cannot adjust to the new rhythms face extinction.[18]

One high-profile impact of the new damscape was the effect on fish populations. Native to the Alpine regions are species that prefer cold, fast, and oxygen-rich waterways. This means minnow, bullhead, barbel, nase, souffie, and above all trout and grayling. The dams, reservoirs, and diversions necessary to extract energy from water altered the conditions in which these organisms thrived. The fate of the huchen (Danube salmon), a species endemic to the Danube catchment and a good indicator of intact aquatic ecosystems, is a case in point. The huchen, which can grow up to 1.5 meters in length, is one of the largest salmonid species worldwide. An apex predator, it is considered the flagship species of Danube tributaries like the Inn, Salzach, and Drau—waterways that we have seen transformed during

Austria's passage through the twentieth century.[19] According to an Austrian study, today the Danube salmon can be found in less than half of its historic habitat. The stretches where it still resides tend to be free-flowing ones not affected by hydraulic modifications.[20]

Most species diminished or disappeared with only local protest, but the vanishing of the iconic Rhine salmon in the twentieth century has provoked widespread debate and some restoration efforts. The Atlantic salmon (*Salmo salar*) that used to populate the Rhine is an anadromous fish species—one that hatches in freshwater rivers, migrates to the ocean, and then returns to its birthplace to spawn. Guided by their sense of smell, these remarkable creatures return to their exact point of origin. Into the late nineteenth century, over 200,000 salmon were harvested annually from the Rhine. But, more than anything else, dredging and dams annihilated the salmon. By the 1930s, the Rhine salmon industry had completely collapsed. The decline was not only due to changes in the Alpine headwaters. In all of its stretches, the Rhine was affected by modern hydraulic engineering to facilitate navigation and pollution by the heavy concentration of industry on its banks. Dam building in the Alps was particularly devastating, as it destroyed the fish's best spawning grounds.

Since the late 1980s, the European Union and other agencies have engaged in efforts to restore the Rhine salmon run. The highest profile of these was Salmon 2000, which sought to identify and protect potential spawning grounds, construct fish ladders to enable migratory species to bypass dams on their runs upstream, and restock the river with eggs from other European rivers. So far, the success has been limited, as evidence suggests that at most 200 fish annually have been able to return to their spawning grounds on the Rhine's lower stretches. No salmon have yet made it to the beginning of the Alpine sections of the Rhine in Basel. But even if the Rhine salmon population were to recover, dams and reservoirs in the Alps would prevent them from reaching their previous spawning grounds in the upper reaches. A mountain range engineered to store hydraulic energy is inimical to migratory species.[21]

As the Rhine salmon and other examples show, there are movements afoot to ameliorate what are seen as the worst excesses of the energy landscape. The catchwords are "revitalization" and "mitigation." While a lot of different activities fall under these rubrics, in general they translate to the removal of engineering works and the restoration of earlier water quality. Terms such as "revitalization" and "renaturation" betray an understanding of previous actions as having deadened waterways and removed them from their natural state. Restorative exertions include the removal of bank fortifications, as along the upper Drau river, to arrest the scouring of the river's bed and allow it to reconnect with its floodplain. Or fish passes along the Isar south of Munich. In western Austria, power companies have changed operations to dampen hydropeaking. In each of these cases, researchers have noted progress in encouraging the return of the Danube salmon.[22] Revitalization efforts are a result of decades of postwar conservation science, which also have informed EU environmental legislation. Policies like the EU's 2000 Water Framework Directive, the most important European water protection legislation, have promoted further actions. In response to it, the Austrian province of Tyrol has launched the largest experiment to mitigate hydropeaking by combining buffer reservoirs and retention basins in conjunction with new facilities. Once again, Europeans are creating new lakes in defiance of hydrological trends. These activities currently being implemented are the most substantial in all of Central Europe.[23]

Centralized electric production also has its environmental impacts. High-voltage transmission lines, for example, intervene substantially into ecosystems. Besides the aesthetic impact, laying transmission lines also requires the clearing of substantial corridors. Since the air between the wires and the ground is intended to function as an additional isolator, vegetation underneath the lines cannot exceed a certain height. Up until the 1980s, herbicides were employed to limit the growth of plants below the lines. On agricultural lands, the impact is restricted to the area around the foundation of the towers. But wherever transmission routes intersect with forests, the consequences multiply. To ensure safe operation, swaths up to seventy meters in width must be cleared through the woods. These lanes make surrounding trees more vulnerable to winds and change the local climate. In Germany, high-voltage wires were also suspected of decimating avian populations—particularly storks—prompting the placement of colorful balls on the lines to catch the birds' attention. It has also recently come to light that the paints used to treat and rustproof the masts contained high quantities of lead that then pollute the ground below. Lead levels in the soil beneath high-voltage towers in RWE's grid, for instance, sometimes contained five times the acceptable amount.[24]

This selective recounting of some of the highest profile changes that have resulted from hydroelectric development can only hint at the magnitude of the impacts wrought in the nineteenth and twentieth centuries. The point is to recognize that increased electricity production can only come at the cost of the environmental changes necessary to redirect energy within this system to new ends. While Alpine dams increased a certain type of energy bounty, they altered dramatically the parameters of a major biome. Europeans created a new landscape of water storage in the Alps and inadvertently—or indifferently—fashioned new ecosystems as well.[25]

The Future of the Mountain Battery

The electric Alps emerged in a particular time period as a result of a specific constellation of social and environmental circumstances. The energy these choices reallocated in turn changed the material basis of European society and enabled even more environmental change. Completion of large dams has waned since the early 1970s, but questions about the future human relationship with the Alps will persist.

There are signs that environmental and political developments might make Alpine water even more significant in coming years. On the environmental side of the ledger, global warming is poised to dramatically alter the hydrology of the mountain range. In fact, the glaciers of the European Alps have experienced accelerated melting since the 1860s. Scientists could not explain this in light of average temperatures, which actually remained relatively low until the 1930s. Recent research has suggested that this melting was a consequence of increased coal soot from the steam engines of the industrial revolution. These black particles absorbed sunlight, melting snow cover and exposing glaciers earlier in the year. In this manner coal-fired industrialization ironically made Alpine water more abundant as an energy source.[26] Nevertheless, the rate of glacial melt has picked up speed in the postwar period. This can be seen in the fate of the Aletsch, the chain's largest glacier. Since 1945 the Aletsch has receded by several kilometers. Whereas local villagers had been holding a papally sanctioned procession against its encroachment since 1678, in 2010 the pope permitted the opposite.[27]

The disappearance of glaciers would mean increased volatility in the flow of Alpine water. As global temperatures continue to rise, scientists predict summers in Europe will be significantly hotter and much drier. Winters, on the other hand, will be somewhat warmer and wetter. The result will be exacerbated drought conditions on Alpine rivers after the snowmelt. This will affect power production for run-of-river dams, but also intensify competition for other uses of water: irrigation, industry, navigation, and drinking water for much of Europe. Water could emerge as *the* critical resource of the Alpine region.[28] Some are already predicting potential water wars among conflicting users—for example, between farmers in the Po Valley and Swiss reservoir operators whose interests are diametrically opposed.[29] Regardless of whether Alpine water becomes most prized for its energy properties, the technical solution to less regular flows would be to find more options for storage. An era of more and taller dams would be a possible outcome.

On the political side, the twenty-first century has been accompanied by discussions of the need for a fundamental energy transition. Much of this discourse has been motivated by the desire to find substitutes for the intensive fossil fuel use that has been causing global warming. Highest on the list of preferable alternatives are so-called green energy sources, primarily wind and solar power. Hydropower is less frequently included this group, due perhaps to a greater familiarity with its negative environmental impacts and the widespread belief that in places like the Alps, there is not much room for growth.

But, as I hope this book has shown, what is considered economical energy development depends very much on prevailing circumstances. These can change dramatically. Higher demand for decarbonized electricity could make increasing rationalization of white coal socially acceptable and profitable. Much of Europe has committed itself to significantly reducing greenhouse emissions through frameworks like the Paris climate agreement in 2015 and the EU's renewable energy directives. Most Alpine states—only France and Slovenia excepted—have also announced their departure from nuclear energy in the wake of the Fukushima disaster (Austria built but never operated its single nuclear power plant). In the non-nuclear Alpine countries, then, the focus will be on renewables, and white coal continues to offer advantages. As I have shown in these pages, modern energy experts have long decried the inconstancy of renewables, and those laments have returned for wind and solar. Hydropower proponents note that presently waterpower makes up some 97 percent of energy storage worldwide, and Alpine reservoirs can once again stabilize a European grid fed by intermittent flows of sun and wind. From a life-cycle perspective, moreover, the durability of dams seems to grant hydroelectricity the lowest carbon footprint of available options—including other renewables.[30]

Global warming, however, is not the only impetus for Europe to wean itself off fossil fuels. Recent geopolitical disruptions have revealed once again some of the same energy insecurities that led to the damming of the Alps in the first place. West-Central Europe has limited fossil fuel supplies, distributed unevenly between numerous states. This is perhaps the major reason that white coal emerged as an alternative source for electricity in the first place. In the postwar period, Europe has been able to meet much of its demand with oil and gas from the Middle East and the Soviet Union/Russia. The Russian invasion of Ukraine in February 2022, however, has called into question the future reliability of a major source of Europe's fossil fuels. Energy independence for West-Central Europe would require more intensive utilization of domestic resources. Here again, the pressure to expand Europe's battery might be irresistible. Looking to the future, one thing is certain: Europeans will continue to decide anew what Alpine water means to them, and how they will coexist with the Alpine landscape.

NOTES

Introduction

1. Marcel Mirande, *Le Comte de Cavour et la Houille Blanche* (Grenoble: Allier Père & Fils, 1927), 5-6. Unless otherwise noted, all translations are my own. The turn of phrase on the waterwheel comes from David Blackbourn, *The Conquest of Nature: Water, Landscape, and the Making of Modern Germany* (New York: W. W. Norton, 2006), 7.

2. Louis André, *Aristide Bergès, une vie d'innovateur: De la papeterie à la houille blanche* (Grenoble: Presses universitaires de Grenoble, 2013), 199-201.

3. A recent consideration of aluminum's role in the twentieth century, however, does make the case for the metal's centrality: Mimi Sheller, *Aluminum Dreams: The Making of Light Modernity* (Cambridge, MA: MIT Press, 2014).

4. Though the term is rarely defined precisely, it commonly refers to the new cluster of industries that revolutionized economies in the latter third of the nineteenth century. Electricity is often grouped with chemicals and steel.

5. Jean-Claude Debeir, Jean-Paul Deléage, and Daniel Hémery, *In the Servitude of Power: Energy and Civilization Through the Ages* (London: Zed Books, 1991); Rolf Peter Sieferle, *The Subterranean Forest: Energy Systems and the Industrial Revolution* (Cambridge, UK: White Horse Press, 2001); Vaclav Smil: *Energy in World History* (Boulder: Westview Press, 1994); James C. Williams, *Energy and the Making of Modern California* (Akron: University of Akron Press, 1997); Alfred W. Crosby, *Children of the Sun: A History of Humanity's Unappeasable Appetite for Energy* (New York: W. W. Norton, 2006); Astrid Kander, Paolo Malanima, and Paul Warde, *Power to the People: Energy in Europe over the Last Five Centuries* (Princeton: Princeton University Press, 2013).

6. Sieferle, *The Subterranean Forest*; Daniel Yergin, *The Prize: The Epic Quest for Oil, Money & Power* (New York: Free Press, 1991); Brian Black, *Petrolia: The Landscape of America's First Oil Boom* (Baltimore: Johns Hopkins University Press, 2000); Alison Fleig Frank,

Oil Empire: Visions of Prosperity in Austrian Galicia (Cambridge, MA: Harvard University Press, 2005); Simon Pirani, *Burning Up: A Global History of Fossil Fuel Consumption* (London: Pluto Press, 2018); Thomas G. Andrews, *Killing for Coal: America's Deadliest Labor War* (Cambridge, MA: Harvard University Press, 2008); William T. Vollmann, *Carbon Ideologies*, 2 vols. (New York: Viking, 2018). A recent spate of studies have focused in particular on coal: Victor Seow, *Carbon Technocracy: Energy Regimes in Modern East Asia* (Chicago: University of Chicago Press, 2021); On Barak, *Powering Empire: How Coal Made the Middle East and Sparked Global Carbonization* (Oakland: University of California Press, 2020); Andreas Malm, *Fossil Capital: The Rise of Steam Power and the Roots of Global Warming* (London: Verso, 2016); Shellen Xiao Wu, *Empires of Coal: Fueling China's Entry into the Modern World Order, 1860–1920* (Stanford, CA: Stanford University Press, 2015).

    7. This is not to say that no research on renewable energy exists. Dams are often explored in histories of water and electrification. But studies focusing on their trajectory as energy sources are rare. Some exceptions are Theodore Steinberg, *Nature Incorporated: Industrialization and the Waters of New England* (Amherst: University of Massachusetts Press, 1994); Matthew Evenden, *Allied Power: Mobilizing Hydro-Electricity During Canada's Second World War* (Toronto: University of Toronto Press, 2015); David Massell, *Amassing Power: J. B. Duke and the Saguenay River, 1897–1927* (Montreal: McGill-Queen's University Press, 2000); Matthias Heymann, *Die Geschichte der Windenergienutzung, 1890–1990* (Frankfurt/Main: Campus, 1995). A recent history of an alternative fuel is Jennifer Eaglin, *Sweet Fuel: A Political and Environmental History of Brazilian Ethanol* (Oxford: Oxford University Press, 2022).

    8. Historians have understood coal to be so unique that its intensive use inaugurated new energy epochs. J. R. McNeill describes the energy regime prior to the Industrial Revolution as "somatic"—coming mostly from bodies. Thereafter he uses the term "fossil fuel age" once the transition to intensive coal use had begun: *Something New Under the Sun: An Environmental History of the Twentieth-Century World* (New York: W. W. Norton, 2000), 10–16. E. A. Wrigley argues that coal brought about an "energy revolution" from an "organic" to an "energy-rich" economy: *Energy and the English Industrial Revolution* (Cambridge, UK: Cambridge University Press, 2010). In his classic study of waterpower in the United States, Louis C. Hunter also chose to frame the development of this energy source against the backdrop of coal's ascendancy: *Waterpower in the Century of the Steam Engine* (Charlottesville: University of Virginia Press, 1979).

    9. Thanks to the importance of storage for energy systems, environmental historians have recently begun exploring the history of batteries: James Morton Turner, *Charged: A History of Batteries and Lessons for a Clean Energy Future* (Seattle: University of Washington Press, 2022).

    10. Richard White influentially advanced the idea of the "organic machine" as the product of human modification of landscapes. In a sense, my work builds on this idea by attempting to characterize the specific type of machine (battery) that Europeans tried to construct in the Alps. Richard White, *The Organic Machine: The Remaking of the Columbia River* (New York: Hill and Wang, 1995). The Alps can also be understood as an "envirotechnical landscape": Sara B. Pritchard, *Confluence: The Nature of Technology and the Remaking of the Rhône* (Cambridge, MA: Harvard University Press, 2011).

11. Brian Black coined the term "petrolia" (*Petrolia*, 2000) to capture the broader "biotic reaction" that occurred thanks to the first oil boom in the United States. Christopher F. Jones referred to "landscapes of intensification" to tease out how the addition of transport infrastructure to an energy source enabled it to be utilized intensively: *Routes of Power: Energy and Modern America* (Cambridge, MA: Harvard University Press, 2014), 6-10.

12. An important theme of this book is that it is in some cases misleading to consider the history of water separately from its surroundings. For this reason, in his "biography" of the Rhine River, author Mark Cioc divides the river into sections based on the prevailing geomorphology: *The Rhine: An Eco-Biography, 1815-2000* (Seattle: University of Washington Press, 2002); rivers are analyzed in connection to cities in Martin Knoll, Uwe Lübken, and Dieter Schott, eds., *Rivers Lost, Rivers Regained: Rethinking City-River Relations* (Pittsburgh: University of Pittsburgh Press, 2017). A recent edited volume on the rivers of the Alps brings together interdisciplinary research on upland waterways and also includes explorations of human interactions with these environments. See, for example, Gertrud Haidvogl, Didier Pont, and Žiga Zwitter, "Geschichte menschlicher Nutzungen und Eingriffe: Alpenflüsse als Ressource und Risiko," in *Flüsse der Alpen: Vielfalt in Natur und Kultur*, eds. Susanna Muhar et al. (Bern: Haupt, 2019), 36-45.

13. Vaclav Smil, *Energy Transitions: History, Requirements, Prospects* (Santa Barbara, CA: Praeger, 2010), 39-42.

14. Vaclav Smil, *Energy at the Crossroads: Global Perspectives and Uncertainties* (Cambridge, MA: MIT Press, 2003), 39.

15. The literature on electrification is vast, and only some key works can be cited here: Thomas P. Hughes, *Networks of Power: Electrification in Western Society, 1880-1930* (Baltimore: Johns Hopkins University Press, 1983); William J. Hausman, Peter Hertner, and Mira Wilkins, *Global Electrification: Multinational Enterprise and International Finance in the History of Light and Power* (Cambridge, UK: Cambridge University Press, 2008); David E. Nye, *Electrifying America: Social Meanings of a New Technology, 1880-1940* (Cambridge, MA: MIT Press, 1990); Andrew Needham, *Power Lines: Phoenix and the Making of the Modern Southwest* (Princeton: Princeton University Press, 2014); Julie A. Cohn, *The Grid: Biography of an American Technology* (Cambridge, MA: MIT Press, 2017); Vincent Lagendijk, *Electrifying Europe: The Power of Europe in the Construction of Electricity Networks* (Amsterdam: Aksant, 2008).

16. Serge Paquier, *Histoire de l'électricité en Suisse: La dynamique d'un petit pays européen 1875-1939*, 2 vols. (Geneva: Ed. Passé Présent, 1998). On France, see *Histoire générale de l'électricité en France*, 3 vols. (Paris: Fayard, 1991-1996). On Italy, see *Storia dell'industria elettrica in Italia*, 5 vols. (Bari: Laterza, 1992-1994).

17. Jones, *Routes of Power*; Casey P. Cater, *Regenerating Dixie: Electric Energy and the Modern South* (Pittsburgh: University of Pittsburgh Press, 2019); Robert D. Lifset, *Power on the Hudson: Storm King Mountain and the Emergence of Modern American Environmentalism* (Pittsburgh: University of Pittsburgh Press, 2014).

18. See, for instance, the special issue on energy transitions: Yves Bouvier and Giovanni Paoloni, eds., "Transitions in Energy History: History in Energy Transitions," special issue, *Journal of Energy History/Revue d'Histoire de l'Énergie*, no. 4 (June 2020).

19. Some scholars challenge the idea of energy transitions in the first place. There is indeed

a lack of clarity about what, precisely, is meant by the term "transitions," and its various uses can lead to confusion. Taken together, these voices are correct that the term should not be adopted uncritically, and that discussing transitions on national or global scales masks local particularities. But some of this fervor is misplaced and seems to rest on the reading that authors using the transition label imply that previous energy sources are abandoned. See, for example, Christophe Bonneuil and Jean-Baptiste Fressoz, *The Shock of the Anthropocene: The Earth, History, and Us*, trans. David Fernbach (London: Verso, 2017), 100–104.

20. Smil, *Energy Transitions*, viii.

21. David Blackbourn, "The Culture and Politics of Energy in Germany: A Historical Perspective," *RCC Perspectives*, no. 4 (2013): 8.

22. Jones, *Routes of Power*, 5.

23. Ian Miller et al., "Forum: The Environmental History of Energy Transitions," *Environmental History* 24, no. 3 (July 2019): 463–533; Blackbourn, "The Culture and Politics of Energy in Germany," 8; Charles-François Mathis and Geneviève Massard-Guilbaud, eds., *Sous le soleil: Systèmes et transitions énergétiques du Moyen Âge à nos jours* (Paris: Sorbonne, 2019).

24. Timothy Mitchell, *Carbon Democracy: Political Power in the Age of Oil* (London: Verso, 2011).

25. Which is not to suggest that only national histories exist. A recent example of a modern European environmental history is Patrick Kupper, *Umweltgeschichte* (Gottingen: Vandenhoeck & Ruprecht, 2021). Historians of technology in the Tensions of Europe network have written transnational European histories through the lens of technology, particularly in their *Making Europe* book series (makingeurope.eu). Per Högselius, Arne Kaijser, and Erik van der Vleuten, *Europe's Infrastructure Transition: Economy, War, Nature* (New York: Palgrave Macmillan, 2016) is especially relevant for hydropower development in the Alps.

26. Georg G. Iggers, *The German Conception of History: The National Tradition of Historical Thought from Herder to the Present* (Middletown, CT: Wesleyan University Press, 1968).

27. On transnational environmental history, see Sterling Evans, "Recent Developments in Transnational Environmental History: Labor, Settler Communities, and Comparative Histories," *Radical History Review* 107 (Spring 2010): 195–208.

28. One aspect of the appeal might be found in their symbolism: Paul Josephson, *Industrialized Nature: Brute Force Technology and the Transformation of the Natural World* (Washington, DC: Island Press, 2002).

29. It should be noted that this history of a transnational landscape of necessity made choices about where to focus. Though the analysis leaves out some national settings—Yugoslavia/Slovenia most notably—I contend that the patterns presented here in general hold true for the entire region.

30. See, for example, David S. Landes, *The Unbound Prometheus: Technological Change and Industrial Development in Western Europe from 1750 to the Present*, 2nd ed. (Cambridge, UK: Cambridge University Press, 2003).

31. Giacomo Parrinello, "Systems of Power: A Spatial Envirotechnical Approach to Water Power and Industrialization in the Po Valley of Italy, ca.1880–1970," *Technology and*

Culture 59, no. 3 (July 2018): 652-688. Parrinello makes this case clearly for Italian industrialization in the Po basin, while also demonstrating its connections to preindustrial waterpower use.

32. Here I borrow a term often utilized in the historiography of Germany and Europe to describe a historical trajectory that diverges from the norm. See, for example, Rolf Peter Sieferle, *Der Europäische Sonderweg. Ursachen und Faktoren* (Stuttgart: Breuninger, 2003).

33. Martin V. Melosi, *Coping with Abundance: Energy and Environment in Industrial America* (New York: Knopf, 1985).

34. Fernand Braudel, *The Mediterranean and the Mediterranean World in the Age of Philip II*, vol. 1 (New York: Harper & Row, 1972), 25-53.

35. Paul Guichonnet, *Histoire et civilisations des Alpes*, 2 vols. (Toulouse: Privat, 1980); Werner Bätzing, *Die Alpen: Geschichte und Zukunft einer europäischen Kulturlandschaft* (Munich: Beck, 2003); Jon Mathieu, *Geschichte der Alpen 1500-1900: Umwelt, Entwicklung, Gesellschaft* (Vienna: Böhlau, 2001). See also the annual journal *Histoire des Alpes—Storia delle Alpi—Geschichte der Alpen* published by the International Association for Alpine History (located at the University of Lugano) since 1995.

36. Marco Armiero, *A Rugged Nation: Mountains and the Making of Modern Italy* (Cambridge, UK: White Horse Press, 2011); Patrick Kupper, *Creating Wilderness: A Transnational History of the Swiss National Park*, trans. Giselle Weiss (New York: Berghahn, 2014); Tait Keller, *Apostles of the Alps: Mountaineering and Nation Building in Germany and Austria, 1860-1939* (Chapel Hill: University of North Carolina Press, 2016); Andrew Denning, *Skiing into Modernity* (Oakland: University of California Press, 2015).

37 Jon Mathieu, *The Alps: An Environmental History*, trans. Rose Hadshar (Boston: Polity, 2019).

38. Leslie Stephen, *The Playground of Europe* (London: Longmans, Green and Co., 1871).

39. Arno J. Mayer, *The Persistence of the Old Regime: Europe to the Great War* (New York: Pantheon, 1981), x.

40. One example is David E. Nye, *Consuming Power: A Social History of American Energies* (Cambridge, MA: MIT Press, 1997).

41. For this reason, a professor at an Austrian mining academy characterized "relief" as a natural resource, no different than iron ore, coal, or oil. Bartel Granigg, *Die Wasserkraftnutzung in Österreich und deren geograpischen Grundlagen* (Vienna: Springer, 1925), 1. The question of the price of hydroelectricity was often advanced as an argument in favor of the source over coal. In general, the most lucrative hydropower sites offered the lowest unit prices for electricity of any energy source. Typically, the main economic disadvantage of white coal was portrayed as its very large upfront costs. By contrast, the operating costs of hydropower facilities were comparatively lower, tipping the scales in favor of hydropower over the long term. For an exemplary discussion of the economics of white coal (and a rare contemporary analysis of hydropower composed by a woman), see Alice Schirmer, "Die schweizerischen Wasserkräfte als wirtschaftliches Gut," PhD dissertation, University of Zurich, 1921.

42. To borrow an influential formulation from E. J. Hobsbawm and T. O. Ranger, *The Invention of Tradition* (Cambridge, UK: Cambridge University Press, 1983). One example of

a history of energy's social and discursive construction is David Gugerli, *Redeströme: Zur Elektrifizierung der Schweiz, 1880-1914* (Zurich: Chronos, 1996).

Chapter 1

1. Terry S. Reynolds, *Stronger Than a Hundred Men: A History of the Vertical Water Wheel* (Baltimore: Johns Hopkins University Press, 1983), 342. On the limitations of waterwheels see Christian Zumbrägel, *"Viele Wenige machen ein Viel": Eine Technik- und Umweltgeschichte der Kleinwasserkraft (1880-1930)*. Paderborn: Ferdinand Schoeningh, 2018), 80-84.

2. Moritz Ruhlman[n], *On Horizontal Water-Wheels, Especially Turbines or Whirl-Wheels . . .* , ed. Robert Kane (Dublin, 1846), 17-18. Quoted in Reynolds, *Stronger*, 342.

3. Anne Dalmasso, "L'ingénieur, la Houille Blanche et les Alpes: Une utopie modernisatrice?" in "Le temps bricolé. Les représentations du Progrès (XIXe-XXe siècles)," Special issue, *Le Monde alpin et rhodanien: Revue régionale d'ethnologie* 29, no. 1-3 (2001): 25-38.

4. Anne Dalmasso, "D'une hydraulique à l'autre: L'évolution des usages industriels de l'eau dans la vallée de l'Arve de la fin du XVIII siècle au début du XX siècle," in *L'eau à Genève et dans la région Rhône-Alpes, XIXe-XXe siècles,* ed. Serge Paquier (Paris: L'Harmattan, 2007).

5. A recent history by Christian Zumbrägel argues convincingly for the remarkable staying power of traditional waterpower technology into the twentieth century. As the author recognizes, however, traditional waterwheels could not compete with turbines in tapping into larger falls. Zumbrägel, *"Viele Wenige machen ein Viel."*

6. Werner Bätzing, *Die Alpen: Geschichte und Zukunft einer europäischen Kulturlandschaft* (Munich: Beck, 2003), 13-24. What separates the Alps from the Apennines, Carpathians, and Balkan chains is another question.

7. Walter Woodburn Hyde, "The Alps in History," *Proceedings of the American Philosophical Society* 75, no. 6 (1935): 432; Jon Mathieu, *Die Alpen: Raum, Kultur, Geschichte* (Stuttgart: Reclam, 2015), 28.

8. Heinz Veit, *Die Alpen: Geoökologie und Landschaftsentwicklung* (Stuttgart: Ulmer, 2002), 16.

9. The next section draws from Bätzing, *Die Alpen*, 25-43; Veit, *Die Alpen*, 16-34; and A. Autran, "Introduction to the Geology of Western and Southern Europe," in *Geology and the Environment in Western Europe: A Coordinated Statement by the Western European Geological Surveys,* ed. G. Innes Lumsden (Oxford: Clarendon Press, 1994), 9-33.

10. Bätzing, *Die Alpen*, 30-31.

11. This section is based on Veit, *Geoökologie*, 73-76, and Bätzing, *Die Alpen*, 190-191.

12. Bätzing, *Die Alpen*, 190.

13. Veit, *Geoökologie*, 76-78.

14. Bätzing, *Die Alpen*, 42.

15. Veit, *Geoökologie*, 82-83.

16. "Niederschrift über die 2. Sitzung des Wasserwirtschaftsrats" (May 28, 1910), Bayerisches Hauptstaatsarchiv (hereafter BayHStA), Landesamt für Wasserwirtschaft Vorl. Nr. 39, 50.

17. On the Isère, see Jacky Girel, "River Diking and Reclamation in the Alpine Piedmont: The Case of the Isère," in *Rivers in History: Perspectives on Waterways in Europe and*

North America, ed. Christof Mauch and Thomas Zeller (Pittsburgh: University of Pittsburgh Press, 2008), 78-88.

18. The previous section and what follows draws from Reynolds, *Stronger*, 9-14, and Vaclav Smil, *Energy in World History* (Boulder: Westview Press, 1994), 103-108.

19. On the geographical basis of waterpower, see Louis C. Hunter, *Waterpower in the Century of the Steam Engine* (Charlottesville: University of Virginia Press, 1979), 114-139; and Raoul Blanchard, "Geographical Conditions of Water Power Development," *Geographical Review* 14, no. 1 (January 1924): 88-90.

20. On the Po River basin, see Fabrizio Frascaroli, Giacomo Parrinello, and Meredith Root-Bernstein "Linking Contemporary River Restoration to Economics, Technology, Politics, and Society: Perspectives from a Historical Case Study of the Po River Basin, Italy," *Ambio* 50 (2021): 493-494.

21. In his discussion of the diffusion of the vertical waterwheel in Europe, Terry S. Reynolds borrows Lynn White's concept of the "Medieval power revolution," where for the first time, complex civilization in Europe rested on non-human energy sources such as waterpower. See Reynolds, *Stronger*, 47-50, 94-95.

22. Reynolds, *Stronger*, 62-63; see also Zumbrägel, "*Viele Wenige machen ein Viel*," 40-50.

23. Peter Kaiser, "Das Wasser der Berge—Bedrohung und Nutzen für die Menschen: Notizen für eine Umweltgeschichte," in *La découverte des Alpes. La scoperta delle Alpi. Die Entdeckung der Alpen*, eds. Jean-François Bergier and Sandro Guzzi (Basel: Schwabe, 1992), 96-97.

24. Alfred W. Crosby, *Children of the Sun: A History of Humanity's Unappeasable Appetite for Energy* (New York: W. W. Norton, 2006), 61-62, 68-70; Smil, *Energy in World History*, 159.

25. Historians have developed conceptual frameworks for understanding energy use in the past. A useful concept for comparing energy use among different societies is the *energy regime*, which simply refers to the collection of energy sources and converters used to get energy into useful form. From a bird's-eye perspective, human history can be broadly divided into two main energy regimes, pre- and post-fossil fuel transition. These epochs have been variously characterized as agrarian/industrial (Sieferle), organic/inorganic (Wrigley), and somatic/exosomatic (McNeill). I prefer the distinction between the fossil fuel age and everything that came before, because it best captures the most important difference about modern, high-energy society: that it has been influenced above all by the extraordinary windfall of coal, gas, and oil. See Rolf Peter Sieferle, *The Subterranean Forest: Energy Systems and the Industrial Revolution* (Cambridge, UK: White Horse Press, 2001), 19-46; J. R. McNeill, *Something New Under the Sun: An Environmental History of the Twentieth-Century World* (New York: W. W. Norton, 2000); E. A. Wrigley, *Energy and the English Industrial Revolution* (Cambridge, UK: Cambridge University Press, 2010), 9-52; Christopher F. Jones, *Routes of Power: Energy and Modern America* (Cambridge, MA: Harvard University Press, 2014), 14-21.

26. Sieferle, *Subterranean Forest*, 94-105.

27. Crosby, *Children of the Sun*, 71-75.

28. The term comes from the seventeenth century *Sylva Subterranea* tract by the German jurist Johann Philipp Bünting. Sieferle, *Subterranean Forest*, 181-184.

29. McNeill, *Something New*, 13.

30. Crosby, *Children of the Sun*, 76-78.

31. J. R. McNeill and Peter Engelke, *The Great Acceleration: An Environmental History of the Anthropocene Since 1945* (Cambridge, MA: Belknap Press, 2014), 10.

32. W. S. Jevons, *The Coal Question* (London: 1865), 145. Quoted in Sieferle, *Subterranean Forest*, 199.

33. Autran, "Introduction," 26.

34. B. Kelk, "Natural Resources in the Geological Environment," in *Geology and the Environment in Western Europe*, 65.

35. Reynolds, *Stronger*, 307; Zumbrägel, "*Viele Wenige machen ein Viel*," 80-84.

36. Blanchard, "Geographical Conditions," 89; Veit, *Geoökologie*, 101.

37. Smil, *Energy in World History*, 108.

38. Jean-Claude Debeir, Jean-Paul Deléage, and Daniel Hémery, *In the Servitude of Power: Energy and Civilization Through the Ages* (London: Zed Books, 1991), 103-105.

39. Hunter, *Waterpower*, 292-342; Reynolds, *Stronger*, 338-349; Smil, *Energy in World History*, 108; Thomas P. Hughes, *Networks of Power: Electrification in Western Society, 1880-1930* (Baltimore: Johns Hopkins University Press, 1983), 263.

40. W. Bernard Carlson, *Tesla: Inventor of the Electrical Age* (Princeton: Princeton University Press, 2013), 41. While Tesla did not design any part of the Niagara plant, he convinced Westinghouse leadership to opt for an alternating current system developed in part to transmit Alpine waterpower (see next chapter).

41. *Elektrotechnische Zeitschrift* 1916, no. 32 (August 10, 1916): 431.

42. Marcel Mirande, *Le Comte de Cavour et la Houille Blanche* (Grenoble: Allier Père & Fils, 1927), 30-31, 33-37. The point of Mirande's essay was indeed to prove that while Cavour had recognized the potential of Alpine waterpower, the count never actually uttered the words "white coal." Mirande determined this burst of creativity remained the merit of his countryman, Aristide Bergès.

43. Louis André, *Aristide Bergès, une vie d'innovateur: De la papeterie à la houille blanche* (Grenoble: Presses universitaires de Grenoble, 2013), 308.

44. R. Avezou, *Petite Histoire du Dauphiné* (Grenoble: B. Arthaud, 1946), 8-10; Andrew Beattie, *The Alps: A Cultural History* (Oxford: Oxford University Press, 2006), 40-41.

45. Reynolds, *Stronger*, 76, 83.

46. Nicholas Shoumatoff and Nina Shoumatoff, *The Alps: Europe's Mountain Heart* (Ann Arbor: University of Michigan Press, 2001), 45; Raoul Blanchard, *Les Alpes françaises* (Paris: Armand Colin, 1952), 35, 146.

47. André, *Aristide Bergès*, 31, 64.

48. "Notes biographiques...," Archives du Musée de la Houille Blanche à Lancey (Conseil Général de l'Isère) (hereafter AMHB), 1A1; André, *Aristide Bergès*, 27-33.

49. "Notes biographiques...," AMHB, 1A1.

50. André, *Aristide Bergès*, 71.

51. As Jones suggests was also the case for energy transitions in the modern United States: *Routes of Power*, 5.

52. André, *Aristide Bergès*, 72.

53. This quote comes from a pamphlet Bergès printed for the white coal exhibit he created for the Exposition Universelle in 1889. AMHB, 1F1.

54. Projet d'association Marmonier-Berges (January 9, 1869), AMHB, 2C38.

55. Marmonier to Bergès (March 5, 1869), AMHB, 2C38.

56. André, *Aristide Bergès*, 94-95.

57. Imbert to Bergès (September 27, 1869), AMHB, 1F2.

58. Marmonier to Bergès (October 30, 1868), AMHB, 2C41.

59. Marmonier to Bergès (January 26, 1870), AMHB, 2C41.

60. F. Trollié, "A Travers la Ville: L'Electricite," *Le Clairon des Alpes* (July 1899).

61. André, *Aristide Bergès*, 99, 319.

62. See the table in Louis André and Serge Benoit, "L'innovation énergétique dans l'industrie papetière, des défibreurs aux hautes chutes," in *Innovation dans la gestion environnementale des territoires de montagne. Actes du 131e Congrès national des sociétés historiques et scientifiques* (Paris: Editions du CTHS, 2009), 44.

63. "La Vallée Lumineuse," *Le Clairon des Alpes* (January 4, 1899).

64. Smil, *Energy in World History*, 181.

65. André, *Aristide Bergès*, 195. Underlined in original document.

66. Luitgard Marschall, *Aluminium—Metall der Moderne* (Munich: Oekom, 2008), 69-83; Walter Wyssling, *Die Entwicklung der schweizerischen Elektrizitätswerke und ihrer Bestandteile in den ersten 50 Jahren* (Zurich: Schweizerischer Elektrotechnischer Verein Fachschriften Verlag, 1946), 215; Cioc, *The Rhine*, 131-132.

67. Trollié, "A Travers la Ville."

68. Trollié, "A Travers la Ville. "

69. For a longer consideration of the social impacts of electrification in one country, see David E. Nye, *Electrifying America: Social Meanings of a New Technology* (Cambridge, MA: MIT Press, 1990).

70. Quotations from G. Clerc, "A. Berges: L'Inventeur de la HOUILLE BLANCHE," *Revue des Alpes: Journal Illustré Hebdomadaire* (December 4, 1892): 391. Found in AMHB, 1E7.

71. F. B., "Actualité," *Le Clairon des Alpes* (November 15, 1897).

72. Syndicat des propriétaires et industriels possédent ou exploitant des forces motrices hydrauliques pour la réunion du Congrès de la houille blanche, *Compte rendu des travaux du congrès, des visites industrielles et des excursions: Congrès de la houille blanche: Grenoble-Annecy-Chamonix, 7-13 Septembre 1902* (St. Cloud: Belin, 1902), 42.

73. AMHB, 1A2.

Chapter 2

1. H. G. Wells, *Anticipations of the Reaction of Mechanical and Scientific Progress upon Human Life and Thought*, 2nd ed. (London: Chapman & Hall, 1902), 239-242, http://www.gutenberg.org/files/19229/19229-h/19229-h.htm

2. For the centrality of transport infrastructure in energy transitions, see Christopher F. Jones, *Routes of Power: Energy and Modern America* (Cambridge, MA: Harvard University Press, 2014).

3. The key example here is Thomas P. Hughes, *Networks of Power: Electrification in*

Western Society, 1880–1930 (Baltimore: Johns Hopkins University Press, 1983). The work starts with Edison and details the history of his illumination systems in Berlin, Chicago, and London before a chapter on hydroelectricity in California.

4. Wilhelm Füßl, *Oskar von Miller 1855–1934: eine Biographie* (Munich: Beck, 2005), 18–19. Much of the following will build off of this excellent biography of Miller.

5. Füßl, *Oskar von Miller*, 40–41.

6. Ludwig Strobel, "Bayerische Geschichte im Donauraum, Wasserbau und Wasserwirtschaft im Lande Bayern—Historische Abrisse, Entwicklungslinien und Wechselbeziehungen," in *Geschichtliche Entwicklung der Wasserwirtschaft und des Wasserbaus in Bayern*, vol. 1, ed. Ludwig Strobel (Munich: Bayerisches Landesamt für Wasserwirtschaft, 1981), 41–42. Indeed in Bavaria and other waterpower regions, the use of traditional waterwheels persisted longer than the historiographical focus on hydroelectricity would suggest. See Christian Zumbrägel, *"Viele Wenige machen ein Viel": Eine Technik-und Umweltgeschichte der Kleinwasserkraft (1880–1930)* (Paderborn: Ferdinand Schöningh, 2018).

7. On Lowell, see Theodore Steinberg, *Nature Incorporated: Industrialization and the Waters of New England* (Amherst: University of Massachusetts Press, 1994).

8. Richard Rhodes, *Energy: A Human History* (New York: Simon & Schuster, 2018), 200.

9. Quote in *Vorträge des Kgl. Direktionsrates Dr. Cassimir "Über die Ausnützung der Wasserkräfte im Auslande und deren Entwicklung nach der technischen, wirtschaftlichen und rechtlichen Seite," ferner "Über den Stand der Ausnützung der Wasserkräfte in Bayern unter Berücksichtigung des Programmes der Verkehrsverwaltung" gehalten am 20. und 27. November 1909 und am 23. April 1910 in den Ministerialbesprechungen des K. Bayerischen Staatsministeriums für Verkehrsangelegenheiten* (Munich: Carl Gerber ?), BayHStA, Ministerium der Finanzen (hereafter MF) 67744.

10. Hughes, *Networks of Power*, 51–52.

11. Oskar von Miller, "Erinnerungen an die Internationale Elektrizitäts-Ausstellung im Glaspalast zu München im Jahre 1882," *Abhandlungen und Berichte des Deutschen Museums*, 4, no. 6 (1932): 154, quoted in Anne-Katrin Ziesak, "Am Vorabend des elektrischen Säkulum. Die Zeit der Ausstellungen 1882–1891," in *Unbedingt modern sein. Elektrizität und Zeitgeist um 1900*, ed. Rolf Spilker (Osnabruck: Rasch, 2001), 26–27.

12. For a company-sponsored history of the AEG, see Peter Strunk, *Die AEG: Aufstieg und Niedergang einer Industrielegende* (Berlin: Nicolai, 1999).

13. The concept of reverse salient comes from Hughes, *Networks of Power*, 33, 83. On the environmental history of the copper mines that provided the critical conductors in the early years of electrification, see Timothy J. Lecain, *Mass Destruction: The Men and Giant Mines That Wired America and Scarred the Planet* (New Brunswick: Rutgers University Press, 2009).

14. According to Richard Hufschmied, the second hydroelectric facility in present-day Austria was built in Vorarlberg under the direction of the industrialist Schindler, who had been living in England and traveled to Paris for the exhibition, coming away fascinated by electric technology. Josef Werndl built the first in March 1880. Richard Hufschmied, "'Weißes Gold' in der Donaumonarchie," in *Wasserkraft. Elektrizität. Gesellschaft. Kraftwerksprojekte ab 1880 im Spannungsfeld*, eds. Oliver Rathkolb et al. (Vienna: Kremayr & Scheriau, 2012), 27–33.

15. The original of the report can be found in the Archiv des Deutschen Museums (DMA), NL 114/Vorl. Nr. 324. In 1932, on the occasion of the fiftieth anniversary of the International Electrical Engineering Exhibition in Munich, the Bavarian minister of the interior presented Miller with a reproduction of the original, which can be found in the library of the Deutsches Museum. Oskar von Miller, *Reise-Bericht des Ingenieur Praktikanten Oscar v. Miller, Theil I*, Archiv des Deutschen Museums (DMA), NL 114/Vorl. Nr. 324.

16. Miller, *Reise-Bericht*, 45-46.

17. Richard Rühlmann, "Marcel Deprez' Versuche über elektrische Kraftübertragung," *Zeitschrift des Vereines deutscher Ingenieure* 29, no. 50 (December 12, 1885): 981.

18. Miller, *Reise-Bericht*, 42, 45-47.

19. For Miller's "social electricity" concept, see Füßl, *Miller*, 140-164.

20. Oskar von Miller, *Die Naturkräfte im Dienste der Elektrotechnik* (Leipzig: F. C. W. Vogel, 1902), 3.

21. Alexander von Urbanitzky, "Neueste Ergebnisse und praktische Fortschritte der Elektrizität: Rückblicke auf die internationale Aufstellung für Elektricität zu München," *Neueste Erfindungen und Erfahrungen auf den Gebieten der praktischen Technik, der Gewerbe, Industrie, Chemie, der Land- und Hauswirtschaft* 10 (1883): 61.

22. Füßl, *Miller*, 52-58.

23. Füßl, *Miller*, 56.

24. Walter Wyssling, *Die Entwicklung der schweizerischen Elektrizitätswerke und ihrer Bestandteile in den ersten 50 Jahren* (Zurich: Schweizerischer Elektrotechnischer Verein Fachschriften Verlag, 1946), ch. 7.

25. Rühlmann, "Marcel Deprez' Versuche," 981; Louis André, *Aristide Bergès, une vie d'innovateur: De la papeterie à la houille blanche* (Grenoble: Presses universitaires de Grenoble, 2013), 153-154.

26. Anna Guagnini, "A Bold Leap into Electric Light: The Creation of the Società Italiana Edison, 1880-1886," *History of Technology* 32 (2014): 167.

27. The biographical notes on Colombo come from Guagnini, "A Bold Leap," 162-164. Edison quote on 170.

28. Guagnini, "A Bold Leap," 159-162.

29. Guagnini, "A Bold Leap," 168. Whether Colombo was at this point aware of the existence of the Italian Edison syndicate is an open historiographical question. Most Italian historians of the Edison company seem to think he was the mastermind; Guagnini suggests that he was not yet in the loop.

30. Vaclav Smil, *Energy and Civilization: A History* (Cambridge, MA: MIT Press, 2018), 263.

31. Thomas Hughes argues that Gaulard and Gibbs should be seen as most responsible for the system that overcame the transmission problem. See Hughes, *Networks of Power*, 86-105. For further agreement on this point, see also Smil, *Energy and Civilization*, 263.

32. "Gaulard and Gibbs' System of Electrical Distribution," *Engineering* 35 (1883): 205-206.

33. Thanks to the Museo Torino whose website includes a section on the 1884 exhibition replete with fully searchable digitized copies of materials related to the fair: www.museotorino.it

34. Torino e l'Esposizione Italiana del 1884, "I Premi Decrati Dal Re," in *Cronaca Illustrate della Esposizione Nazionale-Industriale ed Artistica del 1884* (Turin: Roux E Favale & Fratelli Treves, 1884), 54.

35. Esposizione Generale Italiana in Torino 1884, *Catalogo Ufficiale della Meccanica Agraria, Elettricità e Meccanica di Precisione* (Turin: Unione Tipografico, 1884), 11, 21–23.

36. G. Colombo, "Le Système Gaulard et Gibbs à l'Exposition de Turin," *La Lumière Électrique* 14, no. 41 (October 11, 1884): 44. The demonstration was so important that the main French electrical journal, *La Lumière Électrique*, devoted the first pages of this issue to a discussion less than a month after the exhibition.

37. Hughes, *Networks of Power*, 94.

38. W. Bernard Carlson, *Tesla: Inventor of the Electrical Age* (Princeton: Princeton University Press, 2013), 26, 85–90, 158–175.

39. Giacomo Parrinello, "Systems of Power: A Spatial Envirotechnical Approach to Water Power and Industrialization in the Po Valley of Italy, ca.1880–1970," *Technology and Culture* 59, no. 3 (July 2018): 662–664.

40. Füßl, *Miller*, 120.

41. Füßl, *Miller*, 123.

42. Losses under 30 percent were considered impressive. One could say in this case, the economy at work was that it was cheaper to ruthlessly exploit as much power as possible at a remote Alpine torrent and transmit this—at considerable loss—than to have more steam engines located throughout a city or attempt to transmit direct current through large-diameter cables at distance.

43. Description of the plant from Alfred Lambotte, *Quelques applications de l'életrotechnie dans l'Europe centrale: Notes de voyage par Alfred Lambotte* (Brussels: Ramlot, 1907). See also Parrinello, "Systems of Power," 665.

44. Pierre Leroy-Beaulieu, "La mise en valeur des forces hydro-électriques el les modifications législatives," *Economist français* (May 14, 1904), cited in Theodor Köhn, "Einige allgemeine Betrachtungen über den Ausbau von Wasserkräften," *Die Weisse Kohle* 1, no. 2 (January 25, 1908): 5.

45. Von Liebig, "Über die Bedeutung der Industrie für die Volks- und Staatswirtschaft und die Ausnützung der staatlichen Wasserkräfte Bayerns," *Die Weisse Kohle* 1, no. 2 (January 25, 1908): 5.

46. Jakob Zinßmeister, "Was die neue Zeitschrift will?" *Die Weisse Kohle* 1, no. 1 (January 10, 1908): 1.

47. F. Kreuter, "Bericht zur Sitzung im K. Staatsministerium des Innern am 19. Januar 1907" (January 19, 1907), BayHStA, Landesamt für Wasserwirtschaft Vorl. Nr. 180.

48. Cassimir, *Vorträge*, BayHStA, MF 67744, 23–25.

49. Kretzschmar, "Die Mächtigkeit der weißen Kohle," *Die Weisse Kohle* 1, no. 11 (June 10, 1908): 125.

50. Wilhelm Ostwald, *Energetische Grundlagen der Kulturwissenschaft* (Leipzig: Werner Klinkhardt, 1909), 13, 25; Rolf Peter Sieferle, *The Subterranean Forest: Energy Systems and the Industrial Revolution* (Cambridge, UK: White Horse Press, 2001), ix.

51. See, for instance, Bericht der Generaldirektion der K.B. Staatseisenbahnen (October 12, 1905), BayHStA, MF 677742.

52. Theodor Rehbock, "Der wirtschaftliche Wert der binnenländischen Wasserkräfte, unter besonderer Berücksichtigung des Großherzogtums Baden," *Die Weisse Kohle* 1, no. 3 (February 10, 1908): 3.

53. Cited in K. Oberste Baubehörde, *Die Wasserkräfte Bayerns* (Munich: Piloty & Loehle, 1907), 37-38.

54. This question of the connection between energy systems and political power has been explored by Timothy Mitchell, *Carbon Democracy: Political Power in the Age of Oil* (London: Verso, 2011); Bruce Podobnik, *Global Energy Shifts: Fostering Sustainability in a Turbulent Age* (Philadelphia: Temple University Press, 2006).

55. In France, for example, the distribution of electricity into the homes of craftsmen was promoted to rescue the artisanal textile industry. Philanthropic associations also formed to assist artisans in acquiring electric equipment for their trades. See Cassimir, *Vorträge*, BayHStA, MF 67744, 8.

56. Füßl, *Miller*, 144

57. This discussion of nationalization is based on the chapter on waterpower nationalization in K. Oberste Baubehörde, *Die Wasserkräfte Bayerns*, 54-64.

58. K. Oberste Baubehörde, *Die Wasserkräfte Bayerns*, 59-60.

59. K. Oberste Baubehörde, *Die Wasserkräfte Bayerns*, 60-61.

60. K. Oberste Baubehörde, *Die Wasserkräfte Bayerns*, 56-57.

Chapter 3

1. Reinhard Falter, "Achtzig Jahre 'Wasserkrieg': Das Walchensee-Kraftwerk." In *Von der Bittschrift zur Platzbesetzung: Konflikte um techn. Großprojekte; Laufenburg, Walchensee, Wyhl, Wackersdorf*, ed. Ulrich Linse (Berlin: Dietz, 1988), 66.

2. On the history of batteries, see James Morton Turner, *Charged: A History of Batteries and Lessons for a Clean Energy Future* (Seattle: University of Washington Press, 2022).

3. Raoul Blanchard, "Geographical Conditions of Waterpower Development," *Geographical Review* 14, no. 1 (January 1924): 89-90.

4. This argument refers to the influential viewpoint in James C. Scott, *Seeing Like a State: How Certain Schemes to Improve the Human Condition Have Failed* (New Haven: Yale University Press, 1998). Scott argues that grand technological projects often fail because the states that initiate them make nature "legible" by simplifying it, leading to unintended consequences. In the case of the Walchensee, the Bavarian state's initial moderate approach to the scheme was actually abandoned in the face of public clamor for a superlative project.

5. Since Thomas Hughes's *Networks of Power* (*Networks of Power: Electrification in Western Society, 1880-1930* [Baltimore: Johns Hopkins University Press, 1983]), the focus has been on the evolution of urban illumination systems into regional power grids. The smaller grids that powered electrified rail networks, now the rule in Europe, have not received as much attention. This history of transport electrification will be ever more relevant as the world contemplates decarbonizing this sector.

6. Andreas Malm, "Fleeing the Flowing Commons: The Expansion of Waterpower That Never Happened," in *Fossil Capital: The Rise of Steam Power and the Roots of Global Warming* (New York: Version, 2016), 196-220.

7. This and the following paragraph based on Heinz Veit, *Die Alpen: Geoökologie und Landschaftsentwicklung* (Stuttgart: Ulmer, 2002), 86.

8. Bergès to Isère Prefecture (November 15, 1887), AMHB, 2C.53.

9. Syndicat des propriétaires et industriels possédent ou exploitant des forces motrices hydrauliques pour la réunion du Congrès de la houille blanche, *Compte rendu des travaux du congrès, des visites industrielles et des excursions: Congrès de la houille blanche: Grenoble-Annecy-Chamonix, 7–13 Septembre 1902* (St. Cloud: Belin, 1902), 223.

10. K. Oberste Baubehörde, *Die Wasserkräfte Bayerns* (Munich: Piloty & Loehle, 1907), 119–121.

11. "Das Wasserkraft-Elektrizitätswerk Robbia der Kraftwerke Brusio," *Zeitschrift des Vereines deutscher Ingenieure* 54, no. 50 (December 10, 1910): 2115.

12. Background on the Walchensee from Falter, "Achtzig Jahre," 63, 66.

13. Albert Schmidt, "Das Schicksal und die Zukunft des Walchensees und der Isar," *Münchner Neueste Nachrichten* (December 5, 1907).

14. "Niederschrift über die Beratungen der Kommission zur Prüfung der Wirkungen des Walchenseeprojektes auf Hygiene und Fischerei" (March 4, 1910), BayHStA, Landesamt für Wasserwirtschaft Vorl. Nr. 180.

15. Wilhelm Volkert, *Geschichte Bayerns*, 5th ed. (Munich: Beck, 2017), 13–23.

16. The district construction office and the government forestry department both initially rejected the plan. See the comments in "Wasserkraftgewinnung durch Überleitung der Isar in den Walchensee und Erbauung eines Elektrizitätswerk in Kochel" (January 16, 1905), BayHStA, Minsterium der Finanzen (MF) 67742.

17. "Wasserkraftgewinnung durch Überleitung der Isar in den Walchensee und Erbauung eines Elektrizitätswerk in Kochel" (January 3, 1905), BayHStA, MF 67742.

18. Hensel, "Gutachten des Hydrotechnischen Bureaus" (February 10, 1905), BayHStA, Landesamt für Wasserwirtschaft Vorl. Nr. 188.

19. "Wasserkraftgewinnung durch Überleitung der Isar in den Walchensee und Erbauung eines Elektrizitätswerk in Kochel" (January 3, 1905), BayHStA, MF 67742.

20. Universität Zürich, "Fischer-Reinau, Ludwig," *Matrikeledition*, http://www.matrikel.uzh.ch/active/static/6132.htm

21. Ludwig Fischer-Reinau, *Die Wasserkräfte der bayerischen Alpen* (Munich: Süddeutsche Verlags-Anstalt, 1906), 3.

22. Fischer-Reinau, *Wasserkräfte der bayerischen Alpen*, 4.

23. Artur Budau, "Über hydraulische Akkumulierungsanlagen bei Kraftwerken," *Zeitschrift des österreichischen Ingenieure- und Architekten Vereines*, no. 11 (March 13, 1908): 169.

24. Budau, "Uber hydraulische Akkumulierungsanlagen," 169. Connecting European industry and culture was common among engineers. Cf. Oskar von Miller, *Die Naturkräfte im Dienste der Elektrotechnik* (Leipzig: F. C. W. Vogel, 1902), 11.

25. Fischer-Reinau, *Wasserkräfte der bayerischen Alpen*, 4–5.

26. For a discussion of interurbans, see David E. Nye, *Electrifying America: Social Meanings of a New Technology, 1880–1940* (Cambridge: MIT Press, 1990), 119–122.

27. Wolfgang König, "Bergbahnen in den Alpen (1870–1940): Zwischen Naturer-

schließung und Naturbewahrung," in *Umwelt-Geschichte: Arbeitsfelder-Forschungsansätze-Perspektiven*, eds. Sylvia Hahn and Reinhold Reith (Munich: Oldenbourg, 2001).

28. History of electrification from Emil Huber-Stockar, *Elektrifizierung der Schweizerischen Bundesbahnen bis Ende 1928. Neujahrsblatt der Naturforschenden Gesellschaft in Zürich auf das Jahr 1929* (Zurich: Beer, 1929), 8.

29. "Memorial betreffend die Gründung eines Studien-Komitees für den Elektrischen Betrieb der Schweizerischen Normalbahnen," Schweizerisches Bundesarchiv (hereafter BAR), E8100A#1974/88#54*, Az. 15.01, Drucksachen der Gründung, 1902–1905.

30. Letter Post- and Eisenbahndepartment to Schweizerischer Bundesrat (June, 18, 1904), BAR, E8001B-01#1000/1131#173#, Az. 15.02, Studienkommission für elektrischen Bahnbetrieb.

31. Letter Oskar von Miller to Frauendorfer (November 6, 1903), BayHStA, Verkehrsarchiv (hereafter VArch) 8691.

32. Richard Hufschmied, "'Weißes Gold' der Donaumonarchie," in *Wasserkraft. Elektrizität. Gesellschaft. Kraftwerksprojekte ab 1880 im Spannungsfeld*, eds. Oliver Rathkolb et al. (Vienna: Kremayr & Scheriau, 2012), 51–52; K. Oberste Baubehörde, *Die Wasserkräfte Bayerns*, 111, 144.

33. Miller, *Die Naturkräfte*, 6.

34. Walter Wyssling, ed., *Mitteilungen der Schweizerischen Studienkommission für elektrischen Betrieb*, vol. 1: *Der Kraftbedarf für elektrischen Bahnbetrieb* (Zurich: Rascher & Cie., 1906), 1.

35. C. Hartl, *Bayern auf dem Weg zum Industriestaat: Eine vergleichende volkswirtschaftliche Studie über die Ausnützung der bayer. Wassekräfte, sowie über Staats- und Privatbetrieb in den Industrien der schwarzen und der weißen Kohle. Zugleich ein Beitrag zur Kartellfrage* (Munich: Max Steinebach, 1911), 38.

36. Wyssling, *Der Kraftbedarf*, 22.

37. Wyssling, *Der Kraftbedarf*, 23.

38. Wyssling, *Der Kraftbedarf*, 24.

39. In the Austrian half of the Dual Monarchy, for instance, an "Office for the Preparation of Electrical Operation on State Railways" was added to the Railway Ministry in 1905. Hufschmied, "'Weißes Gold' der Donaumonarchie," 51.

40. Fischer-Reinau, *Wasserkräfte der bayerischen Alpen*, 9.

41. Fischer-Reinau, *Wasserkräfte der bayerischen Alpen*, 16, 23–24.

42. "Bericht der Generaldirektion der K.B. Staatseisenbahnen vom 12. Oktober 1905," BayHStA, MF 67742, 1.

43. "Bericht der Generaldirektion," BayHStA, MF 67742, 1.

44. The influence of Fischer-Reinau is made explicit in Oberste Baubehörde to Hydrotechnisches Bureau (July 11, 1906), Untersuchung über die Ausnützung der staatlichen Wasserkräfte in Bayern, BayHStA, Landesamt für Wasserwirtschaft Vorl. Nr. 27.

45. As argued in Scott, *Seeing Like a State*.

46. Though ideas to drain this area go back to antiquity, a renewed and intense discussion reemerged in the modern period. After Donat claimed to be the inventor of the idea to drain the marshes, the Bavarian legation in Rome looked into the question. It learned

that Donat had a claim to have originated the idea, but his actual proposal was impossible. See "Niederschrift über die am Montag, den 5. März 1906 im K. Staatsministerium des Innern stattgehabte Besprechung über die Ausnützung der staatlichen Wasserkräfte," 41, BayHStA, Landesamt für Wasserwirtschaft Vorl. Nr. 27.

47. See Ludwig Mühlhofer and Carl Reindl, *Das Achensee-Kraftwerk*, Special issue, *Wasserkraft und Wasserwirtschaft* [1928?]. In 1927, the Achensee began serving as a reservoir for a hydropower plant.

48. Fedor Maria von Donat, *Die Kraft der Isar, eine Quelle des Reichtums für Staat und Volk* (Munich: Lindauer, 1906), 4.

49. Donat, *Die Kraft der Isar*, 22.

50. Donat, *Die Kraft der Isar*, 27.

51. David Blackbourn disagrees with historian David Nye's contention that the American reaction to structures such as the Hoover Dam represent another instance of American exceptionalism. See David Blackbourn, *The Conquest of Nature: Water, Landscape, and the Making of Modern Germany* (New York: W. W. Norton, 2006), 195–196.

52. "Der Kampf um die Isar," *Münchner Neueste Nachrichten* (February 27, 1907).

53. On Intze see Blackbourn, *Conquest of Nature*, 198–210.

54. Oberste Baubehörde Evaluation (January 31, 1905), BayHStA, VArch 10543.

55. "Ausnützung staatlicher Wasserkräfte, hier Immediateingabe des K. Majors a.d. von Donat die Kraft der Isar betreffend," K. Staatsministerium des Innern to Königliche Hoheit Prinzen Luitpold (March 13, 1907), BayHStA, MF 67742.

56. K. Oberste Baubehörde, *Die Wasserkräfte Bayerns*, ii.

57. K. Oberste Baubehörde, *Die Wasserkräfte Bayerns*, 501.

58. Thomas M. Lekan, *Imagining the Nation in Nature: Landscape Preservation and German Identity, 1885–1945* (Cambridge, MA: Harvard University Press, 2004).

59. Contained in that peculiarly German notion of *Heimat* or "homeland": Celia Applegate, *A Nation of Provincials: The German Idea of Heimat* (Berkeley: University of California Press, 1990).

60. Falter, "Achtzig Jahre."

61. Schmidt, "Das Schicksal."

62. Report F. Kreuter, Zur Sitzung in K. Staatsministerium des Innern am 19. Januar 1907 Betreffend: die Ausnützung der staatlichen Wasserkräfte in Bayern, hier Ausnützung der oberen Isar mit Zuhilfnahme des Walchensees, BayHStA, Landesamt für Wasserwirtschaft Vorl. Nr. 180.

63. Renzo Boccardi, "Un occhio azzurro si spegne," *Rivista mensile del Touring Club Italiano* 19 (1913): 429. Found in James Sievert, *The Origins of Nature Conservation in Italy* (Bern: Peter Lang, 2000), 95.

64. "Niederschrift über die am Samstag, den 19. October 1907 im K. Staatsministerium des Innern stattgehabte Besprechung über die Ausnützung der staatlichen Wasserkräfte," BayHStA, MF 67742, 36–37.

65. For records on the competition, see the file BayHStA, VArch 10545.

66. "Niederschrift über die Beratungen der Kommission zur Prüfung der Wirkungen des Walchenseeprojektes auf Hygiene und Fischerei" (March 4, 1910), BayHStA, Landesamt für Wasserwirtschaft Vorl. Nr. 180.

67. Jon Mathieu, *Die Alpen: Raum, Kultur, Geschichte* (Stuttgart: Reclam, 2015), 170. Giacomo Bonan portrays this situation in detail for the Piave. See Giacomo Bonan, "An Alpine Energy Transition: The Piave River from Charcoal to 'White Coal,'" *Environmental History* 25, no. 4 (October 2020): 693–696.

68. Niederschrift des Wasserwirtschaftsrates, May 28, 1910, BayHStA, Landesamt für Wasserwirtschaft Vorl. Nr. 39, 20.

69. "Niederschrift über die Versammlungen in Tölz und Kochel am 10. Juli 1910 das Walchenseeprojekt betreffend," BayHStA, VArch 10510.

70. Landesausschuss für Naturpflege to Königliches Staatsministerium des Innern, September 21, 1909, BayHStA, VArch 10547.

71. See for instance the front page of the Sunday morning edition of one of Germany's most important national papers, "Die Opposition gegen das Walchenseeprojekt," *Frankfurter Zeitung und Handelsblatt* Nr. 15 Sonntag (May 31, 1908).

72. Fritz Blaich, *Die Energiepolitik Bayerns, 1900–1921* (Kallmünz/Opf: Lassleben, 1981).

73. E. Auer, *Das Bayernwerk und sein Zusammenhang mit dem Walchenseewerke* (Stuttgart: J. H. W. Dietz, [1918?]), 3. Found in BayHStA, Ministerium für Wirtschaft, Infrastruktur, Verkehr und Technologie (hereafter MWi) 2881.

74. The suggestion was made in a pamphlet published under a pseudonym Wigulè Kreittmayr II, *Ministerium Hertling Weg oder ein Angriff auf's Bayerische Wespennest. Eine hochpolitische Lektüre aus einer kleinen Zauberstube* (Munich: F. Bosch, 1912).

75. Auer, *Das Bayernwerk*, 3.

76. On the debates in the upper chamber, see Bernhard Löffler, *Die bayerische Kammer der Reichsräte 1848 bis 1918: Grundlagen, Zusammensetzung, Politik* (Munich: Beck, 1996); K. Bayerisches Staatsministerium für Verkehrsangelegenheiten, *Bericht über den Stand der Staatseisenbahnverwaltung vorbehaltenen staatlichen Wasserkräfte* (Munich: C. Wolf & Sohn, 1914).

Chapter 4

1. Louis-Jules Arrigon, *La Houille Blanche et l'Avenir Industriel du Sud-Est* (Paris: Attinger, 1918), 5, 8–9.

2. Quoted in Daniel Yergin, *The Prize: The Epic Quest for Oil, Money & Power* (New York: Free Press, 1991), 167.

3. These points made in Dan Tamïr, "Something New Under the Fog of War: The First World War and the Debut of Oil on the Global Stage," in *Environmental Histories of the First World War*, eds. Richard P. Tucker et al. (Cambridge, UK: Cambridge University Press, 2018), 117–135.

4. Walter Wyssling, *Die Entwicklung der Schweizerischen Elektrizitätswerken und ihrer Bestandteile in den ersten 50 Jahren* (Schweizerischer Elektrotechnischer Verein: Zurich, 1946), 510, 533.

5. Some key overview works are Edmund P. Russell, *War and Nature: Fighting Humans and Insects with Chemicals from World War I to Silent Spring* (Cambridge, UK: Cambridge University Press, 2001); Richard P. Tucker and Edmund Russell, eds., *Natural Enemy, Natural Ally: Towards an Environmental History of Warfare* (Corvallis: Oregon State University Press, 2004); Charles E. Closmann, ed., *War and the Environment: Military Destruction*

in the Modern Age (College Station: Texas A&M University Press, 2009); Chris Pearson, Peter Coates, and Tim Cole, eds., *Military Landscapes: From Gettysburg to Salisbury Plain* (London: Continuum, 2010).

6. An excellent, agenda-setting collection of essays is Tucker et al., *Environmental Histories of the First World War*. See also Dorothee Brantz, "Environments of Death: Trench Warfare on the Western Front, 1914-1918," in Closmann, *War and the Environment*. The leading historian of war and the environment in France has considered the impacts of trench warfare and remaking of postwar landscapes but not the Alps: Chris Pearson, *Mobilizing Nature: The Environmental History of War and Militarization in Modern France* (Manchester: Manchester University Press, 2012). An early environmental interpretation of the war is Avner Offer, *The First World War: An Agrarian Interpretation* (New York: Oxford University Press, 1989).

7. Hermann J. W. Kuprian and Oswald Überegger, eds., *Der Erste Weltkrieg im Alpenraum. Erfahrung, Deutung, Erinnerung. La Grande Guerra nell'arco alpino. Esperienze e memoria* (Innsbruck: Wagner, 2006), 11.

8. One exception is Tait Keller, "The Mountains Roar: The Alps During the Great War," *Environmental History* 14 (April 2009): 253-274.

9. Since Thomas Hughes, *Networks of Power*, the First World War has mostly been understood as a turning point within the history of electrification when the idea of "centralization" won out: *Networks of Power: Electrification in Western Society, 1880-1930* (Baltimore: Johns Hopkins University Press, 1983).

10. See the discussion in David Blackbourn, *The Conquest of Nature: Water, Landscape, and the Making of Modern Germany* (New York: W. W. Norton, 2006), 228.

11. Arrigon, *Houille Blanche*, 7.

12. This history draws from Raoul Blanchard, *Les Forces Hydro-Électriques Pendant la Guerre* (New Haven: Yale University Press, 1924), 18-26.

13. Blanchard, *Les Forces*, 7-9.

14. Arrigon, *Houille Blanche*, 36-39, 53-54.

15 Arrigon, *Houille Blanche*, 38.

16. Arrigon, *Houille Blanche*, 31-33.

17. Blanchard, *Les Forces,* 31-35.

18. Blanchard, *Les Forces,* 35.

19. Arrigon, *Houille Blanche*, 62.

20. Blanchard, *Les Forces*, 36-37, 109-110.

21. Blanchard, *Les Forces,* 8-9.

22. Arrigon, *Houille Blanche*, 37, 55-56.

23. Arrigon, *Houille Blanche*, 55.

24. Blanchard, *Les Forces*, 39-40. The reservoir was later augmented by the construction of a dam, begun during the Second World War.

25. On the Rhône, see Sara B. Pritchard, *Confluence: The Nature of Technology and the Remaking of the Rhône* (Cambridge, MA: Harvard University Press, 2011).

26. Historian Anne Dalmasso believes most French, including local populations, supported dam building: "Barrages et développement dans les Alpes françaises de l'entre-deux guerres," *Revue de géographie alpine* 96, no. 1 (2008): 21.

27. For a history of the Bayernwerk enterprise see Manfred Pohl, *Das Bayernwerk, 1921 bis 1996* (Munich: Piper, 1996).

28. Oskar von Miller to Staatsministerium des Innern (October 15, 1915), BayHStA, MF 67746. Miller's biographer argues the engineer sought to use electricity for social purposes: Wilhelm Füßl, *Oskar von Miller, 1855–1934: Eine Biographie* (Munich: Beck, 2005).

29. "Niederschrift über eine Besprechung in der Kriegsrohstoff-Abteilung in Berlin am 29. April 1916," BayHSTA, MWi 2880.

30. Fritz Blaich, *Die Energiepolitik Bayerns, 1900–1921* (Kallmünz/Opf: Lassleben, 1981), 144–146.

31. Report (February 18, 1917), BayHStA, Kriegsministerium (hereafter MKr) 9704.

32. Harald Winkel, *Wirtschaft im Aufbruch: Der Wirtschaftsraum München-Oberbayern und seine Industrie- und Handelskammer im Wandel der Zeit* (Munich: Beck, 1990), 63–64.

33. Miller to Exner (December 21, 1918), Österreichisches Staatsarchiv, Archiv der Republik (hereafter AT-OeStA, AdR), Bundeskanzleramt Inneres Sonderlegungen (hereafter BKA BKA-I SL), Wasser- und Elektrizitätswirtschaftsamt 1919–1931 (hereafter WEWA), 1, P.Z. 44.

34. Such rhetoric supports Timothy Mitchell's thesis that coal-based energy systems were especially vulnerable to disruption because workers were instrumental in the production and distribution of the critical fuel. See Timothy Mitchell, *Carbon Democracy: Political Power in the Age of Oil* (London: Verso, 2011).

35. Walter Köhler, "Zur Einführung!" *Die Wasserkraft* 1, no. 1 (April 1, 1919): 1–2.

36. Georg Schmitz, "Das Walchensee-Kraftwerk und seine Bedeutung für die Energiewirtschaft Deutschlands," *Die Braunschweiger G.N.C. Monatsschrift* (October 1921): 611.

37. "Niederschrift über die Sitzung des Bayerischen Wasserwirtschaftsrates am 7. Juli 1921," BayHStA, Ministerium für Handel, Industrie und Gewerbe (hereafter MHIG) 3128.

38. Johann Hallinger, *Bayerns Wasserkräfte und Wasserwirtschaft: Zum 10jährigen Bestehen des bayer. Wasserwirtschaftsrates* (Diessen vor München: Huber, 1918), 5, 7–8.

39. Blackbourn, *Conquest of Nature*, 227.

40. Miller to Exner (December 21, 1918), AT-OeStA, AdR, BKA BKA-I SL, WEWA, 1, P.Z. 44; Staatsminister des Innern to Oskar von Miller (January 22, 1919), BayHStA, MF 67748.

41. "Niederschrift über die Besprechung der beteiligten Minister und Sachreferenten, betreffend die finanzielle Lage des Walchensees- und Bayernwerkes" (February 9, 1920), BayHStA, MF 67748.

42. "Schuldverschreibungen Walchenseewerk A.-G. und Mittlere Isar A.-G," BayHStA, MF 67748.

43. "Wage and Labor Circumstances at the Walchensee Construction Site" (April 15, 1919), BayHStA, VArch 10546.

44. Knilling to Cuno (September 24, 1923), Bundesarchiv Deutschland (hereafter BArch), R43I/2217.

45. "Vom Tage," *Simplicissimus* 27, no. 49 (March 5, 1923): 687.

46. "Inbetriebsetzung des Walchensee- und des Bayernwerkes," *Münchner Neueste Nachrichten* (January 28, 1924).

Chapter 5

1. Henri Cavaillés, *La houille blanche* (Paris: 1922), 202-203, quoted in Raoul Blanchard, "Geographical Conditions of Water Power Development," *Geographical Review* 14, no. 1 (January 1924): 93.

2. Perhaps the best-known example of utopic hydroelectric visions is the Atlantropa project of Herman Sörgel. Sörgel was the son of a Bavarian state bureaucrat principally involved in the construction of the Walchenseewerk. See Alexander Gall, *Das Atlantropa-Projekt: Die Geschichte einer gescheiterten Vision. Herman Sörgel und die Absenkung des Mittelmeers* (Frankfurt: Campus, 1998) and Philipp Nicolas Lehmann, "Infinite Power to Change the World: Hydroelectricity and Engineered Climate Change in the Atlantropa Project," *American Historical Review* (February 2016): 70-100.

3. See Vincent Lagendijk, "Ideas, Individuals and Institutions: Notion and Practices of a European Electricity System," *Contemporary European History* 27, no. 2 (2018): 204-209.

4. Erik van der Vleuten and Arne Kaijser, eds., *Networking Europe: Transnational Infrastructures and the Shaping of Europe, 1850-2000* (Sagamore Beach, MA: Science History Publications, 2006); Martin Conway and Kiran Klaus Patel, eds., *Europeanization in the Twentieth Century: Historical Approaches* (Basingstoke: Palgrave, 2010); Vincent Lagendijk, *Electrifying Europe: The Power of Europe in the Construction of Electricity Networks* (Amsterdam: Aksant, 2008).

5. Per Högselius, Arne Kaijser, and Erik van der Vleuten, *Europe's Infrastructure Transition: Economy, War, Nature* (New York: Palgrave Macmillan, 2016), 40-46, 70-74.

6. These ideas of course, stemming from an "ideology of circulation," were always contested and should not be adopted uncritically. See Erik van der Vleuten et al., "Europe's System Builders: The Contested Shaping of Transnational Road, Electricity and Rail Networks," *Contemporary European History* 16, no. 3 (August 2007): 322.

7. Tiago Saraiva and M. Norton Wise, "Autarky/Autarchy: Genetics, Food Production and the Building of Fascism," *Historical Studies in the Natural Sciences* 40, no. 4 (Fall 2010): 419-448.

8. See in particular ch. 9 of Nicholas Mulder, *The Economic Weapon: The Rise of Sanctions as a Tool of Modern War* (New Haven: Yale University Press, 2022).

9. See also the discussion of interwar political regimes and environmental policy in Patrick Kupper, *Umweltgeschichte* (Göttingen: Vandenhoeck & Ruprecht, 2021), 144-158. Kupper notes the importance of WWI in all states' prioritization of economic development after the war. The example of interwar US dams, furthermore, was envied by all states, from fascist to communist. Paul Josephson also highlights the general similarities between the capitalist USA and communist USSR regarding enthusiastic promotion of large hydropower dams, while noting cultural differences in *Industrialized Nature: Brute Force Technology and the Transformation of the Natural World* (Washington, DC: Island Press, 2002). The link between water management and political legitimacy is almost as old as civilization itself: J. R. McNeill, *Something New Under the Sun: An Environmental History of the Twentieth-Century World* (New York: W. W. Norton, 2000), 157-159.

10. The most comprehensive environmental history of Italian Fascism posits autarky as foundational to the Fascist "socioecological" project, but the special case of South Tyrol does not receive consideration. Marco Armiero, Roberta Biasillo, and Wilko Graf

von Hardenberg, *Mussolini's Nature: An Environmental History of Italian Fascism*, trans. James Sievert (Cambridge, MA: MIT Press, 2022), 63-81. On hydroelectricity and autarky, see Mulder, *The Economic Weapon*, 238.

11. On the history of dams, see Patrick McCully, *Silenced Rivers: The Ecology and Politics of Large Dams* (London: Zed Books, 1996), 13-14; and David P. Billington, Donald C. Jackson, and Martin V. Melosi, *The History of Large Federal Dams: Planning, Design, and Construction* (Denver: US Department of the Interior Bureau of Reclamation, 2005), 49-50.

12. Schweiz, Departement des Innern/Departement suisse de l'interieur, *Verfügbare Wasserkräfte/Forces hydrauliques disponibles*, vol. 5. *Die Wasserkräfte der Schweiz/Les forces hydrauliques de la Suisse* (Bern: Rösch & Schatzmann, 1916), iv-v. The foreword was written in November 1916.

13. Schweiz, *Verfügbare Wasserkräfte/Forces hydrauliques disponibles*, iv-v.

14. A postwar survey of Alpine reservoirs calculated that up until 1914, only 3 percent of the eventual total Alpine reservoir capacity (as of 1970) had been developed. Reservoir building was a post-WWI phenomenon, with booms in the mid-1920s and the post-WWII period. See Harald Link, "Die Speicherseen der Alpen. Bassins d'accumulation des Alpes," *Cours d'eau et énergie* 62 (October 1970): 255.

15. Anne Dalmasso, "Barrages et développement dans les Alpes françaises de l'entre-deux-guerres," *Revue de géographie alpine* 96, no. 1 (2008): 2, 4-7.

16. Quoted in Dalmasso, "Barrages et développement," 6.

17. Emil Mattern, "Die hochalpinen Wasserkräfte im Rahmen der mitteleuropäischen Stromversorgung," *Elektrotechnische Zeitschrift*, no. 38 (September 22, 1932): 907-911.

18. Vincent Lagendijk, "'To Consolidate Peace'? The International Electro-Technical Community and the Grid for the United States of Europe," *Journal of Contemporary History* 47, no. 2 (2012): 402-226.

19. Anita Ziegerhofer, *Botschafter Europas: Richard Nikolaus Coudenhove-Kalergi und die Paneuropa-Bewegung in den zwanziger und dreißiger Jahren* (Vienna: Böhlau, 2004).

20. Several short historical treatments concerning the World Power Conferences exist. Two of these take the form of small articles in the published proceedings of the 11th Symposium of the International Cooperation in History of Technology Committee (ICOHTEC). See Hans-Joachim Braun, "Die Weltenergiekonferenzen als Beispiel internationaler Kooperation," and Bruce Sinclair, "Regenerating the Future: The First World Power Conference," in *Energie in der Geschichte: zur Aktualität der Technikgeschichte* (Düsseldorf: [?] 1984. See also Ian Fells, *World Energy 1923-1998 and Beyond: A Commemoration of the World Energy Council on Its 75th Anniversary* (London: Atalink, 1998).

21 Sinclair, "Regenerating the Future," 1-2.

22. C. von Gruenewaldt, *Die Wasserkraft: Zeitschrift für die gesamte Wasserwirtschaft* 15, no. 14 (July 15, 1924): 235.

23. The quote comes from a small brochure prepared for the conference. See *The First World Power Conference June 30-July 12, 1924. Under the Presidency of the Rt. Hon. The Earl of Derby, K.G. at the Conference Halls of the British Empire Exhibition Wembley, London* (London: Offices of the WPC, [1924?]), 16.

24. In some cases, as with the High Rhine, there had already been considerable hydropower development by the 1920s. Some at the time also advocated extending major

shipping into the Alps. For example, plans to connect the High Rhine, which permitted only local shipping, to the North Sea also existed. The scheme would have made a town in Austria the highest "seaport" in the world, but it failed for lack of a cost-effective way around the formidable Rhine Falls at Schaffhausen. Mark Cioc, *The Rhine: An Eco-Biography, 1815-2000* (Seattle: University of Washington Press, 2002), 27, 63-64.

25. *The First World Power Conference*, 16-17.

26. *Transactions of the World Power Conference, Basle Sectional Meeting/Compte Rendu, de la Conférence Mondiale de l'Énergie, session spéciale de Bâle 1926/Berichterstattung der Weltkraftkonferenz, Sondertagung Basel 1926*, vol. 1 (Basel: E. Birkhaeuser, 1927), 1192.

27. Gustav Kruck, *Das Kraftwerk Wäggital: Neujahrsblatt der Naturforschenden Gesellschaft in Zürich* (Zurich: Gebr. Fretz, 1925), 5-30.

28. *Transactions*, vol. 1, 1204-1205.

29. *Transactions*, vol. 1, 1199-1201.

30. *Transactions*, vol. 1, 1199-1201.

31. *Transactions*, vol. 1, 1113-1114.

32. *Transactions*, vol. 1, 1170.

33. *Transactions*, vol. 1, 1171.

34. *Transactions*, vol. 1, 1124.

35. *Transactions*, vol. 1, 1163.

36. On the history of RWE, see Dieter Schweer and Wolf Thieme, eds., *"Der gläserne Riese": RWE, ein Konzern wird transparent* (Wiesbaden: Gabler, 1998).

37. What follows is from Hughes, *Networks of Power*, 408-428.

38. For a specialized definition of the concept, see Georg Boll, *Geschichte des Verbundbetriebs: Entstehung und Entwicklung des Verbundbetriebs in der deutschen Elektrizitätswirtschaft bis zum europäischen Verbund: Ein Rückblick zum 20-Jährigen Bestehen der deutschen Verbundgesellschaft* (Frankfurt: VWEW, 1969), 13-16.

39. Hughes, *Networks of Power*, 423.

40. "Oesterreichs Elektrizitäts Wirtschaft, ihre Bedeutung für das Reich: Nutzbarmachung der österreichischen Wasserkräfte als Zukunftsaufgabe," *Völkischer Beobachter* March 22, 1938). Found in BayHStA, MHIG 3235. The switching station in Brauweiler (near Cologne) built to control RWE's transmission lines serves today as the central switchyard for the northern European grid.

41. Oskar Oliven, "Europas Großkraftlinien: Vorschlag eines europäischen Großkraftnetzes," in *Gesamtbericht. Zweite Weltkraftkonferenz. Transactions. Second World Power Conference. Compte Rendu. Deuxième Conférence Mondiale de l'Énergie*, ed. F. Zur Nedden (Berlin: VDI-Verlag, 1930), vol. XIX: 31.

42. Oliven, "Europas Großkraftlinien," 32.

43. Oliven, "Europas Großkraftlinien," 32.

44. Oliven, "Europas Großkraftlinien," 32.

45. Andrea Leonardi, "Energia e sviluppo nell'area trentina e sudtirolese," in *Energia e Sviluppo in Area Alpina: Secoli XIX-XX: atti della 7. sessione del Seminario permanente sulla Storia dell'economia e dell'imprenditorialità nelle Alpi in età moderna e contemporánea*, eds. Andrea Bonoldi and Andrea Leonardi (Milan: FrancoAngeli, 2004), 131-164.

46. What follows draws from Josef Riedmann, *Geschichte Tirols*, 3rd ed. (Vienna: Verlag

für Geschichte und Politik, 2001), 218–220, 267–268; Othmar Parteli, *Geschichte des Landes Tirol: Die Zeit von 1918 bis 1970*, vol. 4 part 1, *Südtirol (1918 bis 1970)* (Bozen: Athesia, 1988), 53–55, 291, 652–653; Rolf Steininger, *Südtirol im 20. Jahrhundert: Vom Leben und Überleben einer Minderheit*, 3rd ed. (Innsbruck: Studien Verlag, 2004), 109–116.

47. The political history of South Tyrol (Südtirol/Alto Adige) makes language and designations for place names and geographic phenomena a complicated issue, particularly when dealing with different time periods. Since South Tyrol's first Autonomy Statute of 1946, both German and Italian have been in official use. In this book I provide the German and Italian names on first use and revert to using the most common form in English, except where this would hamper clarity.

48. In the decade after 1914, hydroelectric capacity doubled and production tripled, with the greatest gains coming from the Alps. W. O. Blanchard, "White Coal in Italian Industry," *Geographical Review* 18, no. 2 (April 1928): 264.

49. Steininger, *Südtirol im 20. Jahrhundert*, 109–110.

50. Steininger, *Südtirol im 20. Jahrhundert*, 109–110, 113–116.

Chapter 6

1. This facility, earlier referred to as the "Tauernkraftwerk," eventually came to be known simply as "Kaprun" after the valley in which the major dams were built. Kaprun would attain immense national significance when it was finished during the early years of the Second Austrian Republic. Austrian historians have been active in exploring the history and "mythology" of Kaprun. See Georg Rigele, "Der Marshall-Plan und Österreichs Alpenwasser-Kräfte: Kaprun," in *"80 Dollar": 50 Jahre ERP-Fonds und Marshall-Plan in Österreich, 1948–1998*, eds. Günter Bischof and Dieter Stiefel (Vienna: Ueberreuter, 1999), 183–216; Georg Rigele, "The Marshall Plan and Austria's Hydroelectric Industry: Kaprun," in *The Marshall Plan in Austria*, vol. 8 of *Contemporary Austrian Studies*, eds. Günter Bischof, Anton Pelinka, and Dieter Stiefel (New Brunswick: Transaction, 2000), 323–356; Georg Rigele, "Das Tauernkraftwerk Glockner-Kaprun—Neue Forschungsergebnisse und offene Fragen," *Blätter für Technikgeschichte* 59 (1997): 55–94.

2. W. G. Jensen, "The Importance of Energy in the First and Second World Wars," *Historical Journal* 11, no. 3 (1968): 538–554; David Edgerton, "Controlling Resources: Coal, Iron and Oil in the Second World War," in *Total War: Economy, Society and Culture at War*, vol. 3 of *Cambridge History of the Second World War*, eds. Michael Geyer and Adam Tooze (Cambridge, UK: Cambridge University Press, 2015), 122–148; Matthew Evenden, *Allied Power: Mobilizing Hydro-Electricity during Canada's Second World War* (Toronto: University of Toronto Press, 2015). For an analysis of Nazi Germany's interest in Austrian waterpower, see Julie Cohn, Matthew Evenden, and Marc Landry, "Water Powers: The Second World War and the Mobilization of Hydroelectricity in Canada, the United States, and Germany," *Journal of Global History* 15, no. 1 (2020): 123–147.

3. Simo Laakkonen, Richard Tucker, and Timo Vuorisalo, eds., *The Long Shadows: A Global Environmental History of the Second World War* (Corvallis: Oregon State University Press, 2017); Chris Pearson, "Environments, States and Societies at War," in *Total War: Economy, Society and Culture at War*, vol. 3 of *Cambridge History of the Second World War*, eds. Michael Geyer and Adam Tooze (Cambridge, UK: Cambridge University Press, 2015),

220–243; Martin Gutmann, "The Nature of Total War: Grasping the Global Environmental Dimensions of World War II," *History Compass* 13, no. 5 (2015): 251–261. Although the severity of these dams' incursions into the Alpine landscape confirm research that the Nazi regime placed mobilization above nature conservation, that will not be the focus of this chapter. See Franz-Josef Brüggemeier, Mark Cioc, and Thomas Zeller, eds., *How Green Were the Nazis? Nature, Environment, and Nation in the Third Reich* (Athens: Ohio University Press, 2005); Frank Uekoetter, *The Green and the Brown: A History of Conservation in Nazi Germany* (Cambridge, UK: Cambridge University Press, 2006).

4. Adam Tooze, *The Wages of Destruction: The Making and Breaking of the Nazi Economy* (New York: Viking, 2006).

5. An important exception is Helmut Maier who has considered Nazi energy policy in relation to its environmental impacts, but who also suggests that the focus on long-term energy projects like dam building that did not result in immediate energy production was misguided and a failure: Helmut Maier, "Kippenlandschaft, 'Wasserkrafttaumel' und 'Kahlschlag': Anspruch und Wirklichkeit nationalsozialistischer Energiepolitik," in *Umweltgeschichte-Methoden, Themen, Potentiale: Tagung des Hamburger Arbeitskreises für Umweltgeschichte*, ed. Günter Bayerl (Muenster: Waxmann, 1996), 266.

6. A standard work on the history of the Habsburg empire is Robert A. Kann, *A History of the Habsburg Empire, 1526–1918* (Berkeley: University of California Press, 1974). A recent reinterpretation that emphasizes the successes and remarkable durability of the imperial dynasty is Pieter M. Judson, *The Habsburg Empire: A New History* (Cambridge, MA: Belknap Press, 2016).

7. On the history of the first Austrian republic see Anton Pelinka, *Die gescheiterte Republik: Kultur und Politik in Österreich 1918–1938* (Vienna: Böhlau, 2017); the earliest years are treated in Walter Rauscher, *Die verzweifelte Republik, Österreich 1918–1922* (Vienna: Kremayr & Scheriau, 2017).

8. Quote in Richard Hufschmied, "'Weißes Gold' in (Deutsch-)Österreich—Kontinuität und Wandel nach dem Epochenjahr 1918," in *Wasserkraft. Elektrizität. Gesellschaft. Kraftwerksprojekte ab 1880 im Spannungsfeld*, eds. Oliver Rathkolb et al. (Vienna: Kremayr & Scheriau, 2012), 84–148.

9. "Protokoll über die 1. Sitzung der Beratenden Kommission Wasserkraft- und Elektrizitätswirtschaftsamtes am 10. und 11. Juli 1919," AT-OeStA, AdR, BKA BKA-I SL, WEWA 1919, Kt. 3, 669/19, 3.

10. Petition in AT-OeStA, AdR, BKA BKA-I SL, WEWA 1919, Kt. 3, P.Z. 40.

11. Letter from Ellenbogen to provincial governments (January 13, 1919), AT-OeStA, AdR, BKA BKA-I SL, WEWA 1919, Kt. 1, P.Z. 32.

12. See the correspondence in AT-OeStA, AdR, BKA BKA-I SL, WEWA 1919, Kt. 1, P.Z. 24.

13. Generaldirektion der Österreichischen Bundesbahnen, *Die österreichischen Eisenbahnen. Gedenkblätter zur Hundertjahrfeier der Eröffnung der ersten österreichischen Dampfeisenbahnen* (Vienna: Verlag der österreichischen Bundesbahnen, 1937), 75. For more on the Austrian electrification program and its Swiss influences, see Peter Staudacher, "'In der Schweiz zum Beispiel': Die Anfänge der österreichischen Elektrowirtschaft und das Schweizer Vorbild," in *Allmächtige Zauberin unserer Zeit: Zur Geschichte der elektrischen Energie in der Schweiz*, ed. David Gugerli (Zurich: Chronos, 1994), 185–198.

14. Wilhelm Ellenbogen, *Anschluß und Energiewirtschaft* (Vienna: Verlag Deutsche Einheit, 1927), 32.

15. Ellenbogen, *Anschluß und Energiewirtschaft*, 32.

16. Erin R. Hochman, *Imagining a Greater Germany: Republican Nationalism and the Idea of the Anschluss* (Ithaca: Cornell University Press, 2016).

17. Ernst Hanisch, *Landschaft und Identität: Versuch einer österreichischen Erfahrungsgeschichte* (Vienna: Böhlau, 2019), 165, 168.

18. Wilhelm Münch, "Das Tauernwerk," *Deutsche Wasserwirtschaft* 26, no. 1 (January 10, 1931): 1–5.

19. Franz Rehrl, "Bericht der Landesregierung über das Projekt einer Verwertung der Wasserkräfte im Bereiche der Tauernkette" (January 29, 1929), Salzburger Landesarchiv (hereafter SLA), RehrlTW 1929/0001.

20. The debate on the project was voluminous: Richard Moro, "Betrachtungen über das Tauernkraftwerks-projekt der AEG," *Die Wasserwirtschaft* 24, no. 8 (1931): 109–117; Thürnau, "Das Tauernwerksprojekt der AEG," *Die Wasserwirtschaft* 24, no. 11 (1931): 157–158; "Stellungnahme der Ingenieurkammer für Tirol und Vorarlberg zu den Projekten der Allgemeinen Elektrizitäts-Gesellschaft Berlin (AEG) und der Österreichischen Kraftwerke A.-G. (Oeka) für das Tauernkraftwerk," *Die Wasserwirtschaft* 24, no. 11 (1931): 158–162; Richard Hofbauer, "Zeitgemäße Betrachtungen über den projektierten Ausbau der österr. Alpenwasserkräfte mit Bezug auf deren künftige Funktion in der Energiewirtschaft Mitteleuropas," *Die Wasserwirtschaft* 24, no. 12 (1931): 169–171; Erich Heller, "Technische Probleme des Tauernwerks," *Die Wasserwirtschaft* 24, no. 13/14 (1931): 181–187; Petersen, "Die Energiewirtschaft des Tauernkraftwerkes," *Die Wasserwirtschaft* 24, no. 13/14 (1931): 188–190; Oskar Vas, "Die Ausnutzung der Tauernwasserkräfte als österreichisches Problem," *Die Wasserwirtschaft* 24, no. 13/14 (1931): 191–201; Hans Dreyer, "Großwasserkräfte im Dienst der deutschen Energiewirtschaft," *Die Wasserkraft und Wasserwirtschaft* 26, no. 15 (August 1, 1931): 177–183; Johann Hallinger, "Das Tauernwerk im Wirtschaftsspiegel," *Wasserkraft und Wasserwirtschaft* 27, no. 2 (January 16, 1932): 13–15.

21. Rigele, "Das Tauernkraftwerk Glockner-Kaprun," 59.

22. SLA RStH V/3 212, "Tauernkraftwerk."

23. A good recent study of this period in Austrian history is Julie Thorpe, *Pan-Germanism and the Austrofascist State, 1933–38* (Manchester: Manchester University Press, 2011).

24. Sidney B. Fay, "Dollfuss: Victim of Nazi Crime," *Current History* 40, no. 6 (September 1934): 733.

25. Albrecht Czimatis, *Energiewirtschaft als Grundlage der Kriegswirtschaft* (Hamburg: Hanseatische Verlagsanstalt, 1936), 8.

26. Czimatis, *Energiewirtshaft*, 9, 14–16.

27. Czimatis, *Energiewirtschaft*, 30–31.

28. Czimatis, *Energiewirtschaft*, 28–29, 34.

29. Dieter Petzina, *Autarkiepolitik im Dritten Reich: Der nationalsozialistische Vierjahresplan* (Stuttgart: Deutsche Verlags-Anstalt, 1968), 60–61.

30. Georg Boll, *Geschichte des Verbundbetriebs: Entstehung und Entwicklung des Verbundbetriebs in der deutschen Elektrizitätswirtschaft bis zum europäischen Verbund: Ein Rückblick zum 20-Jährigen Bestehen der deutschen Verbundgesellschaft* (Frankfurt: VWEW,

1969), 88. The plan for such a line was not new. It had been submitted during the first round of negotiations over the Tauern energy.

31. Norbert Schausberger, "Deutsche Wirtschaftsinteressen in Österreich vor und nach dem März 1938," *Österreich, Deutschland und die Mächte: Internationale und Österreichische Aspekte des "Anschlusses" vom März 1938*, eds. Gerald Stourzh and Birgitta Bader-Zaar (Vienna: Verlag der Österreichischen Akademie der Wissenschaften 1990), 193–196. See also Stefan Karner and Peter Ruggenthaler, ed., *1938: Der „Anschluss" im internationalen Kontext* (Vienna: Leykam, 2020).

32. On the political aspects of the Anschluss, see Gerhard Botz, *Die Eingliederung Österreichs in das Deutsche Reich. Planung und Verwirklichung des politisch-administrativen Anschlusses (1938–1940)*, 3rd ed. (Vienna: Europa Verlag, 1988).

33. Maier, "Kippenlandschaft."

34. Quotation from "Österreichs Energiewirtschaft," in *Wirtschaft, Technik, Verkehr* 14, no. 5 (1939): 5–8. Cited in Maier, "Kippenlandschaft," 260.

35. Maier, "Kippenlandschaft," 266. On RWE and the West Tyrolean water power and nature protection, see also Helmut Maier, "'Unter Wasser und unter die Erde.' Die süddeutschen und alpinen Wasserkraftprojekte des Rheinisch-Westfälischen Elektrizitätswerks (RWE) und der Natur- und Landschaftsschutz während des 'Dritten Reiches,'" in *Die Veränderungen der Kulturlandschaft: Nutzungen-Sichtweisen-Planungen*, ed. Günter Bayerl and Torsten Meyer (Muenster: Waxmann, 2003), 139–175.

36. Reichstelle für Wirtschaftsausbau, "Erste Ermittlungen zur Aufstellung eines Vierjahresplanes für das Land ÖSTERREICH," BArch Berlin, R3112/45: 5, 16.

37. Hermann Göring, "Aufbauprogramm für Österreich," Keesings Archiv 1938, 3488. Found in Norbert Schausberger, *Rüstung in Österreich 1938–1945: Eine Studie über die Wechselwirkung von Wirtschaft, Politik und Kriegsführung* (Vienna: Hollinek, 1970), 186.

38. "Baubeginn des größten Kraftwerks Deutschlands," *Völkischer Beobachter*, no. 137 (May 17, 1938): n.p. Found in BayHStA, MHIG 3235. Göring's insistence on quickly beginning construction on the Tauern project and his rhetoric about action was also likely intended to impress Austrian audiences with the effectiveness of their new leadership. Another article on the project in the same issue entitled "Six Weeks Later the Deed" explicitly denigrated the old regime and its chancellor, Kurt Schuschnigg, a man who "made promises galore," which "lacked only their fulfillment."

39. "Die Stromversorgung Deutsch-Österreich und ihre Eingliederung in das Altreich" (May 11, 1938), AT-OeStA, AdR, 04, "Bürckel"-Materien, Kt. 88, 2155/O Bd. I, 2, 18, 25.

40. AEW was a subsidiary of the Reich-owned industrial giant Vereinigten-Industrieunternehmungen AG (VIAG). On the history of the AEW, see Maria Magdalena Koller "Elektrizitätswirtschaft in Österreich, 1938–1947: von den Alpenelektrowerken zur Verbundgesellschaft," PhD dissertation, Karl-Franzens University of Graz, 1985.

41. Although the report did not explain why Austrian industrialization was important, many Germans viewed Austria strategically as an area less vulnerable to potential air attacks than Germany. Alpen-Elektrowerke AG Wien, "Geschäftsbericht über das erste Geschäftsjahr 1938," AT-OeStA, AdR, 04, "Bürckel"-Materien, Kt. 68, 2155/O: Bd. II-E-wirtschaft in der Ostmark allgemein, 5.

42. "Eigenart," SLA RStH V/3 171, 15–16.

43. "Eigenart," SLA RStH V/3 171, 2-4.

44. Konrad Meyer, "Die deutsche Elektrizitätswirtschaft 1933 bis 1948," *Elektrizitätswirtschaft* 48, no. 2 (February 1949): 35; Bernhard Stier, *Staat und Strom: die politische Steuerung des Elektrizitätssystem in Deutschland, 1890–1950* (Ubstadt-Weiher: Regionalkultur, 1999).

45. Letter from Dillgardt to Bürckel (March 9, 1939), AT-OeStA, AdR, 04, "Bürckel"-Materien 2155/2.

46. His official title was Commissar for the Reunification of Austria with the German Reich. The task of performing the evaluation and the subsequent meeting are described in "Gutachten Thürnau-Schmidt" (April 11, 1939), SLA, RStH V/3 171.

47. Thürnau, "Das Tauernwerksprojekt der AEG," *Die Wasserwirtschaft* 24, no. 11 (1931): 157–158.

48. "Thürnau-Schmidt," SLA, RStH V/3 171, 7-8, 11, 28.

49. "Thürnau-Schmidt," SLA, RStH V/3 171, 33-34.

50. Hermann Grengg, Stellungnahme der Alpen-Elektrowerke A.G. Wien (AEW) zum Gutachten Thürnau-Schmidt vom 11.IV.1939 in Sachen Tauernrkraftwerk," SLA, RStH V/3 171, 10, 14.

51. Handwritten summary of meeting on May 24, 1939. SLA, RStH V/3 171. A vocal current within the National Socialist electrical engineering community opposed centralized electricity production because of its security implications. For an expression of this viewpoint from its leading proponent, see Franz Lawaczeck, *Elektrowirtschaft* (Munich: Lehmann, 1936).

52. Meyer, "Die deutsche Elektrizitätswirtschaft," 36; Boll, *Geschichte des Verbundbetriebs*, 100–101.

53. "Bericht über die Energieversorgung der Ostmark," author unknown (September 1939?), AT-OeStA, AdR, 04, "Bürckel"-Materien 2155, Kt. 88, 5-6.

54. Meyer, "Die deutsche Elektrizitätswirtschaft," 34-36, 39.

55. Report quoted in Boll, *Geschichte des Verbundbetriebs*, 97.

56. Boll, *Geschichte des Verbundbetriebs*, 96-97.

57. "Entwicklung der Energieversorgung im 2. Vierjahresplan und in der späteren Zeit," BArch R3112/38, 25-26.

58. "Plan über den technischen Ausbau der deutschen Elektrizitätsversorgung," BArch R43II/344, 25.

59. Henry Picker, *Hitlers Tischgespräche im Führerhauptquartier, 1941–1942* (Bonn: Athenaeum, 1951), 65-66. Found in Stier, *Staat und Strom*, 486-488. See too the copy of the letter from Bormann to Fiehler (February 5, 1941), "Verhältnisse in der Elektrizitätswirtschaft," BArch R43II/378.

60. Letter from Bormann to Lammers, "Ausbau der Enns" (February 21, 1941), BArch R43II/377, 8.

61. Letter from Bormann to Lammers (February 20, 1941), BArch R43II/377, 7.

62. Letter from Lammers to Funk, "Energiewirtschaft" (March 1, 1941), BArch R43II/377, 40-42. The letter also suggests that the delay in the Enns project (and others as well) stemmed from Todt's own hand in prioritizing "defense" projects over energy ones.

63. Meyer, "Die deutsche Elektrizitätswirtschaft," 37.

64. Oliver Rathkolb, "NS-Erbe, Wiederaufbau, Marshallplan und das "Weiße Gold" in den europäischen Netzwerken," in *Wasserkraft. Elektrizität. Gesellschaft. Kraftwerksprojekte ab 1880 im Spannungsfeld*, eds. Oliver Rathkolb et al. (Vienna: Kremayr & Scheriau, 2012), 187.

65. Boll, *Geschichte des Verbundbetriebes*, 87.

66. Boll, *Geschichte des Verbundbetriebes*, 99. Boll suggests here that the choice to develop hydro to conserve coal may have had a "certain influence" on the initial slow tempo in expanding capacity, particularly before the war.

67. See the comments in Boll, *Geschichte des Verbundbetriebes*, 102.

68. Meyer, "Die deutsche Elektrizitätswirtschaft," 38.

69. Grengg later explained that he utilized the bombing crisis to make the switch to arch dams—a choice that he gradually had come to prefer but feared to make lest he provide attack fodder for the opponents of dam building during wartime. Grengg maintained that Swiss engineers also later opted for arch dams for similar reasons. See Hermann Grengg, *Das Großspeicherwerk Glockner-Kaprun* (Vienna: Springer, 1952), 31.

70. The Austrian nationalized utility, the Verbundgesellschaft, convened a historical commission to investigate the use of forced labor to construct power plants during the Nazi period: Oliver Rathkolb and Florian Freund, eds., *NS-Zwangsarbeit in der Elektrizitätswirtschaft der "Ostmark", 1938-1945* (Vienna: Böhlau, 2002).

71. On the construction history of the Tauern project during the National Socialist period and the use of forced labor, see Margit Reiter, "The "Myth" of Kaprun: Forced Labor at the Tauern Power Plant in Kaprun and How Postwar Austria Dealt with It," in *The Dollfuss/Schuschnigg Era in Austria: A Reassessment*, vol. 11 of *Contemporary Austrian Studies*, eds. Günter Bischof, Anton Pelinka, and Dieter Stiefel (New Brunswick: Transaction, 2003), 260-261; Margit Reiter, "Das Tauernkraftwerk Kaprun," in *NS-Zwangsarbeit in der Elektrizitätswirtschaft der "Ostmark,"* 127-198. Reiter acknowledges that the number of POWs or forced laborers at Kaprun is a "dark figure" as they do not often appear in the records.

72. Helmut Maier, "Systems Connected: IG Auschwitz, Kaprun, and the Building of European Power Grids up to 1945," in *Networking Europe: Transnational Infrastructures and the Shaping of Europe, 1850-2000*, eds. Erik van der Vleuten and Arne Kaijser (Sagamore Beach, MA: Science History Publications, 2006), 129-130; Boll, *Geschichte des Verbundbetriebes*, 91.

73. The numbers come from Norbert Schausberger, "Sieben Jahre deutsche Kriegswirtschaft in Österreich (1938-1945)," in *Jahrbuch 1986 des Dokumentationsarchivs des österreichischen Widerstandes*, ed. Siegwald Ganglmair (Vienna: Österreichischer Bundesverlag, 1986), 53. Found in Hufschmied, "'Weißes Gold' in (Deutsch-)Österreich," 146-147.

74. With respect to energy projects, a case could be made that contemporary Austria's environmental history has been unusually influenced by external actors; whether this holds true in general awaits further study. Though Austria has a very vibrant environmental history community, much of the research has taken a long-term perspective. The contemporary period remains understudied: Verena Winiwarter et al., "Environmental Histories of Contemporary Austria: An Introduction," in *Austrian Environmental History*,

vol. 27 of *Contemporary Austrian Studies,* eds. Marc Landry and Patrick Kupper (New Orleans and Innsbruck: UNO Press/Innsbruck University Press, 2018), 25-48.

Chapter 7

1. Historians have begun considering the environmental history of the Cold War: J. R. McNeill and Corinna R. Unger, eds., *Environmental Histories of the Cold War* (Cambridge, UK: Cambridge University Press, 2010). The Alps must be added to the list of regions where American strategic interests encouraged the spread of high dams. See Richard P. Tucker's chapter in this volume "Containing Communism by Impounding Rivers: American Strategic Interests and the Global Spread of High Dams in the Early Cold War," 139-163.

2. Christophe Bonneuil and Jean-Baptiste Fressoz, *The Shock of the Anthropocene: The Earth, History, and Us,* trans. David Fernbach (London: Verso, 2017), ch. 7, especially 166. Classic works on the Americanization of postwar western Europe are Reinhold Wagnleitner, *Coca-Colonization and the Cold War: The Cultural Mission of the United States in Austria After the Second World War,* trans. Diana Wolf (Chapel Hill: University of North Carolina Press, 1994); Richard Kuisel, *Seducing the French: The Dilemma of Americanization* (Berkeley: University of Californian Press, 1997); Victoria De Grazia, *Irresistible Empire: American's Advance Through Twentieth-Century Europe* (Cambridge, MA: Harvard University Press, 2006).

3. The term "Anthropocene" has become the subject of intense debate within the environmental humanities. Beginning in 2000, it became a popular term to denote that humanity had become the most powerful influencer of global ecology, therefore requiring the naming of a new geologic era. Areas of contention include the specific responsibility for these changes as well their periodization. The record of dam building in the Alps joins research confirming the post-WWII period as transitional. See J. R. McNeill and Peter Engelke, *The Great Acceleration: An Environmental History of the Anthropocene Since 1945* (Cambridge, MA: Belknap Press, 2014). Already in the 1990s, a group of researchers surrounding Swiss climate historian Christian Pfister suggested a phenomenon similar to the "Great Acceleration" that they called the "1950s syndrome" and laid at the feet of cheap Mideast oil: Christian Pfister, ed., *Das 1950er Syndrom: Der Weg in die Konsumgesellschaft* (Bern: Haupt, 1995). As I have argued in previous chapters, to single out oil as the keystone ignores that consumer society would not have been possible without electrified homes and factories. For an application of the concept of "Great Acceleration," see Robert Groß, *Die Beschleunigung der Berge: Eine Umweltgeschichte des Wintertourismus in Vorarlberg/Österreich (1920-2010)* (Vienna: Böhlau, 2019).

4. In fact, the postwar US Strategic Bombing Survey would conclude that had it targeted electric infrastructure from the very beginning, it "would have had a catastrophic effect on Germany's war production." United States Strategic Bombing Survey, *German Electric Utilities Industry Report,* 2nd ed. (Washington, DC: Utilities Division, 1947), 3.

5. Situation as portrayed in Vereinte Nationen Wirtschaftskommission für Europa, "Bericht des Elektro-Energie Komitees, Beilage B: Geschichtliches zum Elektro-Energie Problem," SLA, Präsidial Akten (hereafter Prä) 1948/39, 1-4.

6. John Gillingham, *Coal, Steel, and the Rebirth of Europe: 1945-1955.* Cambridge, UK: Cambridge University Press, 1991.

7. Letters to Walchensee Community Council, February and March 1949, BayHStA, Office of the Military Government for Bavaria (hereafter OMGB) 13/134-2/4.

8. Georg Rigele, "Der Marshall-Plan und Österreichs Alpenwasser-Kräfte: Kaprun," in *"80 Dollar": 50 Jahre ERP-Fonds und Marshall-Plan in Österreich, 1948-1998*, eds. Günter Bischof and Dieter Stiefel (Vienna: Ueberreuter, 1999), 184.

9. Oliver Rathkolb, "Die Zweite Republik (seit 1945)," in *Geschichte Österreichs*, ed. Thomas Winkelbauer (Stuttgart: Reclam, 2015), 525-550; Ernst Hanisch, *Der lange Schatten des Staates. Österreichische Gesellschaftsgeschichte im 20. Jahrhundert (Österreichische Geschichte 1890-1990)* (Vienna: Ueberreuter, 1994), 395-425. A good recent overview of the historiography on the early Austrian Second Republic is Günter Bischof, "Zweite Republik," in *Österreichische Zeitgeschichte—Zeitgeschichte in Österreich: Eine Standortbestimmung in Zeiten des Umbruchs*, eds. Marcus Gräser and Dirk Rupnow (Vienna: Böhlau, 2021), 160-177.

10. Richard Hufschmied, "'Weißes Gold' in (Deutsch-)Österreich—Kontinuität und Wandel nach dem Epochenjahr 1918," in *Wasserkraft. Elektrizität. Gesellschaft. Kraftwerksprojekte ab 1880 im Spannungsfeld*, eds. Oliver Rathkolb et al. (Vienna: Kremayr & Scheriau, 2012), 147.

11. "Beilage 3: Abschrift Gedächtnisprotokoll über die vom Herrn Bundesminister für Energiewirtschaft und Elektrifizierung veranstaltete Pressekonferenz" (January 10, 1946), SLA, Prä 1948/40, 2.

12. "Gedächtnisprotokoll," SLA, Prä 1948/40, 6.

13. "Gedächtnisprotokoll," SLA, Prä 1948/40, 2.

14. On the process in France, see Robert L. Frost, *Alternating Currents: Nationalized Power in France, 1946-1970* (Ithaca: Cornell University Press, 1991).

15. "Resolution, bettrefend Organisation der Elektrizitätswirtschaft in Österreich," SLA, Prä 1948/40, 1. The major exception was Italy, where private companies continued to play an important role in hydroelectric development. This changed with the nationalization of the electricity supply in 1962.

16. "Resolution, bettrefend Organisation der Elektrizitätswirtschaft in Österreich," SLA, Prä 1948/40, 1.

17. Günter Bischof and Hans Petschar, *The Marshall Plan: Saving Europe, Rebuilding Austria* (New Orleans: University of New Orleans Press, 2017), 66-68.

18. Oskar Vas, *Der Anteil Österreichs an der elektrizitätswirtschaftlichen Gemeinschaftsplanung in Europa* (Vienna: Springer, 1948), 7.

19. Vas, *Der Anteil Österreichs*, 27.

20. United States Department of State, *Committee of European Economic Co-operation: Volume I: General Report* (Washington, DC: Division of Publications, 1947) 38, 86-87.

21. United Nations Economic Commission for Europe, *ECE: The First Ten Years, 1947-1957* (Geneva: (?), 1957), II-5.

22. Vereinte Nationen Wirtschaftskommission für Europa, "Bericht des Elektro-Energie Komitees," SLA, Prä 1948/39, 1.

23. Vereinte Nationen Wirtschaftskommission für Europa, "Beilage A," SLA, Prä 1948/39, 1.

24. Vereinte Nationen Wirtschaftskommission für Europa, "Bericht des Elektro-Energie Komitees," SLA, Prä 1948/39, 2.

25. Vereinte Nationen Wirtschaftskommission für Europa, "Bericht des Elektro-Energie Komitees," SLA, Prä 1948/39, 4, 7.

26. Oskar Vas, *Probleme der Kraftwirtschaft in Mitteleuropa* (Vienna: Springer, 1952), 6.

27. The total investments were 2,961 million schillings. Rathkolb, "NS-Erbe, Wiederaufbau, Marshallplan und das "Weiße Gold" in den europäischen Netzwerken," 200.

28. Rigele, "Der Marshall-Plan," 202.

29. Rigele, "Der Marshall-Plan," 196–216.

30. On Kaprun's function as a stamp motif, see Christian Rohr, "The Austrian Environment *en Miniature*: 'Official' Perceptions of the Austrian Landscape Reflected Through Postal Stamps Since 1945," in *Austrian Environmental History*, vol. 27 of *Contemporary Austrian Studies*, eds. Marc Landry and Patrick Kupper (New Orleans and Innsbruck: UNO Press/Innsbruck University Press, 2018), 175–176.

31. A good recent summary of the Kaprun myth and its role in Austrian identity is Ernst Hanisch, *Landschaft und Identität: Versuch einer österreichischen Erfahrungsgeschichte* (Vienna: Böhlau, 2019), 180–186. Kaprun is described as one of the most important chapters in the environmental history of modern Austria in Martin Schmid and Ortrun Veichtlbauer, *Vom Naturschutz zur Ökologiebewegung: Umweltgeschichte Österreichs in der Zweiten Republik* (Vienna: Studienverlag, 2006), 28–35. Nobel Prize-winning Austrian author Elfriede Jelinek wrote a play entitled *The Works* that thematized Kaprun at the turn of the millennium.

32. J. R. McNeill, *Something New Under the Sun: An Environmental History of the Twentieth-Century World* (New York: W. W. Norton, 2000), 157–159.

33. Bischof and Petschar, *The Marshall Plan*, 13.

34. "Der Bundeskanzler über Südtirol," *Wiener Kurier* (January 28, 1946).

35. Rolf Steininger, *Südtirol im 20. Jahrhundert: Vom Leben und Überleben einer Minderheit*, 3rd ed. (Innsbruck: Studien Verlag, 1997), 110, 225–232.

36. On the history of the Reschensee and South Tyrolean hydroelectricity, see Andrea Bonoldi, "Technologien, Kapitalien und Kontrolle der Ressourcen: die regionale Elektrizitätswirtschaft," in *Die Region Trentino-Südtirol im 20. Jahrhundert*, vol. 2, ed. Andrea Leonardi (Trent: Museo Storico in Trento, 2009), 235–252; Philipp Tolloi, "'Alpen unter Strom': Zur Geschichte der Elektrifizierung Südtirols," *Tiroler Chronist*, no. 2 (2015): 5–8. In 1966, Montecatini would fuse with Edison, the original company to exploit white coal in Italy.

37. The following sections are based on Harald Link, "Die Speicherseen der Alpen. Bassins d'accumulation des Alpes," *Cours d'eau et énergie* 62 (October 1970): 279–281; Patrick McCully, *Silenced Rivers: The Ecology and Politics of Large Dams* (London: Zed Books, 1996), 101–122.

38. Marco Armiero, *A Rugged Nation: Mountains and the Making of Modern Italy* (Cambridge, UK: White Horse Press, 2011), 173–194.

39. The preceding caveat is necessary to gain some perspective on global dam disasters in the twentieth century. In the worst dam catastrophe in history, up to 200,000 Chinese

may have been killed by a dam failure in August 1975. See McNeill and Engelke, *The Great Acceleration*, 33.

40. Link, "Bassins d'accumulation," 247.

41 Alessandro Botteri Balli, *Wasserkraftwerke der Schweiz: Architektur und Technik* (Zurich: Offizin, 2003), 18-20. French filmmaker Jean-Luc Godard actually worked as a laborer on the dam and made one of his earliest films, a documentary called *Opération béton*, on the subject of the complex.

42. A recent history of the final large power reservoir project in the Swiss Alps also locates the 1970s as the end of the large dam-building era. Interestingly, the author finds that there was little environmental resistance to the project, in part because it was understood to be the end of the era. Sebastian De Pretto, "Un espace sans conflit? Structures de pouvoir et path dependencies autour du lac d'Émosson, 1953-1975," *Geschichte der Alpen—Histoire des Alpes—Storia delle Alpi* 27 (2022): 173-189.

43. Patrick Kupper, "Verkanntes unternehmerisches Risiko. Der übereilte Einstieg der schweizerischen Elektrizitätswirtschaft in die Atomtechnologie: der Fall Motor-Columbus 1961-1966," *Zeitschrift für Unternehmensgeschichte* 47, no. 1 (2002): 51-54.

44. France incorporated nuclear power into its national energy mix more than any other European country. See Gabrielle Hecht, *The Radiance of France, Nuclear Power and National Identity after World War II* (Cambridge, MA: MIT Press, 2000).

Conclusion

1. Peter Matt, Otto Pirker, and Martin Schletterer, "Wasserkraft im Wandel der Zeit: Energiewirtschaftliche Bedeutung der Alpenflüsse," in *Flüsse der Alpen: Vielfalt in Natur und Kultur*, eds. Susanna Muhar et al. (Bern: Haupt, 2019), 256.

2. Werner Bätzing, *Die Alpen: Geschichte und Zukunft einer europäischen Kulturlandschaft* (Munich: Beck, 2003), 195-196.

3. Alessandro Botteri Balli, *Wasserkraftwerke der Schweiz: Architektur und Technik* (Zurich: Offizin, 2003), 19. Itaipú supplies almost all of Paraguay's electricity and about one-fifth of Brazil's.

4. Vaclav Smil, *Energy at the Crossroads: Global Perspectives and Uncertainties* (Cambridge, MA: MIT Press, 2003), 251.

5. Daniel L. Vischer, "Wasserbau und Elektrifizierung," in *Allmächtige Zauberin unserer Zeit: Zur Geschichte der elektrischen Energie in der Schweiz*, ed. David Gugerli (Zurich: Chronos, 1994), 117-130.

6. In the meantime, in part due to the diversification of the European energy mix, Alpine dams are enlisted in the battle against floods as well. The hydropower section of the French Electricity Board (EDF), for instance, now plays a major role in civil defense through "contingency plans to counter the effects of floods on their structures or reservoir management programmes." See Charles Obled and Patrick Tourasse, "Uncertainty in Flood Forecasting: A French Case Study," in *Coping with Floods*, eds. Giuseppi Rossi, Nilgun B. Harmanciogammalu, and V. Yevjevich (Dordrecht: Kluwer, 1994), 474.

7. For a multifaceted look at the connection between dams and political economy, See Patrick McCully, *Silenced Rivers: The Ecology and Politics of Large Dams* (London: Zed Books, 1996), ch. 9.

8. Romed Aschwanden, *Politisierung der Alpen: Umweltbewegungen in der Ära der Europäischen Integration (1970-2000)* (Vienna: Böhlau, 2021), 88-94; Jon Mathieu, *Die Alpen: Raum, Kultur, Geschichte* (Stuttgart: Reclam, 2015), 198-201.

9. The distinction between the environmental impacts of run-of-river and storage dam are from the Alpine geographer Werner Bätzing, *Die Alpen*, 197-198.

10. Severin Hohensinner et al., "What Remains Today of Pre-Industrial Alpine Rivers? Census of Historical and Current Channel Patterns in the Alps," *River Research and Applications* 37, no. 2 (February 2021): 147.

11. Susanna Muhar et al., "Zustand und Schutz der Fließgewässer: Ein alpenweiter Überblick," in *Flüsse der Alpen: Vielfalt in Natur und Kultur*, eds. Susanna Muhar et al. (Bern: Haupt, 2019), 307. Many, though not all, of these obstacles are related to energy.

12. Jean-Paul Bravard et al., "River Incision in South-East France: Morphological Phenomena and Ecological Effects," *River Research and Applications* 13, no 1 (January 1997): 75-90.

13. Severin Hohensinner et al., "Morphologie: Die vielfältige Gestalt der Alpenflüsse," in *Flüsse der Alpen*, 109-110.

14. Muhar et al., "Zustand," 307.

15. Aude Zingraff-Hamed and Gregory Egger, "Isar," in *Flüsse der Alpen*, 408-411.

16. P. Parasiewicz et al., "The Effect of Managed Hydropower Peaking on the Physical Habitat, Benthos and Fish Fauna in the River Bregenzerach in Austria," *Fisheries Management and Ecology* 5 (1998): 403-417.

17. Lukas Denzler, "Der Spöl blüht wieder auf," *Neue Zürcher Zeitung* (June 18, 2014).

18. Bätzing, *Die Alpen*, 198.

19. Günter Unfer, Andreas Meraner, and Didier Pont, "Bedrohte aquatische Biodiversität in den Alpen," in *Flüsse der Alpen*, 139-140.

20. Muhar et al., "Zustand," 311.

21. Mark Cioc, *The Rhine: An Eco-Biography, 1815-2000* (Seattle: University of Washington Press, 2002), 158-171, 185-192

22. Susanna Muhar et al., "Revitalisierung: Neues Leben für die Alpen Flüsse," in *Flüsse der Alpen*, 320-343.

23. Robert Reindl, Johan Neuner, and Martin Schletterer, "Increased Hydropower Production and Hydropeaking Mitigation Along the Upper Inn River (Tyrol, Austria) with a Combination of Buffer Reservoirs, Diversion Hydropower Plants and Retention Basins," *River Research and Applications* 39, no. 3 (2023): 602-609.

24. Wiebke Rögener, "Gift unter Strommasten: Bleihaltige Rostschutzfarbe verseucht das Erdreich—noch immer wird in Deutschland wenig dagegen getan," *Süddeutsche Zeitung* (February 6, 2009). This article also notes the continuing suspicions that the electromagnetic fields created by high-voltage electric transmission can cause cancer in humans.

25. This analysis has focused primarily on the environmental changes required to produce and transport hydroelectricity. A complete account would also consider the environmental impacts of the population growth and industry made possible by the new energy. For an example of how societies became aware of environmental problems associated with white coal, see the file at the Archivio provinciale di Bolzano/Südtiroler Landesar-

chiv Bozen concerning the growing problems (and protection measures) for the Etsch/Adige River in northern Italy during the 1990s: Archiv der grünen Fraktion im Südtiroler Landtag 70.

26. Thomas H. Painter et al., "End of the Little Ice Age in the Alps Forced by Industrial Black Carbon," *Proceedings of the National Academy of Sciences* 110, no. 38 (September 17, 2013): 15216–15221.

27. Jon Mathieu, *Geschichte der Alpen, 1500–1900: Umwelt, Entwicklung, Gesellschaft*, 2nd ed. (Vienna: Böhlau, 2001), 214.

28. Werner Bätzing, *Zwischen Wildnis und Freizeitpark: Eine Streitschrift zur Zukunft der Alpen*, 2nd ed. (Zurich: Rotpunktverlag, 2015), 69–74. Bätzing, one of the foremost Alpine scholars, sees certain sections of the Alps as having gone to seed thanks to the prevalence of a neoliberal outlook that prioritizes capitalist profitmaking over nature protection. He predicts the future of the Alps will be "terrible"—the continuation of the destruction of traditional forms of life and increased marginalization as a European periphery—if the trends of the postwar period persist. See especially his afterword, 139–145. This pessimistic view is a response to historian Jon Mathieu's formulation that in the future, the Alps will remain a dynamic, heterogeneous, and livable space: Mathieu, *Geschichte*, 215.

29. Jamey Keaten, "The Battle of the Alps? Water Woes Loom amid Climate Change," *Seattle Times* (October 27, 2022): https://www.seattletimes.com/business/battle-of-the-alps-water-woes-loom-amid-climate-change/

30. Matt et al.,"Wasserkraft im Wandel der Zeit," 254–258.

BIBLIOGRAPHY

Archives
BERLIN, GERMANY
Bundesarchiv Deutschland, Berlin-Lichterfelde (BArch)
R2, Reichsfinanzministerium, R2/69, R2/70, R2/111, R2/17634, R2/17801, R2/17806,
    R2/17812, R2/17813a, R2/21481, R2/21709
R8IV, Reichsstelle für Elektrizitätswirtschaft, R8IV/1, R8IV/8, R8IV/16, R8IV/19
R12II, Reichgsgruppe Energiewirtschaft
R13XVI/2
R43/598a, R43/599
R43II/344, R43II/345, R43II/346, R43II/377
R26I, Beauftragter für den Vierjahresplan, R26I/11, R26I/12, R26I/29
R3101, Reichswirtschaftsministerium, R3101/1787, R3101/1788, R3101/1791, R3101/1792,
    R3101/1793, R3101/1794, R3101/1795, R3101/20512
R3102/17, R3102/21, R3102/22, R3102/24, R3102/985, R3102/2997, R3102/3133, R3102/5690,
    R3102/5726, R3102/6236, R3102/10017, R3102/10021, R3102/22, R3102/10024
R3112/1, R3112/11, R3112/21, R3112/29, R3112/38, R3112/42, R3112/45, 3112/53, R3112/55,
    3112/87, 3112/92, R3112/93 3112/94, 3112/96, 3112/98, 3112/169
R4604, Generalinspektor für Wasser und Energie, R4604/1, R4604/2, R4604/6,
    R4604/7, R4604/12, R4604/36, R4604/37, R4604/68, R4604/91, R4604/94, R4604/235,
    R4604/236, R4604/512, R4604/519

BERN, SWITZERLAND
Schweizerisches Bundesarchiv Bern (BAR)
E27, Landesverteidigung, Sicherung des elektrischen Bahnbetriebes im Kriegsfall
E53, Eisenbahnwesen, Elektrifizierung des gesamten Streckennetzes
E56, Wasserkraftwerke, Ritomsee/Ritóm, Lago

E2200.53-03, Elektrifizierung der österreichischen Bundesbahnen
E3215-03, Bundesamt für Wasser und Geologie Kraftwerk Ritom, Konzessionserneuerung
E3270A, Abteilung Forstwesen, Jagd und Fischerei, Kraftwerk Massaboden
E3270C, Eidgenössische Forstdirektion—Zentrale Ablage
E8001B-01, Abteilungs Rechtswesen und Sekretariat im Post- und Eisenbahndepartment
E8100A, Eidgenössisches Amt für Verkehr
E8100B, Eidgenössisches Amt für Verkehr
E8170D, Eidgenössisches Amt für Wasserwirtschaft

BOLZANO/BOZEN, ITALY
Archivio provinciale di Bolzano/Südtiroler Landesarchiv Bozen (SLB)
Archiv der Etschwerke
Archiv der grünen Fraktion im Südtiroler Landtag 70

LANCEY, FRANCE
Archives du Musée de la Houille Blanche à Lancey (Conseil Général de l'Isère) (AMHB)

MUNICH, GERMANY
Archiv des Deutschen Museums (DMA)
NL 114

Bayerisches Hauptstaatsarchiv (BayHStA)
Generalstab
Kriegsministerium
Ministerium des Kgl. Hauses und des Äußern (MA), 93011, 93012
Landesamt für Wasserwirtschaft, Vorlage Nr., 26, 27, 31, 33, 39, 40, 41, 152, 180, 181, 188, 189, 190-192
Landesstelle für Naturschutz
Ministerium der Finanzen (MF), 67742-67749
Ministerium des Innern (MInn), 74023
Ministerium für Handel, Industrie und Gewerbe (MHIG), 3128, 3160-3165, 3168, 3169
Ministerium für Land- und Forstwirtschaft (ML), 1856
Ministerium für Unterricht und Kultus (MK), 18262, 51193
Ministerium für Verkehrsangelegenheiten (MV1)
Ministerium für Wirtschaft, Infrastruktur, Verkehr und Technologie (MWi), 2848-2849, 2880-2884, 9499
Oberste Baubehörde (OBB), 9903, 9904, 12812, 15507, 15574, 15575, 15582
Verkehrsarchiv (VArch), 7776, 8690, 8691, 8824-8832, 8838-8840, 8851, 10510, 10525, 10535, 10543-10545, 10546, 10547

SALZBURG, AUSTRIA
Salzburger Landesarchiv (SLA)
Landesregierungsakten 1850-1938
Plakatsammlung 0601

Präsidial-Akten 1948-1950
Rehrl-Akten
   Rehrl-Briefe, 1925-1933
   RehrlTW 1928-1931
Reichsstatthalter (RStH), V/3 171-232, V/d 80-99

VIENNA, AUSTRIA
Österreichisches Staatsarchiv, Archiv der Republik (OeStA/AdR)
Reichskommissar für die Wiedervereinigung Österreichs mit dem Deutschen Reich, 1938-1940 ("Bürckel"-Materien), Kt. 87, Kt. 88
Wasser- und Elektrizitätswirtschaftsamt (WEWA), Kt. 1, Kt. 2

Journals Consulted
Deutsche Technik: Technopolitische Zeitschrift der Architekten, Chemiker, Ingenieure, und Techniker
Elektrizitätswirtschaft
Elektrotechnische Zeitschrift
Wasserkraft: Zeitschrift für Wasserbau und Wasserwirtschaft
Wasserkraftjahrbuch
Wasserwirtschaft
Weisse Kohle
Zeitschrift des Vereines deutscher Ingenieure
Zeitschrift für die gesamte Wasserwirtschaft

Conference Proceedings
Gesamtbericht, Zweite Weltkraftkonferenz. Transactions. Second World Power Conference. *Compte Rendu, Deuxième Conférence Mondiale de l'Énergie.* Edited by F. Zur Nedden and Carl Theodor Kromer. Berlin: VDI-Verlag, 1930.
Syndicat des propriétaires et industriels possédent ou exploitant des forces motrices hydrauliques. *Compte rendu des travaux du Congrès, des visites industrielles et des excursions: Congrès de la Houille Blanche: Grenoble-Annecy-Chamonix 7-13 septembre 1902.* 2 vols. St. Cloud: Belin Frères, 1902.
The Transactions of the First World Power Conference, London, June 30th to July 12th, 1924. 5 vols. London: Percy Lund, Humphries & Co., 1925.
Transactions of the World Power Conference, Basle Sectional Meeting. Compte rendu, de la Conférence Mondiale de l'Énergie, session spéciale de Bâle 1926/Berichterstattung der Weltkraftkonferenz, Sondertagung Basel 1926. 2 vols. Basel: E. Birkhaeuser, 1927.

Published Primary Sources
Adam, Georg. "Der Wert der Wasserkräfte im Kriege." *Zeitschrift für die gesamte Wasserwirtschaft, für Wassertechnik und Wasserrecht* IX, no. 19/20 (October 20, 1914): 271-272.
Arrigon, Louis-Jules. *La Houille Blanche et l'Avenir Industriel du Sud-Est.* Paris: Attinger, 1918.

Austrian Federal Railways. *Over the Arlberg by Electricity: On the Occasion of the Opening of the Electric Traffic on the Line Innsbruck-Bludenz in Spring 1925.* Vienna: Christoph Reisser's Söhne, 1925.

Avezou, R. *Petite Histoire du Dauphiné.* Grenoble: B. Arthaud, 1946.

Blanchard, Raoul. *Les Alpes françaises.* Paris: Librairie Armand Colin, 1952.

———. *Les Forces Hydro-Électriques Pendant la Guerre.* New Haven: Yale University Press, 1924.

———. "Geographical Conditions of Water Power Development." *Geographical Review* 14, no. 1 (January 1924): 88-100.

Blanchard, W. O. "White Coal in Italian Industry." *Geographical Review* 18, no. 2 (April 1928): 261-273.

Budau, Artur. "Über hydraulische Akkumulierungsanlagen bei Kraftwerken." *Zeitschrift des österreichischen Ingenieur- und Architekten-Vereins*, no. 11 (March 13, 1908): 169-174, 185-190.

"Der Bundeskanzler über Südtirol." *Wiener Kurier.* January 28, 1946.

Clerc, G. "A. Berges: L'Inventeur de la HOUILLE BLANCHE." *Revue des Alpes: Journal Illustré Hebdomadaire* (December 4, 1892): 390-391.

Colombo, G. "Le Système Gaulard et Gibbs à l'Exposition de Turin. " *La Lumière Électrique* 14, no. 41 (October 11, 1884): 43-46.

Czimatis, Albrecht. *Energiewirtschaft als Grundlage der Kriegswirtschaft.* Hamburg: Hanseatische Verlagsanstalt, 1936.

Donat, Fedor Maria von. *Die Kraft der Isar, eine Quelle des Reichtums für Staat und Volk. Vortrag gehalten am 13. Dezember 1905, mit Ergänzungen.* Munich: Lindauer, 1906.

Dreyer, Hans. "Großwasserkräfte im Dienst der deutschen Energiewirtschaft." *Die Wasserkraft und Wasserwirtschaft* 26, no. 15 (August 1, 1931): 177-183.

Ellenbogen, Wilhelm. *Anschluss und Energiewirtschaft.* Vienna: Verlag Deutsche Einheit, 1927.

Esposizione Generale Italiana in Torino 1884. *Catalago ufficiale della meccanica agraria, elettricità e meccanica di precisione.* Turin: Unione tipografico-editrice, 1884.

Fay, Sidney B. "Dollfuss: Victim of Nazi Crime," *Current History* 40, no. 6 (September 1934): 733.

The First World Power Conference June 30-July 12, 1924. Under the Presidency of the Rt. Hon. the Earl of Derby, K.G. at the Conference Halls of the British Empire Exhibition Wembley, London. London: Offices of the WPC, [1924?].

Fischer, K. *Die Wahrheit über das staatliche Walchenseeprojekt. Gemeinverständlich dargestellt vom Verfasser des Wettbewerb-Entwurfs Tölz.* [Gustavsburg?]: Selbstverlag, [1912?].

Fischer-Reinau, Ludwig. *Die Wasserkräfte der bayerischen Alpen.* Munich: Süddeutsche Verlags-Anstalt, 1906.

"Gaulard and Gibbs' System of Electrical Distribution," *Engineering* 35 (1883): 205-206.

Generaldirektion der Österreichischen Bundesbahnen. *Die Österreichische Eisenbahnen, 1837-1937. Gedenkblätter zur Hundertjahrfeier der Eröffnung der ersten österreichischen Dampfeisenbahnen.* Vienna: Selbstverlag der Generaldirektion der Österreichischen Bundesbahnen.

Götz, J., ed. *Die Hauptstufe des Tauernwerks Glockner-Kaprun: Festschrift Herausgegeben anlässlich der Fertigstellung der zum Krafthaus Kaprun-Hauptstufe gehörenden Anlagen September 1951.* Vienna: Ueberreuter, 1951.

Granigg, Bartel. *Die Wasserkraftnutzung in Österreich und deren geographischen Grundlagen.* Vienna: Springer, 1925.

Grasmann, M. "Volkswirtschaftliche Bedeutung der bayerischen Wasserkräfte und bayerische Energiewirtschaftspolitik." In *Geschichte der bayerischen Industrie,* edited by Alfred Kuhlo. Munich: Bayerische Druckerei und Verlagsanstalt, 1926.

Grengg, Hermann. *Das Großspeicherwerk Glockner-Kaprun.* Vienna: Springer, 1952.

Gruenewaldt, C. Von. *Die Wasserkraft: Zeitschrift für die gesamte Wasserwirtschaft* 15, no. 14 (July 15, 1924): 235.

Grünhut, Karl, ed. *Elektrizitätswirtschaft und Wasserkraftnutzung: Wechselrede, gehalten in den Fachgruppen Bau- und Eisenbahn-Ingenieure und für Elektrotechnik des Oesterr. Ingenieur und Architekten-Vereines.* Berlin: Urban & Schwarzberg, 1919.

Hallinger, Johann. *Bayerns Wasserkräfte und Wasserwirtschaft: Zum 10jährigen Bestehen des Bayer. Wasserwirtschaftrates.* Diessen vor München: Jos. C. Huber, 1918.

———. "Das Tauernwerk im Wirtschaftsspiegel." *Wasserkraft und Wasserwirtschaft* 27, no. 2 (January 16, 1932): 13-15.

———. *Die großen staatlichen Niederdruckwasserkräfte in Südbayern, deren Erschließung und Verwertung nach den Grundsätzen der größten Wirtschaftlichkeit und des kleinsten Massenaufwandes.* Diessen vor München: Jos. C. Huber, 1916.

Hamann, Fritz. "Das Achenseewerk." In *Jenbacher Buch: Beiträge zur Heimatkunde von Jenbach und Umgebung.* Schlern-Schriften 101. Innsbruck: Wagner, 1953.

Hartl, C. *Bayern auf dem Weg zum Industriestaat: Eine vergleichende volkswirtschaftliche Studie über die Ausnützung der bayer. Wasserkräfte, sowie über Staats- und Privatbetrieb in den Industrien der schwarzen und der weißen Kohle.* Munich: Max Steinebach, 1911.

Heller, Erich. "Technische Probleme des Tauernwerks." *Die Wasserwirtschaft* 24, no. 13/14 (1931): 181-187.

Hofbauer, Richard. "Zeitgemäße Betrachtungen über den projektierten Ausbau der österr. Alpenwasserkräfte mit Bezug auf deren künftige Funktion in der Energiewirtschaft Mitteleuropas." *Die Wasserwirtschaft* 24, no. 12 (1931): 169-171.

Huber-Stockar, Emil. *Die Elektrifizierung der Schweizerischen Bundesbahnen bis Ende 1928. Neujahrsblatt der Naturforschenden Gesellschaft in Zürich auf das Jahr 1929.* Zurich: Beer, 1929.

Kammerer. "Ausnutzung der Wasserläufe im bayerischen Hochlande für elektrische Energieverteilung." *Zeitschrift des Vereines deutscher Ingenieure* 41, no. 30 (1897): 864-867.

"Der Kampf um die Isar." *Münchner Neueste Nachrichten.* February 27, 1907.

K. Bayerisches Staatsministerium für Verkehrsangelegenheiten. *Bericht über den Stand der Ausnützung der Staatseisenbahnverwaltung vorbehaltenen staatlichen Wasserkräfte.* Munich: C. Wolf & Sohn, 1914.

———. *Denkschrift über die Einführung des elektrischen Betriebes auf den bayerischen Staatseisenbahnen.* Munich: Staatsministerium für Verkehrsangelegenheiten, 1908.

Klebelsberg, R. "Das Becken von Längenfeld im Ötztal: Ein Beispiel für: Geologie und Kraftwerksplanung." In *Tiroler Wirtschaft in Vergangenheit und Gegenwart,* edited by

Hermann Gerhardinger and Adolf Günther. Innsbruck: Universitäts-Verlag Wagner, 1951.

K. Oberste Baubehörde. *Die Wasserkräfte Bayerns*. Munich: Piloty & Loehle, 1907.

Köhler, Walter. "Zur Einführung!" *Die Wasserkraft* 1, no. 1 (April 1, 1919): 1-2.

Köhn, Theodor. "Einige allgemeine Betrachtungen über den Ausbau von Wasserkräften. *Die Weisse Kohle* 1, no. 2 (January 25, 1908): 5.

Kollmann, Franz. *Wunderwerke der Technik*, 3rd ed. Stuttgart: Union Deutsche Verlagsgesellschaft, 1931.

Kreittmayr, Wigulè II [pseud.]. *Ministerium Hertling Weg—oder Ein Angriff auf's Bayerische Wespennest: Eine hochpolitische Lektüre aus einer kleinen Zauberstube*. Munich: F. Bosch, [1912?].

Kretzschmar, "Die Mächtigkeit der weißen Kohle." *Die Weisse Kohle* 1, no. 11 (June 10, 1908): 125

Kruck, Gustav. *Das Kraftwerk Wäggital: Neujahrsblatt der Naturforschenden Gesellschaft in Zürich*. Zurich: Gebr. Fretz, 1925.

Lambotte, Alfred. *Quelques applications de l'életrotechnie dans l'Europe centrale: Notes de voyage par Alfred Lambotte*. Brussels: Ramlot, 1907.

Länderrat des Amerikanisches Besatzungsgebiets, ed. *Statistisches Handbuch von Deutschland, 1928-1944*. Munich: Franz Ehrenwirth, 1949.

Lawaczeck, Franz. *Elektrowirtschaft*. Munich: Lehmann, 1936.

Link, Harald. "Die Speicherseen der Alpen. Bassins d'accumulation des Alpes." *Cours d'eau et énergie* 62 (October 1970): 241-357.

———. *Die Speicherseen der Alpen: Bestand und Planung*. Zurich: Schweizerischer Wasserwirtschaftverbands, 1953.

Mattern, Emil. *Die Ausnützung der Wasserkräfte, technische und wirtschaftliche Grundlagen, neuere Bestrebungen der Kulturländer*. Leipzig: W. Engelmann, 1921.

———. "Die hochalpinen Wasserkräfte im Rahmen der mitteleuropäischen Stromversorgung." *Elektrotechnische Zeitschrift*, no. 38 (September 22, 1932): 907-911.

Meyer, Konrad. "Die deutsche Elektrizitätswirtschaft 1933 bis 1948." *Elektrizitätswirtschaft* 48, no. 2 (February 1949): 34-40.

Miller, Oskar von. "Bayerns Wasserkräfte." *Süddeutsche Bauzeitung* 17, no. 30 (July 27, 1907): 233-237.

———. *Die Naturkräfte im Dienste der Elektrotechnik*. Leipzig: F. C. W. Vogel, 1902.

———. *Die Wasserkräfte am Nordabhange der Alpen*. Berlin, 1903.

Mirande, Marcel. *Le Comte de Cavour et la Houille Blanche*. Grenoble: Allier Père & Fils, 1927.

Moro, Richard. "Betrachtungen über das Tauernkraftwerks-projekt der AEG." *Die Wasserwirtschaft* 24, no. 8 (1931): 109-117.

Mühlhofer, Ludwig, and Carl Reindl. *Das Achensee-Kraftwerk*. Special issue, *Wasserkraft und Wasserwirtschaft*. [1928?].

Müller, Hans. *Gesetz zur Förderung der Energiewirtschaft vom 13. Dezember 1935 (RGBL. I S. 1451): Erläutert von Dr. Hans Müller Rechtsanwalt und Notar*. Berlin: Verlag Franz Vahlen, 1936.

Münch, Wilhelm "Das Tauernwerk." *Deutsche Wasserwirtschaft* 26, no. 1 (January 10, 1931): 1-5.

Oliven, Oskar. "Europas Großkraftlinien: Vorschlag eines europäischen Großkraftnetzes." In *Gesamtbericht, Zweite Weltkraftkonferenz/Transactions, Second World Power Conference, Compte rendu, Deuxième Conférence Mondiale de l'Énergie*, vol. XIX, edited by F. zur Nedden, 30-39. Berlin: VDI-Verlag, 1930.

"Die Opposition gegen das Walchenseeprojekt." *Frankfurter Zeitung und Handelsblatt* Nr. 15 Sonntag. May 31, 1908.

Österreichisch-Deutsche Arbeitsgemeinschaft. *Das Österreichische Wirtschaftsproblem. Denkschrift der österr.-deutschen Arbeitsgemeinschaft.* Vienna: Hölder/Pichler/Tempsky, 1925.

Ostwald, Wilhelm. *Energetische Grundlagen der Kulturwissenschaft.* Leipzig: Werner Klinkhardt, 1909.

Petersen, "Die Energiewirtschaft des Tauernkraftwerkes." *Die Wasserwirtschaft* 24, no. 13/14 (1931): 188-190.

Ratzel, Friedrich. *Die Alpen inmitten der geschichtlichen Bewegungen.* [Graz?]: [Lindauer in Komm.?], 1896.

Rehbock, Theodor. "Der wirtschaftliche Wert der binnenländischen Wasserkräfte, unter besonderer Berücksichtigung des Großherzogtums Baden." *Die Weisse Kohle* 1, no. 3 (February 10, 1908): 1-4.

Reichsverkehrsministerium. *Hundert Jahre deutsche Eisenbahnen. Jubiläumsschrift zum hundertjährigen Bestehen der deutschen Eisenbahnen*, 2nd ed. Leipzig: Verkehrswissenschaftliche Lehrmittelgesellschaft, 1938.

Rühlmann, Richard. "Marcel Deprez' Versuche über elektrische Kraftübertragung." *Zeitschrift des Vereines deutscher Ingenieure* 29, no. 50 (December 12, 1885): 981-982.

Schirmer, Alice. "Die schweizerischen Wasserkräfte als volkswirtschaftliches Gut." PhD dissertation, University of Zurich, 1921.

Schmidt, Albert. "Das Schicksal und die Zukunft des Walchensees und der Isar." *Münchner Neueste Nachrichten.* December 5, 1907.

Schmidt-Stölting, Hans. *Das Problem des Ausbaues der deutschen Wasserkräfte.* Berlin: Georg Stilke, 1930.

Schmitz, Georg. "Das Walchensee-Kraftwerk und seine Bedeutung für die Energiewirtschaft Deutschlands." *Die Braunschweiger G.N.C. Monatsschrift* (October 1921): 611-615.

Schweiz, Departement des Innern/Departement suisse de l'interieur. *Verfügbare Wasserkräfte/Forces hydrauliques disponibles.* Vol. 5 of *Die Wasserkräfte der Schweiz/Les forces hydrauliques de la suisse.* Bern: Rösch & Schatzmann, 1916.

Senger, Max. *Wie die Schweizer Alpen erobert wurden.* Zurich: Büchergilde Gutenberg, 1945.

"Stellungnahme der Ingenieurkammer für Tirol und Vorarlberg zu den Projekten der Allgemeinen Elektrizitäts-Gesellschaft Berlin (AEG) und der Österreichischen Kraftwerke A.-G. (Oeka) für das Tauernkraftwerk." *Die Wasserwirtschaft* 24, no. 11 (1931): 158-162.

Stephen, Leslie. *The Playground of Europe.* London: Longmans, Green and Co., 1871.

Thierbach, Bruno. "Die Kraftwerke Oberhasli. " *Elektrotechnische Zeitschrift*, no. 40 (October 6, 1932): 955–958.
Thürnau. "Das Tauernwerksprojekt der AEG." *Die Wasserwirtschaft* 24, no. 11 (1931): 157–158.
Torino e L'Esposizione Italiana del 1884. *Cronaca Illustrata della Esposizione Nazionale Industriale ed Artística del 1884*. Turin: Roux e Favale & Fratelli Treves, 1884.
Trollié, F. "A Travers la Ville: L'Electricite." *Le Clairon des Alpes*. July 1899.
United Nations Economic Commission for Europe. *ECE: The First Ten Years, 1947–1957*. Geneva, 1957.
United States Department of State. *Committee of European Economic Co-operation: Volume I: General Report*. Washington, DC: Division of Publications, 1947.
United States Strategic Bombing Survey. *German Electrical Utilities Industry Report*, 2nd ed. Washington, DC: Utilities Division, 1947.
Urbanitzky, Alexander von. "Neueste Ergebnisse und praktische Fortschritte der Elektrizität: Rückblicke auf die internationale Aufstellung für Electricität zu München." *Neueste Erfindungen und Erfahrungen auf den Gebieten der praktischen Technik, der Gewerbe, Industrie, Chemie, der Land- und Hauswirtschaft* 10 (1883): 50–62.
Vas, Oskar. *Der Anteil Österreichs an der elektrizitäts-wirtschaftlichen Gemeinschaftsplanung in Europa*. Vienna: Springer, 1948.
———. "Die Ausnutzung der Tauernwasserkräfte als österreichisches Problem." *Die Wasserwirtschaft* 24, no. 13/14 (1931): 191–201.
———, ed. *Grundlagen und Entwicklung der Energiewirtschaft Österreichs: Offizieller Bericht des österreichischen Nationalkomitees der Weltkraftkonferenz*. Vienna: Springer, 1930.
———. *Probleme der Kraftwirtschaft in Mitteleuropa*. Vienna: Springer, 1952.
"Vom Tage." *Simplicissimus* 27, no. 49 (March 5, 1923): 687.
Von Liebig. "Über die Bedeutung der Industrie für die Volks- und Staatswirtschaft und die Ausnützung der staatlichen Wasserkräfte Bayerns." *Die Weisse Kohle* 1, no. 2 (January 25, 1908): 5.
Wells, H. G. *Anticipations of the Reaction of Mechanical and Scientific Progress upon Human Life and Thought*, 2nd ed. London: Chapman & Hall, 1902.
Wyssling, Walter. *Die Entwicklung der schweizerischen Elektrizitätswerke und ihrer Bestandteile in den ersten 50 Jahren*. Zurich: Schweizerischer Elektrotechnischer Verein Fachschriften Verlag, 1946.
———, ed. *Mitteilungen der Schweizerischen Studienkommission für elektrischen Bahnbetrieb*. Vol. 1: *Der Kraftbedarf für elektrischen Bahnbetrieb*. Zurich: Rascher & Cie., 1906.
Zinßmeister, Jakob. "Was die neue Zeitschrift will?" *Die Weiße Kohle* 1, no. 1 (January 10, 1908): 1

Secondary Books, Articles, and Websites
Allen, Robert C. *The British Industrial Revolution in Global Perspective*. Cambridge, UK: Cambridge University Press, 2009.
André, Louis. *Aristide Bergès, une vie d'innovateur: De la papeterie à la houille blanche*. Grenoble: Presses universitaires de Grenoble, 2013.
André, Louis, and Serge Benoit. "L'innovation énergétique dans l'industrie papetière, des

défibreurs aux hautes chutes. " In *Innovation dans la gestion environnementale des territoires de montagne. Actes du 131e Congrès national des sociétés historiques et scientifiques*, 37–48. Paris: Editions du CTHS, 2009.

Andrews, Thomas G. *Killing for Coal: America's Deadliest Labor War*. Cambridge, MA: Harvard University Press, 2008.

Applegate, Celia. *A Nation of Provincials: The German Idea of Heimat*. Berkeley: University of California Press, 1990.

Armiero, Marco. *A Rugged Nation: Mountains and the Making of Modern Italy*. Cambridge, UK: White Horse Press, 2011.

Armiero, Marco, Roberta Biasillo, and Wilko Graf von Hardenberg. *Mussolini's Nature: An Environmental History of Italian Fascism*. Translated by James Sievert. Cambridge, MA: MIT Press, 2022.

Aschwanden, Romed. *Politisierung der Alpen: Umweltbewegungen in der Ära der Europäischen Integration (1970–2000)*. Vienna: Böhlau, 2021.

Autran, A. "Introduction to the Geology of Western and Southern Europe." In *Geology and the Environment in Western Europe: A Coordinated Statement by the Western European Geological Surveys*, edited by G. Innes Lumsden, 9–33. Oxford: Clarendon Press, 1994.

Barak, On. *Powering Empire: How Coal Made the Middle East and Sparked Global Carbonization*. Oakland: University of California Press, 2020.

Bätzing, Werner. *Die Alpen: Geschichte und Zukunft einer europäischen Kulturlandschaft*, 3rd ed. Munich: Beck, 2005.

——. *Zwischen Wildnis und Freizeitpark: Eine Streitschrift zur Zukunft der Alpen*. Zurich: Rotpunktverlag, 2015.

Bayerl, Günter, ed. *Umweltgeschichte-Methoden, Themen, Potentiale: Tagung des Hamburger Arbeitskreises für Umweltgeschichte*. Muenster: Waxmann, 1996.

Bayerl, Günter, and Torsten Meyer, eds. *Die Veränderungen der Kulturlandschaft: Nutzungen-Sichtweisen-Planungen*. Muenster: Waxmann, 2003.

Beattie, Andrew. *The Alps: A Cultural History*. Oxford: Oxford University Press, 2006.

Bergier, Jean-François, and Sandro Guzzi, eds. *La découverte des Alpes. La scoperta delle Alpi. Die Entdeckung der Alpen*. Basel: Schwabe, 1992.

Billington, David P., and Donald C. Jackson. *Big Dams of the New Deal Era*. Norman: University of Oklahoma Press, 2006.

Billington, David P. Donald C. Jackson, and Martin V. Melosi. *The History of Large Federal Dams: Planning, Design, and Construction*. Denver: US Department of the Interior, Bureau of Reclamation, 2005.

Bischof, Günter. "Zweite Republik." In *Österreichische Zeitgeschichte—Zeitgeschichte in Österreich: Eine Standortbestimmung in Zeiten des Umbruchs*, edited by Marcus Gräser and Dirk Rupnow, 160–177. Vienna: Böhlau, 2021.

Bischof, Günter, and Hans Petschar. *The Marshall Plan: Saving Europe, Rebuilding Austria*. New Orleans: University of New Orleans Press, 2017.

Black, Brian. *Petrolia: The Landscape of America's First Oil Boom*. Baltimore: Johns Hopkins University Press, 2000.

Blackbourn, David. *The Conquest of Nature: Water, Landscape, and the Making of Modern Germany*. New York: W. W. Norton, 2006.

———. "The Culture and Politics of Energy in Germany: A Historical Perspective." *RCC Perspectives*, no. 4 (2013): 3-31.

Blaich, Fritz. *Die Energiepolitik Bayerns, 1900–1921.* Kallmünz/Opf: Lassleben, 1981.

Boll, Georg. *Geschichte des Verbundbetriebs: Entstehung und Entwicklung des Verbundbetriebs in der deutschen Elektrizitätswirtschaft bis zum europäischen Verbund: Ein Rückblick zum 20-Jährigen Bestehen der deutschen Verbundgesellschaft.* Frankfurt: VWEW, 1969.

Bonan, Giacomo. "An Alpine Energy Transition: The Piave River from Charcoal to 'White Coal.'" *Environmental History* 25, no. 4 (October 2020): 687–710.

Bonneuil, Christophe, and Jean-Baptiste Fressoz. *The Shock of the Anthropocene: The Earth, History, and Us.* Translated by David Fernbach. Brooklyn: Verso, 2016.

Bonoldi, Andrea. "Technologien, Kapitalien und Kontrolle der Ressourcen: die regionale Elektrizitätswirtschaft." In *Die Region Trentino-Südtirol im 20. Jahrhundert*, vol. 2, edited by Andrea Leonardi, 235–252. Trent: Museo Storico in Trento, 2009.

Bonoldi, Andrea, and Andrea Leonardi, eds. *Energia e Sviluppo in Area Alpina: secoli XIX–XX: atti della VII sessione del seminario permanente sulla storia dell'economie e dell'imprenditorialità nelle Alpi in età moderna e contemporanea.* Milan: FrancoAngeli, 2004.

Borsdorf, Axel, Johann Stotter, and Eric Veulliet, eds. *Managing Alpine Future: Proceedings of the Innsbruck Conference October 15–17, 2007.* IGF Forschungsberichte, vol. 2. Vienna: Österreichische Akademie der Wissenschaften, 2008.

Botteri Balli, Alessandro. *Wasserkraftwerke der Schweiz: Architektur und Technik.* Zurich: Offizin, 2003.

Botz, Gerhard. *Die Eingliederung Österreichs in das Deutsche Reich. Planung und Verwirklichung des politisch-administrativen Anschlusses (1938–1940)*, 3rd ed. Vienna: Europa Verlag, 1988.

Bouneau, Christophe. *Modernisation et Territoire: L'Électrification du Grand Sud-Ouest de la Fin du XIXe Siècle à 1946.* Bordeaux: Fédération Historique du Sud-Ouest, 1997.

Bouvier, Yves, and Giovanni Paoloni, eds. "Transitions in Energy History: History in Energy Transitions." Special issue, *Journal of Energy History/Revue d'Histoire de l'Énergie*, no. 4 (June 2020).

Brantz, Dorothee. "Environments of Death: Trench Warfare on the Western Front, 1914–1918." In *War and the Environment: Military Destruction in the Modern Age*, edited by Charles E. Closmann, 68–91. College Station: Texas A&M University Press, 2009.

Braudel, Fernand. *The Mediterranean and the Mediterranean World in the Age of Philip II.* Vol. 1. New York: Harper & Row, 1972.

Braun, Hans-Joachim. "Die Weltenergiekonferenzen als Beispiel internationaler Kooperation." In *Energie in der Geschichte: zur Aktualität der Technikgeschichte.* Düsseldorf: [?], 1984.

Bravard, Jean-Paul et al. "River Incision in South-East France: Morphological Phenomena and Ecological Effects." *River Research and Applications* 13, no 1. (January 1997): 75–90.

Brüggemeier, Franz-Josef, Mark Cioc, and Thomas Zeller, eds. *How Green Were the Nazis? Nature, Environment, and Nation in the Third Reich.* Athens: Ohio University Press, 2005.

Burke, Edmund, III. "The Big Story: Human History, Energy Regimes, and the Environment." In *The Environment and World History*, edited by Edmund Burke III and Kenneth Pomeranz. Berkeley: University of California Press, 2009.

Burns, Robert K., Jr. "The Circum-Alpine Culture Area." *Anthropological Quarterly* 36, no. 3 (July 1963): 130–155.

Carlson, W. Bernard. *Tesla: Inventor of the Electrical Age*. Princeton: Princeton University Press, 2013.

Cater, Casey P. *Regenerating Dixie: Electric Energy and the Modern South*. Pittsburgh: University of Pittsburgh Press, 2019.

Cioc, Mark. *The Rhine: An Eco-Biography, 1815–2000*. Seattle: University of Washington Press, 2002.

CIPRA International. *Nous les Alpes!: Des femmes et des hommes façonnent l'avenir. 3ème Rapport sur l'etat des Alpes*. [Gap]: Y. Michel, 2007.

Clark, John Garretson. *The Political Economy of World Energy: A Twentieth Century Perspective*. Chapel Hill: University of North Carolina Press, 1991.

Cleveland, Cutler J., and Christopher G. Morris, eds. *Handbook of Energy*. 2 vols. Oxford: Elsevier Science, 2013.

Closmann, Charles E. *War and the Environment: Military Destruction in the Modern Age*. College Station: Texas A&M University Press, 2009.

Cohn, Julie A. *The Grid: Biography of an American Technology*. Cambridge, MA: MIT Press, 2017.

Cohn, Julie, Matthew Evenden, and Marc Landry. "Water Powers: The Second World War and the Mobilization of Hydroelectricity in Canada, the United States, and Germany." *Journal of Global History* 15, no. 1 (2020): 123–147.

Cole, John W., and Eric R. Wolf. *The Hidden Frontier: Ecology and Ethnicity in an Alpine Valley*. New York: Academic Press, 1974.

Conte, Christopher A. *Highland Sanctuary: Environmental History in Tanzania's Usambara Mountains*. Athens: Ohio University Press, 2004.

Conway, Martin, and Kiran Klaus Patel, eds. *Europeanization in the Twentieth Century: Historical Approaches*. Basingstoke: Palgrave, 2010.

Crosby, Alfred W. *Children of the Sun: A History of Humanity's Unappeasable Appetite for Energy*. New York: W. W. Norton, 2006.

Cunliffe, Barry. *Europe Between the Oceans: 9000 BC–AD 1000*. New Haven: Yale University Press, 2008.

Dalmasso, Anne. "Barrages et développement dans les Alpes françaises de l'entre-deux-guerres."*Revue de géographie alpine* 96, no. 1 (2008): 45–54.

———. "D'une hydraulique à l'autre: L'évolution des usages industriels de l'eau dans la vallée de l'Arve de la fin du XVIII siècle au début du XX siècle." In *L'eau à Genève et dans la région Rhône-Alpes, XIXe-XXe siècles*, edited by Serge Paquier, 43–56. Paris: L'Harmattan, 2007.

———. "L'ingénieur, la Houille Blanche et les Alpes: Une utopie modernisatrice?" In "Le temps bricolé. Les représentations du Progrès (XIXe-XXe siècles)." Special issue, *Le Monde alpin et rhodanien: Revue régionale d'ethnologie* 29, no. 1–3 (2001): 25–38.

Debeir, Jean-Claude, Jean-Paul Deléage, and Daniel Hémery. *In the Servitude of Power: Energy and Civilization Through the Ages*. London: Zed Books, 1991.

De Grazia, Victoria. *Irresistible Empire: America's Advance Through Twentieth-Century Europe.* Cambridge, MA: Harvard University Press, 2006.

Denning, Andrew. *Skiing into Modernity: A Cultural and Environmental History.* Oakland: University of California Press, 2015.

Denzler, Lukas. "Der Spöl blüht wieder auf." *Neue Zürcher Zeitung.* June 18, 2014.

De Pretto, Sebastian. "Un espace sans conflict? Structures de pouvoir et path dependencies autour du lac d'Émosson, 1953–1975." *Geschichte der Alpen—Histoire des Alpes—Storia delle Alpi* 27 (2022): 173–189.

Dominick, Raymond H. *The Environmental Movement in Germany: Prophets & Pioneers, 1871–1971.* Bloomington: Indiana University Press, 1991.

Eaglin, Jennifer. *Sweet Fuel: A Political and Environmental History of Brazilian Ethanol.* Oxford: Oxford University Press, 2022.

Edgerton, David. "Controlling Resources: Coal, Iron and Oil in the Second World War." In *Total War: Economy, Society and Culture at War,* vol. 3 of *Cambridge History of the Second World War,* edited by Michael Geyer and Adam Tooze, 122–148. Cambridge, UK: Cambridge University Press, 2015.

Evans, Sterling. "Recent Developments in Transnational Environmental History: Labor, Settler Communities, and Comparative Histories." *Radical History Review* 107 (Spring 2010): 195–208.

Evenden, Matthew. *Allied Power: Mobilizing Hydro-Electricity During Canada's Second World War.* Toronto: University of Toronto Press, 2015.

Falter, Reinhard. "Achtzig Jahre 'Wasserkrieg': Das Walchensee-Kraftwerk." In *Von der Bittschrift zur Platzbesetzung: Konflikte um techn. Großprojekte; Laufenburg, Walchensee, Wyhl, Wackersdorf,* edited by Ulrich Linse. Berlin: Dietz, 1988.

Fells, Ian. *World Energy 1923–1998 and Beyond: A Commemoration of the World Energy Council on Its 75th Anniversary.* London: Atalink, 1998.

Fischer, Wolfram. *Die Geschichte der Stromversorgung.* Frankfurt/Main: Verl.- und Wirtschaftsges. der Elektrizitätswerke, 1992.

Frank, Alison Fleig. *Oil Empire: Visions of Prosperity in Austrian Galicia.* Cambridge, MA: Harvard University Press, 2005.

Frank, Wilhelm. "Zur Geschichte der Energieplanung in Österreich." *Wirtschaft und Gesellschaft* 8, no. 2 (1982): 235–270.

Frascaroli, Fabrizio, Giacomo Parrinello, and Meredith Root-Bernstein. "Linking Contemporary River Restoration to Economics, Technology, Politics, and Society: Perspectives from a Historical Case Study of the Po River Basin, Italy." *Ambio* 50 (2021): 492–504.

Frost, Robert L. *Alternating Currents: Nationalized Power in France, 1946–1970.* Ithaca: Cornell University Press, 1991.

Füßl, Wilhelm. *Oskar von Miller, 1855–1934: Eine Biographie.* Munich: Beck, 2005.

Gall, Alexander. *Das Atlantropa-Projekt: Die Geschichte einer gescheiterten Vision. Herman Sörgel und die Absenkung des Mittelmeers.* Frankfurt: Campus, 1998.

Gerber, Sophie. *Küche, Kühlschrank, Kilowatt. Zur Geschichte des privaten Energiekonsums in Deutschland, 1945–1990.* Bielefeld: Transcript, 2015.

Geyer, Michael, and Adam Tooze. *Total War: Economy, Society and Culture at War.* Vol. 3

of *Cambridge History of the Second World War*. Cambridge, UK: Cambridge University Press, 2015.

Gillingham, John. *Coal, Steel, and the Rebirth of Europe, 1945-1955: The Germans and French from Ruhr Conflict to Economic Community*. Cambridge, UK: Cambridge University Press, 1991.

Gilson, Norbert. *Konzepte von Elektrizitätsversorgung und Elektrizitätswirtschaft: Die Entstehung eines neuen Fachgebietes der Technikwissenschaften zwischen 1880 und 1945*. Diepholz: GNT Verlag, 1994.

Girel, Jacky. "River Diking and Reclamation in the Alpine Piedmont: The Case of the Isère." In *Rivers in History: Perspectives on Waterways in Europe and North America*, edited by Christof Mauch and Thomas Zeller, 78-88. Pittsburgh: University of Pittsburgh Press, 2008.

Gräser, Marcus, and Dirk Rupnow. *Österreichische Zeitgeschichte—Zeitgeschichte in Österreich: Eine Standortbestimmung in Zeiten des Umbruchs*. Vienna: Böhlau, 2021.

Groß, Robert. *Die Beschleunigung der Berge: Eine Umweltgeschichte des Wintertourismus in Vorarlberg/Österreich (1920-2010)*. Vienna: Böhlau, 2019.

Guagnini, Anna. "A Bold Leap into Electric Light: The Creation of the Società Italiana Edison, 1880-1886." *History of Technology* 32 (2014): 155-189.

Gugerli, David, ed. *Allmächtige Zauberin unserer Zeit: Zur Geschichte der elektrischen Energie in der Schweiz*. Zurich: Chronos, 1994.

———. *Redeströme: Zur Elektrifizierung der Schweiz, 1880-1914*. Zurich: Chronos, 1996.

Guichonnet, Paul. *Histoire et civilisations des Alpes*. 2 vols. Toulouse: Privat, 1980.

Guichonnet, Paul, and Emanuele Kanceff, eds. *Alpi, laghi e letterature. Les Alpes, les lacs, les lettres*. Cahiers de civilization Alpine 7. Geneva: Slatkine, 1988.

Gutmann, Martin. "The Nature of Total War: Grasping the Global Environmental Dimensions of World War II." *History Compass* 13, no. 5 (2015): 251-261.

Haag, Erich. *Grenzen der Technik: Der Widerstand gegen das Kraftwerkprojekt Urseren*. Zurich: Chronos, 2004.

Hahn, Sylvia, and Reinhold Reith, eds. *Umwelt-Geschichte: Arbeitsfelder-Forschungsansätze- Perspektiven*. Munich: Oldenbourg, 2001.

Haidvogl, Gertrud, Didier Pont, and Žiga Zwitter, "Geschichte menschlicher Nutzungen und Einriffe: Alpenflüsse als Ressource und Risiko." In *Flüsse der Alpen: Vielfalt in Natur und Kultur*, edited by Susanna Muhar et al., 36-45. Bern: Haupt, 2019.

Hanisch, Ernst. *Landschaft und Identität: Versuch einer österreichischen Erfahrungsgeschichte*. Vienna: Böhlau, 2019.

———. *Der lange Schatten des Staates. Österreichische Gesellschaftsgeschichte im 20. Jahrhundert (Österreichische Geschichte 1890-1990)*. Vienna: Ueberreuter, 1994.

Hausman, William J., Peter Hertner, and Mira Wilkins. *Global Electrification: Multinational Enterprise and International Finance in the History of Light and Power, 1878-2007*. Cambridge, UK: Cambridge University Press, 2008.

Hecht, Gabrielle. *The Radiance of France, Nuclear Power and National Identity After World War II*. Cambridge, MA: MIT Press, 2000.

Heymann, Matthias. *Die Geschichte der Windenergienutzung, 1890-1990*. Frankfurt/Main: Campus, 1995.

Hirt, Paul. *The Wired Northwest: The History of Electric Power, 1870s to 1970s.* Lawrence: University of Kansas Press, 2012.

Histoire génerale de l'électricité en France. 3 vols. Paris: Fayard, 1991–1996.

Hobsbawm, E. J., and T. O. Ranger. *The Invention of Tradition.* Cambridge, UK: Cambridge University Press, 1983.

Hochgrassl, Peter. "Die Wasserkräfte Frankreichs und ihre Nutzung im Rahmen der französischen Elektrizitätswirtschaft." PhD dissertation, University of Munich, 1957.

Hochman, Erin R. *Imagining a Greater Germany: Republican Nationalism and the Idea of Anschluss.* Ithaca: Cornell University Press, 2016.

Högselius, Per, Arne Kaijser, and Erik van der Vleuten. *Europe's Infrastructure Transition: Economy, War, Nature.* New York: Palgrave Macmillan, 2016.

Hohensinner, Severin et al. "Die Morphologie: Die vielfältige Gestalt der Alpenflüsse." In *Flüsse der Alpen: Vielfalt in Natur und Kultur,* edited by Susanna Muhar et al., 86–113. Bern: Haupt, 2019.

———. "What Remains Today of Pre-Industrial Alpine Rivers? Census of Historical and Current Channel Patterns in the Alps." *River Research and Applications* 37, no. 2 (February 2021): 128–149.

Hollenstein, Roman, ed. *Kreativität: Herausgegeben zum 75jährigen Jubiläum der Nordostschweizerischen Kraftwerke AG.* Frauenfeld: Huber, 1989.

Hufschmied, Richard. "'Weißes Gold' in (Deutsch-)Österreich—Kontinuität und Wandel nach dem Epochenjahr 1918." In *Wasserkraft. Elektrizität. Gesellschaft. Kraftwerksprojekte ab 1880 im Spannungsfeld,* edited by Oliver Rathkolb, Richard Hufschmied, Andreas Kuchler, and Hannes Leidinger, 84–148. Vienna: Kremayr & Scheriau, 2012.

———. "'Weißes Gold' in der Donaumonarchie." In *Wasserkraft. Elektrizität. Gesellschaft. Kraftwerksprojekte ab 1880 im Spannungsfeld,* edited by Oliver Rathkolb, Richard Hufschmied, Andreas Kuchler, and Hannes Leidinger, 27–56. Vienna: Kremayr & Scheriau, 2012.

Hughes, Thomas P. *Networks of Power: Electrification in Western Society, 1880–1930.* Baltimore: Johns Hopkins University Press, 1983.

Hunter, Louis C. *Waterpower in the Century of the Steam Engine.* Charlottesville: University of Virginia Press, 1979.

Hyde, Walter Woodburn. "The Alps in History." *Proceedings of the American Philosophical Society* 75, no. 6 (1935): 431–442.

Iggers, Georg C. *The German Conception of History: The National Tradition of Historical Thought from Herder to the Present.* Middletown, CT: Wesleyan University Press, 1968.

Jakobsson, Eva. "Industrialization of Rivers: A Water System Approach to Hydropower Development." *Knowledge, Technology, & Policy* 14, no. 4 (Winter 2002): 41–56.

Jensen, W. G. "The Importance of Energy in the First and Second World Wars." *Historical Journal* 11, no. 3 (1968): 538–554.

Jones, Christopher F. "A Landscape of Energy Abundance: Anthracite Coal Canals and the Roots of American Fossil Fuel Dependence, 1820–1860." *Environmental History* 15 (July 2010): 449–484.

———. *Routes of Power: Energy and Modern America.* Cambridge, MA: Harvard University Press, 2014.

Jonnes, Jill. *Eiffel's Tower: And the World's Fair Where Buffalo Bill Beguiled Paris, the Artists Quarreled, and Thomas Edison Became a Count*. New York: Viking, 2009.

Josephson, Paul. *Industrialized Nature: Brute Force Technology and the Transformation of the Natural World*. Washington, DC: Island Press, 2002.

Judson, Pieter M. *The Habsburg Empire: A New History*. Cambridge, MA: Belknap Press, 2016.

Kaiser, Peter. "Das Wasser der Berge—Bedrohung und Nutzen für die Menschen: Notizen für eine Umweltgeschichte." In *La découverte des Alpes. La scoperta delle Alpi. Die Entdeckung der Alpen*, edited by Jean-François Bergier and Sandro Guzzi. Basel: Schwabe, 1992.

Kander, Astrid, Paolo Malanima, and Paul Warde. *Power to the People: Energy in Europe over the Last Five Centuries*. Princeton: Princeton University Press, 2013.

Kann, Robert A. *A History of the Habsburg Empire, 1526–1918*. Berkeley: University of California Press, 1974.

Karner, Stefan, and Peter Ruggenthaler, eds. *1938: Der "Anschluss" im internationalen Kontext*. Vienna: Leykam, 2020.

Keaten, Jamey. "The Battle of the Alps? Water Woes Loom amid Climate Change." *Seattle Times*. October 27, 2022. https://www.seattletimes.com/business/battle-of-the-alps-water-woes-loom-amid-climate-change/

Kelk, B. "Natural Resources in the Geological Environment." In *Geology and the Environment in Western Europe*, edited by G. Innes Lumsden. Oxford: Clarendon Press, 1994.

Keller, Tait. *Apostles of the Alps: Mountaineering and Nation Building in Germany and Austria, 1860–1939*. Chapel Hill: University of North Carolina Press, 2016.

———. "Eternal Mountains, Eternal Germany: The Alpine Association and the Ideology of Alpinism, 1909–1939." PhD dissertation, Georgetown University, 2006.

———. "The Mountains Roar: The Alps During the Great War." *Environmental History* 14, no. 2 (April 2009): 253–274.

Knoll, Martin, Uwe Lübken, and Dieter Schott, eds. *Rivers Lost, Rivers Regained: Rethinking City-River Relations*. Pittsburgh: University of Pittsburgh Press, 2017.

Koller, Maria Magdalena. "Elektrizitätswirtschaft in Österreich, 1938–1947: Von den Alpenelektrowerken zur Verbundgesellschaft." PhD dissertation, Karl-Franzens University of Graz, 1985.

König, Wolfgang. "Bergbahnen in den Alpen (1870–1940): Zwischen Naturerschließung und Naturbewahrung." In *Umwelt-Geschichte: Arbeitsfelder-Forschungsansätze-Perspektiven* edited by Sylvia Hahn and Reinhold Reith. Munich: Oldenbourg, 2001.

Kuisel, Richard. *Seducing the French: The Dilemma of Americanization*. Berkeley: University of Californian Press, 1997.

Kupper, Patrick. *Creating Wilderness: A Transnational History of the Swiss National Park*. Translated by Giselle Weiss. New York: Berghahn, 2014.

———. *Umweltgeschichte*. Goettingen: Vandenhoeck & Ruprecht, 2021.

———. "Verkanntes unternehmerisches Risiko. Der übereilte Einstieg der schweizerischen Elektrizitätswirtschaft in die Atomtechnologie: der Fall Motor Columbus 1961–1966." *Zeitschrift für Unternehmensgeschichte* 47, no. 1 (2002): 48–71.

Kupper, Patrick, and Irene Pallua. *Energieregime in der Schweiz seit 1800*. Bern: Bundesamt für Engergie, 2016.

Kupper, Patrick, and Tobias Wildi. *Motor-Columbus: From 1895 to 2006*. Baden-Dätwill: buag, 2006.

Kuprian, Hermann J. W., and Oswald Überegger, eds. *Der Erste Weltkrieg im Alpenraum Erfahrung, Deutung, Erinnerung. La Grande Guerra nell'arco alpino: Esperienze e memoria*. Innsbruck: Wagner, 2006.

Laakkonen, Simo, Richard Tucker, and Timo Vuorisalo, eds. *The Long Shadows: A Global Environmental History of the Second World War*. Corvallis: Oregon State University Press, 2017.

Lagendijk, Vincent. *Electrifying Europe: The Power of Europe in the Construction of Electricity Networks*. Amsterdam: Aksant, 2008.

———. "Ideas, Individuals and Institutions: Notion and Practices of a European Electricity System." *Contemporary European History* 27, no. 2 (2018): 202–220.

———. "'To Consolidate Peace'? The International Electro-Technical Community and the Grid for the United States of Europe." *Journal of Contemporary History* 47, no. 2 (2012): 402–226.

Landes, David. S. *The Unbound Prometheus: Technological Change and Industrial Development in Western Europe from 1750 to the Present*, 2nd ed. Cambridge, UK: Cambridge University Press, 2003.

Lang, Norbert, and Roland Mosimann, eds. *Faszination Wasserkraft: Technikgeschichte und Maschinenästhetik*. Baden: hier + jetzt, 2003.

Lecain, Timothy J. *Mass Destruction: The Men and Giant Mines That Wired America and Scarred the Planet*. New Brunswick: Rutgers University Press, 2009.

Lehmann, Philipp Nicolas. "Infinite Power to Change the World: Hydroelectricity and Engineered Climate Change in the Atlantropa Project." *American Historical Review* 121, no. 1 (February 2016): 70–100.

Lekan, Thomas M. *Imagining the Nation in Nature: Landscape Preservation and German Identity, 1885–1945*. Cambridge, MA: Harvard University Press, 2004.

Leonardi, Andrea. "Energia e sviluppo nell'area trentina e sudtirolese." In *Energia e Sviluppo in Area Alpina: Secoli XIX–XX: atti della 7. sessione del Seminario permanente sulla Storia dell'economia e dell'imprenditorialità nelle Alpi in età moderna e contemporanea*, edited by Andrea Bonoldi and Andrea Leonardi, 131–164. Milan: FrancoAngeli, 2004.

———, ed. *Die Region Trentino-Südtirol im 20. Jahrhundert*. Vol. 2. Trent: Museo Storico in Trento, 2009.

Lifset, Robert D. *Power on the Hudson: Storm King Mountain and the Emergence of Modern American Environmentalism*. Pittsburgh: University of Pittsburgh Press, 2014.

Linse, Ulrich, ed. *Von der Bittschrift zur Platzbesetzung: Konflikte um techn. Großprojekte; Laufenburg, Walchensee, Wyhl, Wackersdorf*. Berlin: Dietz, 1988.

Löffler, Bernhard. *Die bayerische Kammer der Reichsräte 1848 bis 1918: Grundlagen, Zusammensetzung, Politik*. Munich: Beck, 1996.

Luckin, Bill. *Questions of Power: Electricity and the Environment in Inter-War Britain*. Manchester: Manchester University Press, 1990.

Luger, Kurt, and Franz Rest, eds. *Alpenreisen: Erlebnis—Raumtransformationen—Imagination*. Innsbruck: Studienverlag, 2017.

Lumsden, G. Innes, ed. *Geology and the Environment in Western Europe*. Oxford: Clarendon Press, 1994.

Macfarlane, Daniel. *Negotiating a River: Canada, the US, and the Creation of the St. Lawrence Seaway*. Vancouver: University of British Columbia Press, 2014.

Maier, Helmut. "Kippenlandschaft, 'Wasserkrafttaumel' und 'Kahlschlag': Anspruch und Wirklichkeit nationalsozialistischer Energiepolitik." In *Umweltgeschichte-Methoden, Themen, Potentiale: Tagung des Hamburger Arbeitskreises für Umweltgeschichte*, edited by Günter Bayerl, 247–266. Muenster: Waxmann, 1996.

———. "Systems Connected: IG Auschwitz, Kaprun, and the Building of European Power Grids up to 1945." In *Networking Europe: Transnational Infrastructures and the Shaping of Europe, 1850–2000*, edited by Erik van der Vleuten and Arne Kaijser, 129–158. Sagamore Beach, MA: Science History Publications, 2006.

———. "'Unter Wasser und unter die Erde': Die süddeutschen und alpinen Wasserkraftprojekte des Rheinisch-Westfälischen Elektrizitätswerks (RWE) und der Natur- und Landschaftschutz während des 'Dritten Reiches.'" In *Die Veränderung der Kulturlandschaft: Nutzungen-Sichtweisen-Planungen*, edited by Günter Bayerl and Torsten Meyer. Muenster: Waxmann, 2003.

Malm, Andreas. *Fossil Capital: The Rise of Steam Power and the Roots of Global Warming*. London and New York: Verso, 2016.

Manganiello, Christopher J. *Southern Water, Southern Power: How the Politics of Cheap Energy and Water Scarcity Shaped a Region*. Chapel Hill: University of North Carolina Press, 2015.

Marschall, Luitgard. *Aluminium—Metall der Moderne*. Munich: Oekom, 2008.

Massell, David. *Amassing Power: J. B. Duke and the Saguenay River, 1897–1927*. Montreal: McGill-Queen's University Press, 2000.

———. *Quebec Hydropolitics: The Peribonka Concessions of the Second World War*. Montreal: McGill-Queen's University Press, 2011.

Mathieu, Jon. *Die Alpen: Raum, Kultur, Geschichte*. Stuttgart: Reclam, 2015.

———. *The Alps: An Environmental History*. Translated by Rose Hadshar. Boston: Polity, 2019.

———. *Geschichte der Alpen 1500–1900: Umwelt, Entwicklung, Gesellschaft*. Vienna: Böhlau, 2001.

Mathieu, Jon, and Simona Boscani Leoni, eds. *Die Alpen!: zur europäischen Wahrnehmungsgeschichte seit der Renaissance = Les Alpes!: pour une histoire de la perception européenne depuis la Renaissance*. Bern: Peter Lang, 2005.

Mathis, Charles-François, and Geneviève Massard-Guilbaud, eds. *Sous le soleil: Systèmes et transitions énergétiques du Moyen Âge à nos jours*. Paris: Sorbonne, 2019.

Matt, Peter, Otto Pirker, and Martin Schletterer. "Wasserkraft im Wandel der Zeit: Energiewirtschaftliche Bedeutung der Alpenflüsse." In *Flüsse der Alpen: Vielfalt in Natur und Kultur*, edited by Susanna Muhar et al., 248–261. Bern: Haupt, 2019.

Mauch, Christof, and Thomas Zeller, eds. *Rivers in History: Perspectives on Waterways in Europe and North America*. Pittsburgh: University of Pittsburgh Press, 2008.

Mayer, Arno J. *The Persistence of the Old Regime: Europe to the Great War*. New York: Pantheon, 1981.

McCully, Patrick. *Silenced Rivers: The Ecology and Politics of Large Dams.* London: Zed Books, 1996.

Melosi, Martin V. *Coping with Abundance: Energy and Environment in Industrial America.* New York: Knopf, 1985.

Melosi, Martin V., and Joseph A. Pratt, eds. *Energy Metropolis: An Environmental History of Houston and the Gulf Coast.* Pittsburgh: University of Pittsburgh Press, 2007.

McNeill, J. R. *The Mountains of the Mediterranean World: An Environmental History.* Cambridge, UK: Cambridge University Press, 1992.

———. *Something New Under the Sun: An Environmental History of the Twentieth-Century World.* New York: W. W. Norton, 2000.

McNeill, J. R., and Peter Engelke. *The Great Acceleration: An Environmental History of the Anthropocene Since 1945.* Cambridge, MA: Belknap Press, 2014.

McNeill, J. R., and Corinna R. Unger, eds. *Environmental Histories of the Cold War.* Cambridge, UK: Cambridge University Press, 2010.

Miller, Ian, et al. "Forum: The Environmental History of Energy Transitions." *Environmental History* 24 (2019): 463-533.

Millward, Robert. *Private and Public Enterprise in Europe: Energy, Telecommunications, and Transport, 1830-1990.* Cambridge, UK: Cambridge University Press, 2005.

Milward, Alan S. *The German Economy at War.* London: Athlone Press, 1965.

Mitchell, Timothy. *Carbon Democracy: Political Power in the Age of Oil.* London: Verso, 2011.

Muhar, Susanna, et al. "Revitalisierung: Neues Leben für die Alpenflüsse." In *Flüsse der Alpen: Vielfalt in Natur und Kultur,* edited by Susanna Muhar et al., 320-345. Bern: Haupt, 2019.

———. "Zustand und Schutz der Fließgewässer: Ein alpenweiter Überblick." In *Flüsse der Alpen: Vielfalt in Natur und Kultur,* edited by Susanna Muhar et al., 302-319. Bern: Haupt, 2019.

Muhar, Susanna, Andreas Muhar, Gregory Egger, and Dom Siegrist, eds. *Flüsse der Alpen: Vielfalt in Natur und Kultur.* Bern: Haupt, 2019.

Mulder, Nicholas. *The Economic Weapon: The Rise of Sanctions as a Tool of Modern War.* New Haven: Yale University Press, 2022.

Needham, Andrew. *Power Lines: Phoenix and the Making of the Modern Southwest.* Princeton: Princeton University Press, 2014.

Netting, Robert McC. *Balancing on an Alp: Ecological Change and Continuity in a Swiss Mountain Community.* Cambridge, UK: Cambridge University Press, 1981.

Nye, David E. *Consuming Power: A Social History of American Energies.* Cambridge, MA: MIT Press, 1997.

———. *Electrifying America: Social Meanings of a New Technology, 1880-1940.* Cambridge, MA: MIT Press, 1990.

Obled, Charles, and Patrick Tourasse, "Uncertainty in Flood Forecasting: A French Case Study." In *Coping with Floods,* edited by Giuseppi Rossi, Nilgun B. Harmanciogammalu, and V. Yevjevich, 473-501. Dordrecht: Kluwer, 1994.

Odum, Howard T. *Environment, Power, and Society for the Twenty-First Century: The Hierarchy of Energy.* New York: Columbia University Press, 2007.

Offer, Avner. *The First World War: An Agrarian Interpretation*. New York: Oxford University Press, 1989.
Ott, Hugo, ed. *Historische Energiestatistik von Deutschland*. Vol. 1: *Statistik der öffentlichen Elektrizitätsversorgung Deutschlands, 1890-1913*. St. Katharinen: Scripta Mercaturae, 1986.
Ozenda, Paul. *The Vegetation of the Alps*. Croton, NY: Manhattan Publishing, 1983.
Painter, David S. "The Marshall Plan and Oil." *Cold War History* 9, no. 2 (May 2009): 159-175.
Painter, Thomas H., et al. "End of the Little Ice Age in the Alps Forced by Industrial Black Carbon." *Proceedings of the National Academy of Sciences* 110, no. 38 (September 17, 2013): 15216-15221.
Paquier, Serge, ed. *L'eau à Genève et dans la région Rhône-Alpes, XIXe-XXe siècles*. Paris: L'Harmattan, 2007.
——. *Histoire de l'électricité en Suisse: La dynamique d'un petit pays européen 1875-1939*. 2 vols. Geneva: Ed. Passé Présent, 1998.
Parasiewicz, P., et al. "The Effect of Managed Hydropower Peaking on the Physical Habitat, Benthos, and Fish Fauna in the River Bregenzerach in Austria." *Fisheries Management and Ecology* 5 (1998): 403-417.
Parrinello, Giacomo. "Charting the Flow: Water Science and State Hydrography in the Po Watershed, 1872-1917." *Environment and History* 23, no. 1 (February 2017): 65-96.
——. "Systems of Power: A Spatial Envirotechnical Approach to Water Power and Industrialization in the Po Valley of Italy, ca. 1880-1970." *Technology and Culture* 59, no. 3 (July 2018): 652-688.
Parteli, Othmar. *Geschichte des Landes Tirol: Die Zeit von 1918 bis 1970*, vol. 4 part 1, *Südtirol (1918 bis 1970)*. Bozen: Athesia, 1988.
Pauli, Ludwig. *The Alps: Archaeology and Early History*. London: Thames and Hudson, 1984.
Pearson, Chris. "Environments, States and Societies at War." In *Total War: Economy, Society and Culture at War*, vol. 3 of *Cambridge History of the Second World War*, edited by Michael Geyer and Adam Tooze, 220-243. Cambridge, UK: Cambridge University Press, 2015.
——. *Mobilizing Nature: The Environmental History of War and Militarization in Modern France*. Manchester: Manchester University Press, 2012.
Pearson, Chris, Peter Coates, and Tim Cole, eds. *Military Landscapes: From Gettysburg to Salisbury Plain*. London: Continuum, 2010.
Pelinka, Anton. *Die gescheiterte Republik: Kultur und Politik in Österreich 1918-1938*. Vienna: Böhlau, 2017.
Petzina, Dieter. *Autarkiepolitik im Dritten Reich: Der nationalsozialistische Vierjahresplan*. Stuttgart: Deutsche Verlags-Anstalt, 1968.
Pfister, Christian, ed. *Das 1950er Syndrom: Der Weg in die Konsumgesellschaft*. Bern: Haupt, 1995.
Pirani, Simon. *Burning Up: A Global History of Fossil Fuel Consumption*. London: Pluto Press, 2018.
Podobnik, Bruce. *Global Energy Shifts: Fostering Sustainability in a Turbulent Age*. Philadelphia: Temple University Press, 2006.

Pohl, Manfred. *Das Bayernwerk, 1921 bis 1996*. Munich: Piper, 1996.
Price, Larry W. *Mountains and Man: A Study of Process and Environment*. Berkeley: University of California Press, 1981.
Prinoth, Gabriela. "Die Elektrizitätswirtschaft Nord- und Osttirols von den Anfängen bis zum Jahre 1938." PhD dissertation, University of Innsbruck, 1983.
Pritchard, Sara B. *Confluence: The Nature of Technology and the Remaking of the Rhône*. Cambridge, MA: Harvard University Press, 2011.
Radkau, Joachim. *Nature and Power: A Global History of the Environment*. New York: Cambridge University Press, 2002.
Rathkolb, Oliver. "Die Zweite Republik (seit 1945)." In *Geschichte Österreichs*, edited by Thomas Winkelbauer, 525-594. Stuttgart: Reclam, 2015.
———. "NS-Erbe, Wiederaufbau, Marshallplan und das "Weiße Gold" in den europäischen Netzwerken." In *Wasserkraft. Elektrizität. Gesellschaft. Kraftwerksprojekte ab 1880 im Spannungsfeld*, edited by Oliver Rathkolb et al., 187-206. Vienna: Kremayr & Scheriau, 2012.
Rathkolb, Oliver, and Florian Freund, eds. *NS-Zwangsarbeit in der Elektrizitätswirtschaft der "Ostmark", 1938-1945*. Vienna: Böhlau, 2002.
Rathkolb, Oliver, Richard Hufschmied, Andreas Kuchler, and Hannes Leidinger, eds. *Wasserkraft. Elektrizität. Gesellschaft. Kraftwerksprojekte ab 1880 im Spannungsfeld*. Vienna: Kremayr & Scheriau, 2012.
Rauscher, Walter. *Die verzweifelte Republik, Österreich 1918-1922*. Vienna: Kremayr & Scheriau, 2017.
Reindl, Robert, Johan Neuner, and Martin Schletterer. "Increased Hydropower Production and Hydropeaking Mitigation Along the Upper Inn River (Tyrol, Austria) with a Combination of Buffer Reservoirs, Diversion Hydropower Plants and Retention Basins." *River Research and Applications* 39, no. 3 (2023): 602-609.
Reisner, Marc. *Cadillac Desert: The American West and Its Disappearing Water*. New York: Penguin Press, 1993.
Reiter, Margit. "The "Myth" of Kaprun: Forced Labor at the Tauern Power Plant in Kaprun and How Postwar Austria Dealt with It." In *The Dollfuss/Schuschnigg Era in Austria: A Reassessment*, vol. 11 of *Contemporary Austrian Studies*, eds. Günter Bischof, Anton Pelinka, and Dieter Stiefel, 258-266. New Brunswick: Transaction, 2003.
———. "Das Tauernkraftwerk Kaprun." In *NS-Zwangsarbeit in der Elektrizitätswirtschaft der "Ostmark", 1938-1945*, edited by Oliver Rathkolb and Florian Freund, 127-198. Vienna: Böhlau, 2002.
Reynolds, Terry S. *Stronger Than a Hundred Men: A History of the Vertical Water Wheel*. Baltimore: Johns Hopkins University Press, 1983.
Rhodes, Richard. *Energy: A Human History*. New York: Simon & Schuster, 2018.
Riedmann, Josef. *Geschichte Tirols*, 3rd ed. Vienna: Verlag für Geschichte und Politik, 2001.
Rigele, Georg. "The Marshall Plan and Austria's Hydroelectric Industry: Kaprun." In *The Marshall Plan in Austria*, vol. 8 of *Contemporary Austrian Studies*, edited by Günter Bischof, Anton Pelinka, and Dieter Stiefel, 323-356. New Brunswick: Transaction, 2000.
———. "Der Marshall-Plan und Österreichs Alpenwasser-Kräfte: Kaprun." In *"80 Dollar":*

50 Jahre ERP-Fonds und Marshall-Plan in Österreich, 1948–1998, edited by Günter Bischof and Dieter Stiefel, 183–216. Vienna: Ueberreuter, 1999.

———. "Das Tauernkraftwerk Glockner-Kaprun—Neue Forschungsergebnisse und offene Fragen." *Blätter für Technikgeschichte* 59 (1997): 55–94.

Ritvo, Harriet. *The Dawn of Green: Manchester, Thirlmere, and Modern Environmentalism*. Chicago: University of Chicago Press, 2009.

Rögener, Wiebke. "Gift unter Strommasten: Bleihaltige Rostschutzfarbe verseucht das Erdreich—noch immer wird in Deutschland wenig dagegen getan." *Süddeutsche Zeitung*. February 6, 2009.

Rohr, Christian. "The Austrian Environment *en Miniature*: 'Official' Perceptions of the Austrian Landscape Reflected Through Postal Stamps Since 1945." In *Austrian Environmental History*, vol. 27 of *Contemporary Austrian Studies*, edited by Marc Landry and Patrick Kupper, 155–182. New Orleans and Innsbruck: UNO Press/Innsbruck University Press, 2018.

Rossi, Giuseppi, Nilgun B. Harmanciogammalu, and V. Yevjevich, eds. *Coping with Floods*. Dordrecht: Kluwer, 1994.

Russell, Edmund. *War and Nature: Fighting Humans and Insects with Chemicals from World War I to Silent Spring*. Cambridge, UK: Cambridge University Press, 2001.

Saraiva, Tiago, and M. Norton Wise. "Autarky/Autarchy: Genetics, Food Production and the Building of Fascism." *Historical Studies in the Natural Sciences* 40, no. 4 (Fall 2010): 419–448.

Savoy, Monique. *Lumières sur la ville: Introduction et promotion de l'électricité en Suisse, l'éclairage lausannois, 1881–1921*. Lausanne: Section d'histoire Université de Lausanne, 1987.

Schaumann, Caroline. *Peak Pursuits: The Emergence of Mountaineering in the Nineteenth Century*. New Haven: Yale University Press, 2020.

Schausberger, Norbert. "Deutsche Wirtschaftsinteressen in Österreich vor und nach dem März 1938." In *Österreich, Deutschland und die Mächte: Internationale und Österreichische Aspekte des "Anschlusses" vom März 1938*, edited by Gerald Stourzh and Birgitta Bader-Zaar, 177–213. Vienna: Verlag der Österreichischen Akademie der Wissenschaften, 1990.

———. *Rüstung in Österreich 1938–1945: Eine Studie über die Wechselwirkung von Wirtschaft, Politik und Kriegsführung*. Vienna: Hollinek, 1970.

Schmid, Martin, and Ortrun Veichtlbauer. *Vom Naturschutz zur Ökologiebewegung: Umweltgeschichte in der zweiten Republik*. Innsbruck: Studienverlag, 2006.

Schnitter, Niklaus. *Die Geschichte des Wasserbaus in der Schweiz*. Oberbözberg: Olynthus, 1992.

Schweer, Dieter and Wolf Thieme, eds. *"Der gläserne Riese": RWE, ein Konzern wird transparent*. Wiesbaden: Gabler, 1998.

Scott, James C. *Seeing Like a State: How Certain Schemes to Improve the Human Condition Have Failed*. New Haven: Yale University Press, 1998.

Seow, Victor. *Carbon Technocracy: Energy Regimes in Modern East Asia*. Chicago: University of Chicago Press, 2021.

Sheller, Mimi. *Aluminum Dreams: The Making of Light Modernity*. Cambridge, MA: MIT Press, 2014.

Shoumatoff, Nicholas, and Nina Shoumatoff. *The Alps: Europe's Mountain Heart*. Ann Arbor: University of Michigan Press, 2001.

Sieferle, Rolf Peter. *Der europäische Sonderweg. Ursachen und Faktoren*. Stuttgart: Breuninger, 2003.

———. *The Subterranean Forest: Energy Systems and the Industrial Revolution*. Cambridge, UK: White Horse Press, 2001.

Sievert, James. *The Origins of Nature Conservation in Italy*. Bern: Peter Lang, 2000.

Silver, Timothy. *Mount Mitchell and the Black Mountains: An Environmental History of the Highest Peaks in Eastern America*. Chapel Hill: University of North Carolina Press, 2003.

Sinclair, Bruce. "Regenerating the Future: The First World Power Conference." In *Energie in der Geschichte: zur Aktualität der Technikgeschichte*. Düsseldorf: [?], 1984.

Smil, Vaclav. *Energy and Civilization: A History*. Cambridge, MA: MIT Press, 2018.

———. *Energy at the Crossroads: Global Perspectives and Uncertainties*. Cambridge, MA: MIT Press, 2003.

———. *Energy Transitions: History, Requirements, Prospects*. Santa Barbara: Praeger, 2010.

———. *Energy in World History*. Boulder: Westview Press, 1994.

Spilker, Rolf, ed. *Unbedingt modern sein. Elektrizität und Zeitgeist um 1900*. Osnabruck: Rasch, 2001.

Staudacher, Peter. "'In der Schweiz zum Beispiel': Die Anfänge der österreichischen Elektrowirtschaft und das Schweizer Vorbild." In *Allmächtige Zauberin unserer Zeit: Zur Geschichte der elektrischen Energie in der Schweiz*, edited by David Gugerli, 185-198. Zurich: Chronos, 1994.

Steinberg, Theodore. *Nature Incorporated: Industrialization and the Waters of New England*. Amherst: University of Massachusetts Press, 1994.

Steininger, Rolf. *Südtirol im 20. Jahrhundert: Vom Leben und Überleben einer Minderheit*, 3rd ed. Innsbruck: Studien Verlag, 2004.

Stier, Bernhard. *Staat und Strom: Die politische Steuerung des Elektrizitätssystems in Deutschland, 1890-1950*. Ubstadt-Weiher: Verlag Regionalkultur, 1999.

Storey, William Kelleher. *The First World War: A Concise Global History*. Lanham: Rowman & Littlefield, 2009.

Storia dell'industria elettrica in Italia. 5 vols. Bari: Laterza, 1992-1994.

Stourzh, Gerald, and Birgitta Bader-Zaar. *Österreich, Deutschland und die Mächte: Internationale und Österreichische Aspekte des "Anschlusses" vom März 1938*. Vienna: Verlag der Österreichischen Akademie der Wissenschaften, 1990.

Strobel, Ludwig. "Bayerische Geschichte im Donauraum, Wasserbau und Wasserwirtschaft im Lande Bayern—Historische Abrisse, Entwicklungslinien und Wechselbeziehungen." In *Geschichtliche Entwicklung der Wasserwirtschaft und des Wasserbaus in Bayern*, vol. 1, edited by Ludwig Strobel, 13-62. Munich: Bayerisches Landesamt für Wasserwirtschaft, 1981.

———, ed. *Geschichtliche Entwicklung der Wasserwirtschaft und des Wasserbaus in Bayern*, vol. 1. Munich: Bayerisches Landesamt für Wasserwirtschaft, 1981.

Strunk, Peter. *Die AEG: Aufstieg und Niedergang einer Industrielegende*. Berlin: Nicolai, 1999.

Tamïr, Dan. "Something New Under the Fog of War: The First World War and the Debut of Oil on the Global Stage." In *Environmental Histories of the First World War*, edited by Richard P. Tucker et al., 117-135. Cambridge, UK: Cambridge University Press, 2018.

Tauernkraftwerke AG. *Die Tauernkraftwerke Aktiengesellschaft*. Salzburg: TKWAG, 1991.

Thorpe, Julie. *Pan-Germanism and the Austrofascist State, 1933-38*. Manchester: Manchester University Press, 2011.

Tolloi, Philipp. "'Alpen unter Strom': Zur Geschichte der Elektrifizierung Südtirols." *Tiroler Chronist*, no. 2 (2015): 2-8.

Tooze, Adam. *The Wages of Destruction: The Making and Breaking of the Nazi Economy*. New York: Viking, 2006.

Tucker, Richard P. "Containing Communism by Impounding Rivers: American Strategic Interests and the Global Spread of High Dams in the Early Cold War." In *Environmental Histories of the Cold War*, edited by J. R. McNeill and Corinna R. Unger, 139-163. Cambridge, UK: Cambridge University Press, 2010.

Tucker, Richard P., Tait Keller, J. R. McNeill, and Martin Schmid, eds. *Environmental Histories of the First World War*. Cambridge, UK: Cambridge University Press, 2018.

Tucker, Richard P., and Edmund Russell. *Natural Enemy, Natural Ally: Toward an Environmental History of Warfare*. Corvallis: Oregon State University Press, 2004.

Turner, James Morton. *Charged: A History of Batteries and Lessons for a Clean Energy Future*. Seattle: University of Washington Press, 2022.

Uekoetter, Frank. *The Green and the Brown: A History of Conservation in Nazi Germany*. Cambridge, UK: Cambridge University Press, 2006.

Unfer, Günter, Andreas Meraner, and Didier Pont. "Bedrohte aquatische Biodiversität in den Alpen." In *Flüsse der Alpen: Vielfalt in Natur und Kultur*, edited by Susanna Muhar et al., 126-145. Bern: Haupt, 2019.

Universität Zürich. "Fischer-Reinau, Ludwig." *Matrikeledition*. http://www.matrikel.uzh.ch/active/static/6132.htm

van der Vleuten, Erik, and Arne Kaijser, eds. *Networking Europe: Transnational Infrastructures and the Shaping of Europe, 1850-2000*. Sagamore Beach, MA: Science History Publications, 2006.

van der Vleuten, Erik, et al. "Europe's System Builders: The Contested Shaping of Transnational Road, Electricity and Rail Networks." *Contemporary European History* 16, no. 3 (August 2007): 321-347.

Veit, Heinz. *Die Alpen: Geoökologie und Landschaftsentwicklung*. Stuttgart: Ulmer, 2002.

Viazzo, Pier Paolo. *Upland Communities: Environment, Population and Social Structure in the Alps Since the Sixteenth Century*. Cambridge, UK: Cambridge University Press, 1989.

Vischer, Daniel L. "Wasserbau und Elektrifizierung." In *Allmächtige Zauberin unserer Zeit: Zur Geschichte der elektrischen Energie in der Schweiz*, edited by David Gugerli, 117-130. Zurich: Chronos, 1994.

Volkert, Wilhelm. *Geschichte Bayerns*, 5th ed. Munich: Beck, 2017.

Vollmann, William T. *Carbon Ideologies*. 2 vols. New York: Viking, 2018.

Wagnleitner, Reinhold. *Coca-Colonization and the Cold War: The Cultural Mission of the United States in Austria After the Second World War*. Translated by Diana Wolf. Chapel Hill: University of North Carolina Press, 1994.

Waterbury, John. *Hydropolitics of the Nile Valley*. Syracuse: Syracuse University Press, 1979.

White, Richard. *The Organic Machine: The Remaking of the Columbia River*. New York: Hill and Wang, 1995.

Whited, Tamara L., et al. *Northern Europe: An Environmental History*. Santa Barbara: ABC-CLIO, 2005.

Williams, James C. *Energy and the Making of Modern California*. Akron: University of Akron Press, 1997.

Winiwarter, Verena, et al. "Environmental Histories of Contemporary Austria: An Introduction." In *Austrian Environmental History*, vol. 27 of *Contemporary Austrian Studies*, edited by Marc Landry and Patrick Kupper, 25–48. New Orleans and Innsbruck: UNO Press/Innsbruck University Press, 2018.

Winkel, Harald. *Wirtschaft im Aufbruch: Der Wirtschaftsraum München-Oberbayern und seine Industrie- und Handelskammer im Wandel der Zeit*. Munich: Beck, 1990.

Winkelbauer, Thomas, ed. *Geschichte Österreichs*. Stuttgart: Reclam, 2015.

Worster, Donald. *Rivers of Empire: Water, Aridity and the Growth of the American West*. New York: Oxford University Press, 1985.

Wrigley, E. A. *Energy and the English Industrial Revolution*. Cambridge, UK: Cambridge University Press, 2010.

Wu, Shellen Xiao. *Empires of Coal: Fueling China's Entry into the Modern World Order, 1860–1920*. Stanford, CA: Stanford University Press 2015.

Yergin, Daniel. *The Prize: The Epic Quest for Oil, Money & Power*. New York: Free Press, 1991.

———. *The Quest: Energy, Security, and the Remaking of the Modern World*, rev. and updated ed. New York: Penguin, 2012.

Zeller, Thomas. *Driving Germany: The Landscape of the German Autobahn*. Oxford: Berghahn, 2007.

Ziegerhofer, Anita. *Botschafter Europas: Richard Nikolaus Coudenhove-Kalergi und die Paneuropa-Bewegung in den zwanziger und dreißiger Jahren*. Vienna: Böhlau, 2004.

Zingraff-Hamed, Aude, and Gregory Egger. "Isar." In *Flüsse der Alpen: Vielfalt in Natur und Kultur*, edited by Susanna Muhar et al., 408–411. Bern: Haupt, 2019.

Zumbrägel, Christian. *"Viele Wenige machen ein Viel": Eine Technik- und Umweltgeschichte der Kleinwasserkraft (1880–1930)*. Paderborn: Ferdinand Schoeningh, 2018.

INDEX

Figures are designated with italicized page numbers

Achenseewerk, 172
Adda River, 66-67, 70, 84
Adige River, 157-58, 204
Adriatic Electric Company (SADE), 207-8
agriculture, 25, 108, 127, 221; Alpine, 179, 183, 214; Po Valley, 223
Allgemeine Elektrizitäts-Gesellschaft (AEG), 54, 59, 169-72, 176, 178-79, 182, 184
Alpen-Elektrowerke AG (AEW), 178-79, 181, 184, 188-91, 197, 201, 250n40
Alpine Convention, 215
Alps, 2-8, 10-11, 16, 91, 199; Aristotle's terminology for the, 19; Austrian, 11, 151-52, 161-92, 201; Bavarian, 36, 99, 105, 116; as center of the modern industries associated with the second industrial revolution, 4; eastern, 126-33, 167, 175; as Europe's battery, 136, 161, 214-15; as Europe's water tower, 20; fossil fuels in the, 30; French, 2, 17, 31, 34, 44, 46, 114, 116-25, *118*, 140; glacial retreat in the, 181; hanging valleys of the, 20, 31; Herodotus on the, 18-19; hydrological cycle of the, 18, 211-13, 215; map of large dams and reservoirs of the, *216*; mountain-building processes of the, 10-11, 18-19, 21, 99; northern, 117, 121; postwar landscape of drained lakes and reservoirs in the, 195; as "rain catcher," 20; Savoy, 64; southern, 121; specialness of the, 12, 139; Swiss, 139, 187, 256n42; traces of high culture in the, 12; waterfalls in the, 14. *See also* energy landscapes; glaciers; mountains
Altmann, Karl, 196-98
aluminum, 3, 43-44, 225n3; Austrian industry of, 203; German industry of, 186; Swiss factory for the mass-production of, 44
Anschluss, 163, 165-72, 175, 177, 187; electrical, 192, 197; political aspects of the, 250n32. *See also* Austria; Germany
Anthropocene, 194, 253n3
Armiero, Marco, 208
Arrigon, Louis-Jules, 113, 119-22, 124-26

283

Augsburg, 89
Austria, 18, 72, 86; Alpine waterpower as a question of state survival for, 133, 167; energy map of, *185*; First Austrian Republic, 165-72, 197, 203; high-altitude waterpower exploitation in, 141, 191; hydroelectricity to power the railroads of, 95, 167-68; hydropower cadaster of the imperial government of, 95; incorporation into the German electricity supply of, 162-93; industrialization of, 250n41; lending money for hydro development abroad by, 214; Ministry for Energy Supply and Electrification of, 196; nationalization of waterpower in, 76-78, 168; Nazi adherents and sympathizers in, 171; postwar hydropower development of, 195-203; Salzburg province of, 164, 170, 178, 196; Second Austrian Republic, 195, 202-3, 247n1; second hydroelectric facility in, 234n14; Social Democrats of, 168; special mission in the European electricity grid for, 198-99; Tyrol province of, 221. See also *Anschluss*; Austria-Hungary; Hohe Tauern; Kaprun/Kaprun Valley; Water and Electricity Supply Office (Wasser- und Elektrizitätswirtschaftsamt)
Austria-Hungary, 115, 126-33; dissolution of, 165-66; pulp exports of, 120. See also Austria
Austrian Engineers and Architects Association, 78
autarky, 134-61; Alps as the key to economic, 161-92; development of hydropower as a means of achieving national, 156, 160-92; technological-chemical, 174

barreurs de chutes, 122
Basel, 44, 146-47, 150-51, 220
batteries, 81, 94; environmental consequences of mountain, 214-22; future of mountain, 222-24; history of, 226n9; mountains as natural, 100, 135-36, 161, 187, 194, 214. See also waterpower
Bauer, Otto, 166
Bavaria, 11, 51-57, 82, 84-89; aesthetic importance of the Alpine lakes of, 105; Alpine waterpower of, 93-95, 99-111, 126-33; coal scarcity in, 100, 111, 130; energy problems of, 102; postwar (First World War) political turmoil in, 128; postwar (Second World War) power shortage in, 195; support for state-led electrification by the Social Democrats of, 110-11; war economy in, 127-28. See also Bayernwerk; Germany; Munich; Walchenseewerk
Bavarian Supreme Construction Authority (Oberste Baubehörde), 89-90
Bayernwerk, 126-28, 130-33. See also Bavaria
Bergès, Aristide, 1-6, 9, 13, 17, 34-35, *37*, 44-47, 59, 232n42; electricity system of, 42-43; paper mill of, 3, 36-41, 84, 117, 119-20, 124, 209; "White Coal" exhibit of, 2, 3
Bergès, M. Maurice: "White Coal Coming to the Aid of Black Coal," 123
Berlin, 154
Bern, 25
Bernstein, Eduard, 58
biodiversity, 22; of Alpine ecosystems, 218; of riparian ecosystems, 217. See also environment
biomass fuels, 26
Bismarck, Otto von, 29, 71, 86, 103
Blake, William: "Gnomic Verses," 49
Blanchard, Raoul, 122, 124
Blitzkrieg, 163
Bolzano, 157, 159; industrial zone of, 159-60
Braudel, Fernand, 11-12
Bréda River, 119
Bregenzerach River, 219
Brembo River, 70
Britain, 26, 30, 198; coal industry of, 28, 173; dam building in, 137; as global empire,

145, 164; oil resources of, 172; textile industry of, 28. *See also* England
British Electrical and Allied Manufacturers' Association, 144
Budau, Artur, 92-93, 99
Bürckel, Josef, 182, 184, 251n46
Burgdorf-Thun railroad, 94

canals, 14, 52, 57; network of hydraulic, 53; slope, 169-71, *170*, *171*, 183-84. See also *Hangkanäle*
capitalism, 164-65, 188; profitmaking over nature protection in, 258n28
Caucasus, 19
Cavaillés, Henri, 134
Cavour, Count Camillo Benso di, 34, 232n42
Central Europe, 136, 221; high-Alpine waterpower in, *142*; waterpower of the lower mountain ranges of, 154. *See also* Europe
Central Powers, 126-33
channelization, 216-17. *See also* rivers
chemical industry, 47-48, 86, 120-21; German, 176, 186, 188-89
China, 11, 29; dam catastrophe in, 255n39; Three Gorges Dam of, 159
cirques, 83. *See also* glaciers
climate change, 222-24. *See also* global warming
coal, 3-10, 17-18, 25-26, 30, 45, 86, 111; anthracite, 151-52, 173, 186, 190, 213; Appalachian, 43; bituminous, 186; Bohemia production of, 130, 166; conservation of supplies of, 100, 128-29, 162-65, 174, 178, 183, 187, 252n66; dependence of railroads of Alpine states on, 96; and economic might, 29, 71-73, 129; French imports of, 114, 117, 119; Germany as leading producer of, 172, 213; history of, 79, 82, 93, 195, 210, 226n8, 243n34; independence from earlier geographic and seasonal restrictions on energy use through, 82, 92; Italian imports of, 63-64, 114; lignite, 152, 173; mobility of, 50, 53; "On the Coal Emergency: A deputation of the Trade Association of German Thieves and Burglars presents the nomination for honorary membership to a coal baron" (illustration), 97; reserves of, 52, 93; rising prices for, 104; Ruhr production of, 72, 128-30, 151; steam engine used in mining for, 27-28; stockpiling of, 41; Swiss imports of, 95-96, 115; of Upper Silesia, 129. *See also* fossil fuels; mining; steam engines; steel
Cobden, Richard, 33-34
Cold War, 194, 204, 209
Cologne, 151-52
Colombo, Giuseppe, 60-61, 64, 66, 70, 78, 235n29
Commission Internationale pour la Protection des Régions Alpines (CIPRA), 215
communism, 196-97, 203-4. *See also* Soviet Union
Compagnie National du Rhone, 140
compressed air, 34, 53, 76. *See also* transmission
Congrès de la Houille Blanche, 47
conservationists, 105-7. *See also* environment; popular movements
consumerism, 194, 253n3
Corinth, Lovis, 85
Croatia, 33
Curzon, Foreign Secretary Lord, 114
Czechoslovakia, 130, 166; industrial centers in, 200
Czech Republic, 86
Czimatis, Albrecht, 173-75, 177-78, 183; *Energy Supply as Foundation of the War Economy*, 173-74

dams, 14, 70, 84, 102-5; Alpine, 112, 125, 135, 137-43, 161, 193-95, 199, 222, 256n6; American building of, 244n9; arch, 190, 252n69; Austrian building of, 169, dams (cont.) 176–77, 181, 184, 192, 194, 198; Chinese building of, 159, 213; developing world building of, 210; Egyptian building of, 104; environmental effects of large, 176, 210; failure of, 206–8; French building of, 116, 121–22, 125, 140, 242n26; German building of, 104, 127, 188–89; high-elevation, 141; Italian building of, 136, 160, 194, 204–8; run-of-river, 138, 201, 215, 223; Soviet building of, 244n9; storage, 138, 201–2, 215, 217; Swiss building of, 146, 190, 209, 256n42; wartime boom in the building of, 116, 121, 143. *See also* hydroelectricity; rivers

Danube River, 20, 52, 86–87, 146, 167, 212; fish species that are endemic to the catchment of the, 219–20; hydroelectric dam on the, 176, 184, 189; riparian states of the, 199; shipping on the, 180

Dauphiné, 35, 59, 113, 122; Italian workers in the, 124; paper industry in the, 36–39, 120. *See also* France

Davidson, Robert, 94

d'Azeglio, Marquis Massimo, 33

democratization, 9–10

Deprez, Marcel, 55, 57–60, 63

Der Blaue Reiter, 85

Die Wasserkraft, 145

Dillgardt, Just, 182, 184

Dollfuss, Chancellor Engelbert, 171

Donat, Fedor Maria von, 101–4, 239n46

Drac River, 31, 125, 140–41

Drau River, 184, 219, 221

Dunlop, D. N., 143–44, 148

Durance River, 121, 125

Dusaugey, Ernest, 140–41

dynamos, 55, 212. *See also* electricity; turbines

earthquakes, 207–8. *See also* seismicity

École centrale des arts et manufactures (Paris), 36–37

École Libre des Sciences Politiques (Paris), 71

Economic Commission for Europe (ECE): Electric Power Commission, 199–200, 200

Edison, Thomas, 8, 42, 53–54, 60–61, 66–67, 79

Edward VIII, King, 145

Egypt, 104

Eiffel, Gustave, 1, 36

Eiffel Tower, 1, 3

Eisner, Kurt, 128

electrical engineering, 71

electricity, 3–4, 8; as alternating current, 9, 50, 62–71, 79, 232n40; as commodity, 8; as direct current, 43, 50, 54, 62, 65, 67, 236n42; as energy carrier, 78–79, 92; fossil fuels used to generate, 7, 99, 143, 152, 186; German mobilization and wartime industrial consumption of, 186–87, 189; German wartime synchronous grid for the transmission of, 186–87, 189; high-voltage alternating, 9, 62–64, 67–68, 70, 126, 152; high-voltage direct, 55, 57, 62; history of the emergence and spread of, 7, 35, 51–70, 79; international exchanges of, 148–49, 152; as medium for transmitting flows of natural energy over long distances, 55–70, 212; storage of, 92. *See also* dynamos; energy; generator; hydroelectricity; infrastructure; lighting; power plants; transformers; transmission; waterpower

electric trains, 82, 84, 92, 94–97; Bavarian, 95; electric load demands of, 96–97, 98; Swiss, 95. *See also* locomotives; railroads

electro-chemistry, 112, 117, 158

electrolytic process, 43

electro-metallurgy, 112, 117, 158, 178

Ellenbogen, Wilhelm, 167

Energie de l'Ouest-Suisse (EOS), 148–49

energy: environmental implications of the use of, 6, 74, 90, 214–22; history of, 11, 227n11, 231n25; kinetic, 16, 27; malleable nature of, 13; natural flows of, 27, 55, 81–82; regimes of, 231n25; risky Alpine, 206–8; traditional systems of, 27; transitions of, 8–10, 15, 17, 25–34, 36, 38, 43, 83, 92, 227n19, 231n25; transmission of flows of natural, 55–59, 227n11; transportation and distribution of mechanical, 60. *See also* electricity; energy landscapes; fossil fuels; nuclear power; renewables; solar power; wind power; waterpower

energy landscapes, 6–13, 17, 30, 47, 112, 115, 133; European major, 213; making of the Alpine, 211–14; maturation of, 193. *See also* Alps; energy

Engels, Friedrich, 58–59

England, 5, 8, 17, 34, 72; fossil fuel transition in, 83, 92. *See also* Britain

Enlightenment, 12

Enns River, 86, 188–89

environment: aesthetic concern for the, 83, 106, 215; consequences for the, 214–22; history of war and the French, 242n6; waterpower considered as less damaging than fossil fuels to the, 73–74. *See also* biodiversity; conservationists; hydropeaking; pollution

environmental history, 8, 105, 226n9; Austrian, 252n74, 255n31; Cold War, 253n1; Italian, 137, 208; of warfare, 115, 164

environmental movements, 210, 220. *See also* popular movements

Esposizione Generale Italiana (Turin, 1884), 62–65, 68; Gallery of Electricity, 64, 65

Etsch River, 70, 157, 204

Europe: fascist states of Central, 136, 162–92; geographic distribution of energy sources of, *150*; postwar economic recovery of, 195–202; postwar energy demand in northwestern, 198–99; "superpower grid" (*Großkraftnetz*) for, 154, 198; as united by technological infrastructure, 135–36; vast geographic diversity of energy resources in, 150. *See also* Central Europe

European Coal and Steel Community, 143

European Union, 143, 195, 213, 220; renewable energy directives of the, 223; Water Framework Directive (2000) of the, 221

Evelyn, John, 26

Exposition Universelle. *See* World's Fair (Paris, 1889)

factories: coal as energy supply for, 93, 100, 186, 188–89; steam-powered, 33, 83; water-powered, 37–44, 84, 160. *See also* industrialization

Ferraris, Galileo, 59–60

fertilizer industry, 160

Fier River, 121

Figl, Chancellor Leopold, 196, 204

Finck, Wilhelm Peter von, 75

First World War, 3–4, 9–11, 25, 113–17, 150–51, 165, 204; changing political climate in Bavaria during the, 127; construction of dams during the, 116, 121, 143; demobilizing German soldiers at the end of the, 128; disruptions to fossil fuel supplies in the, 96, 114–15, 133; energy and the, 114–16, 161; hydroelectric development in the mountains during the, 116–33; prioritization of economic development after the, 244n9; resource implications of the, 136; as turning point in the history of electrification, 242n9

Fischer-Reinau, Ludwig, 90–94, 99–100, 239n44

fishing/fisheries, 83, 85, 107–8, 218–21

floodplain environments, 22; restoration of connections of rivers to, 221; widespread alteration of, 215, 218. *See also* rivers fossil fuels, 4–5; expansion of the use of, 8, 17, 25–34, 36, 92; German/Austrian wartime (Second World War) supplies of, 164, 185–87, 190; history of, 4, 7, 74, 83; Italy's imports of, 63–64; motive power of, 81; as non-renewable, 72–74, 90, 93; substitutes for, 223–24; Switzerland's imports of, 95–96; wartime (First World War) supplies of, 96, 114–15. *See also* coal; energy; gas; industrial revolution; oil; steam engines

Fourneyron, Benoît, 16, 32–33, 37, 40

France, 11, 18, 49–50, 57, 126, 133, 198, 210; aluminum industry of, 44; coal production in, 71, 114, 116–19; distribution of electricity into the homes of craftsmen in, 237n55; high-altitude waterpower exploitation in, 141; nationalization of waterpower in, 76; northeastern, 44; northern, 24; nuclear power in, 213, 223, 256n44; pulp imports of, 120; southeastern, 24; southern, 23; transformation of the countryside of, 113–14; Vichy regime in, 141. *See also* Dauphiné; Savoy

Frankfurt, 67

Frei-Land Society, 76–77. *See also* Switzerland

French Revolution, 35, 86, 93

frost, 20

Ganz & Company, 62

gas, 5, 7, 74; North Sea natural, 213. *See also* fossil fuels

gas engines, 55

Gaulard, Lucien, 62–66, 78

generator, 9, 57, 60, 81; alternating-current, 62–63. *See also* electricity

George, Henry, 103

German League of Land Reformers (Bund deutscher Bodenreformer), 103

Germany, 15, 18–19, 44, 49–50, 109, 126–33, 150–54, 210; aerial bombardment of, 190; airplane and tank production as priority in wartime, 189; cartelization of the coal industry of, 96; "coal crisis" in, 110, 130–31, 166; coal resources of, 115, 128–30, 186–87, 190; energy policy of Nazi, 172–90, 192, 247n3, 248n5, 251n51; high-voltage grid of, 152, *153*, 154; industrial centers in western, 200; industrialization of, 75, 86, 127; mobilization of national production in peacetime of, 164–65, 247n3; nationalization of waterpower in the state of Baden in, 76; National Socialist economic planning of, 164, 172–90, 192; Nazi control of, 4, 11, 136, 162–92; postwar political turmoil in, 128; pulp exports of, 120; Ruhr region of, 72, 128–30, 151–52, 161, 193–94; Saar region of, 129–30; southern, 23–24, 51–52, 56–59, 72, 95; Weimar Republic in, 128–30, 163. *See also Anschluss*; Bavaria

Gibbs, John Dixon, 62–66, 78

Girod, Paul, 125

glaciers, 14, 20, 38, 44–45, 180–81; erosion of, 31, 83; as natural equalizers, 134; as reservoirs of potential energy, 181; retreat of, 19, 21, 31, 35–36, 39, 71, 179, 181, 211, 222–23; storage of water in, 21, 84, 181, 211; watersheds of, 21. *See also* Alps; cirques; moraines; reservoirs; water; waterpower

global warming, 4, 8, 223–24. *See also* climate change

Goethe, Johann Wolfgang von: *Faust*, 47; *Wilhelm Meister*, 1

Göring, Hermann, 176–77, *177*, 182, 250n38

Great Depression, 140, 160–61, 198

Grengg, Hermann, 171, 178–81, 183–84, 190, 252n69

Grenoble, 2, 34–36, 39, 45, 47, 59, 84, 121, 140

Haas, Robert, 149–51

Hall, Charles Martin, 43

Hangkanäle, 169. *See also* canals

hautes chutes, 17, 41–42, 44–45, 59

heating, 92
Heim, Albert, 134
herbicides, 221
Héroult, Paul, 43–44
historiography, 10, 12; of Germany and Europe, 229n32; Nazi, 165
Hitler, Adolf, 162, 164, 172, 175, 182, 188–90, 192; Romantic views of waterpower of, 188
Hohe Tauern, 164; foreign and forced laborers in the, 191, 203, 252n71; layout of the Tauern waterpower plant, *170*; rivers in the catchment of, 219; waterpower of the, 169–71, 174, 176, 178–79, 181–84, 190, 198–203. *See also* Austria; Tauernwerk
Holy Roman Empire, 86, 165
Hoover, President Herbert, 145
hydraulic engineering, 101, 135; large-scale projects of, 101, 104–12; and political considerations, 143. *See also* hydroelectricity
hydraulic motors, 55
hydroelectricity, 7–8, 14, 34, 42–59, 71–75, 79–83, 125–26; coupling turbines to generators in the rise of, 9, 78–79; economic costs of, 229n41; European supply of, 212–13; expansion of, 139–43, 207, 209; export market for Austrian, 198; industrial virtues of, 74, 75; mountain, 81, 83, 122, 135, 213, 215; utopian visions of, 135, 244n2; "waste current" of, 143. *See also* dams; electricity; hydraulic engineering; hydroplants; transmission; waterpower
hydropeaking, 218–19; mitigation efforts against, 221. *See also* environment
hydroplants, 4, 14, 67–70, 75, 88, 93, 152; Adamello, 84; Austrian, 169–84; Bavarian, 102, 127; Brusio, 84; French, 117–19, *118*, 121, 123; German, 68, 75; Itaipú, 213; Italian, 67, 70, 159; Kochel, 88; operating costs of, 229n41; run-of-river, 215–18; South Tyrolean, *205*; storage, 217–18;

Swiss, 88, 97, 146. *See also* hydroelectricity; power plants; waterpower
hydropower. *See* waterpower

IG Farben, 191–92
Ill River, 152, 189
incision, 217. *See also* rivers
India, 29
industrialists, 83, 93; Austrian, 53, 72; French, 42, 47, 120–21, 123, 125
industrialization, 108; Bavarian, 86; coal-fired, 222; early American, 53; German, 75, 86; Swiss traditional waterpower, 52. *See also* factories; industrial revolution
industrial revolution, 3–6, 11–12, 17, 25–29, 36, 75, 86, 137; importance of coal in the, 5–6, 226n8; melting of glaciers as consequence of the, 222. *See also* fossil fuels; industrialization; second industrial revolution; steam engines; textile industry
infrastructure: Austrian nationalization of electrical, 198; French Resistance sabotage of electrical, 194; postwar manufacture of electrical equipment for German electrical, 195; transport, 78, 135, 227n11, 233n2; wartime destruction of electrical, 194, 196, 253n4. *See also* electricity; transmission
Inn River, 20, 87, 217, 219
internal combustion engine, 51
International Committee on Large Dams, 208
international cooperation, 143–51, 156, 161, 193–94
International Electrical Exhibition (Frankfurt, 1891), 67–70
International Electrical Exhibition (Munich, 1882), 56–59, 68; artificial waterfall at the, 57, *58*
International Electrical Exhibition (Paris, 1881), 42, 53–56, 59
Intze, Otto, 104

290 Index

iron, 3, 45; energy required for smelting, 43; Han dynasty production of, 26; industry of, 25; Nazi wartime rationing of, 184; Song dynasty production of, 26
Isarco River, 157–59
Isar River, 70, 75, 87–90, 102–5, 108–9, 218, 221
Isère River, 22, 35, 39, 119, 121
Italy, 11, 18, 59–67, 84, 95, 114, 126, 133, 213; as coal poor country, 173; draining of the Pontine Marshes in, 101; failure of the Gleno Dam in the Lombard Alps in, 206, 208; as fascist state, 136, 157–61, 204, 244n10; high-altitude waterpower exploitation in, 141, 247n48; lack of heavy industry in, 158; landslide into the reservoir of the Vajont Dam in the Alps north of Venice in, 207–8; nationalization of the electricity supply in, 254n15; northern, 23–24, 59–60, 63, 67, 198; northwest, 30, 33. *See also* Piedmont; South Tyrol

Japan, 29
Jevons, W. Stanley: *The Coal Question*, 29
Jordan, 137

Kaprun/Kaprun Valley, 169, 171, 176–84, 247n1, 255n31; completion of dams in the, 201–3, 209, 247n1; Nazi-era storage dams in the, 190, 192, 197, 201. *See also* Austria; Tauernwerk
Keppler, Wilhelm, 175
Kreuter, Professor Franz, 105–6

labor, 14; Alpine conditions for, 122–24, 191; disruption of, 9, 75, 128, 131–32; Nazi shortages of, 191; Nazi use of forced, 191, 203, 252n70; northern Italian migrant, 159–60; political power of, 10, 75; wartime expansion of women in the ranks of French, 124
Lac de Caillaouas, 84
Lac de la Girotte, 84, 125

Lac du Crozet, 84
Lago di Antillone, 106
Lake Achensee, 101–2
Lake Como, 84, 94, 200
Lake Constance, 83–84, 167, 219
Lake Geneva, 83–84
Lake Kochelsee, 85, 88–89, 102
Lake Lugano, 94
Lake Reschensee, 204, 209
lakes, 20; Alpine, 82–84, 105, 112; Bavarian mountain, 80–82, 84–89, 98, 99–101, 104–5; and electric trains, 83, 98–100; finger, 83; as *houille bleu*, 84; as reservoirs of waterpower, 80–89, 98–102, 107, 112, 116, 124–25, 135, 139, 149, 172, 204–5; storage of water in, 41, 83–84, 98, 212; upland, 22. *See also* mountains; reservoirs; water
Lake Walchensee, 80–90, 99–107, 130; diversion of the upper Isar River to fill up the, 88–90, 218; popular sentiment to protect the, 105, 107; postwar lowering of the level of, 195; storage capabilities of the, 104, 106; as tourist attraction, 106. *See also* Walchenseewerk
Landry, Jean, 148–49
landslides, 20, 207–8; dynamics of, 207–8
last glacial period, 19
League of Nations, 136, 145
Lech River, 87, 89
Leroy-Beaulieu, Pierre, 71
liberal internationalists, 11
Liechtenstein, 18
lighting: electric, 46, 54, 60–64, 68, 78, 81, 92, 96; gas and petroleum, 46; urban electric, 43, 67, 70, 79, 237n5. *See also* electricity
Linz, 188
locomotives, 28, 81; electric traction, 94–96; steam, 94–96, 100. *See also* electric trains; railroads
Loisach River, 87
London, 63, 144–45
Luitpold, Prince, 104

Lycophron, 19
Lyon, 18

machine construction industry, 124
Marc, Franz, 85
Märjelensee, 22
Marmonnier, Dr. Joseph Melchior, 39–41
Marshall, Secretary of State George C., 198–99
Marshall Plan, 194, 198–99, 201–3, 209
Marx, Karl, 58
mechanical engineering, 86
media. *See* press
Medieval power revolution, 24, 35
Merlin, Tina, 208
Merrill, O. C., 145
metallurgy, 47–48, 52, 60, 63–64; French, 117, 140
Middle East, 224
Milan, 18, 60–61, 70, 84, 94
Miller, Oskar von, 51–60, 67–70, 78, 95–96, 99, 126–30, 149; concept of "social electricity" of, 56–57, 67–68, 75; "Overview Map of the Available and Usable Water Power on the Northern Slope of the Alps in Bavaria" (1903), 87
mining, 52; coal, 27–28, 74–75, 90, 131; environmental history of copper, 234n13. *See also* coal
Montecatini, 204; electric infrastructure of the Montecatini chemical concern, 205
moraines, 31, 83. *See also* glaciers
Moselle River, 44
mountains, 12, 82; building of dams closer to the peaks of Alpine, 161; construction projects as more difficult in the, 14, 122–24; electric traction in the, 95–100; and formation of national identities, 12, 99; Swiss reservoirs in the, 139; transformation of a shared landscape of, 13, 101, 220. *See also* Alps; lakes; relief; waterfalls
Münchner Allgemeine Zeitung, 74

Münchner Neueste Nachrichten, 103
Munich, 18, 51, 56–59, 70–75, 80, 85–90, 105, 108, 221. *See also* Bavaria
Mussolini, Benito, 204

nationalism, 14, 91, 116, 143; French, 34; German, 72, 93, 99, 103, 105; Italian, 158; rise of, 165; self-defeating logic of European, 143
nationalization: of Alpine waterpower, 109; of Austrian energy supply, 197–98, 201; of Austrian railways, 197; of Bavarian railways, 90, 94; of Italian electricity supply, 254n15; of Swiss mainline railways, 94–95; of Swiss waterpower, 75–78. *See also* railroads; waterpower
National Socialists. *See* Germany
Newcomen, Thomas, 27
newspapers. *See* press
New Zealand, 214
Niagara Falls, 31, 33, 53, 66, 102. *See also* United States
Norway, 187
nuclear power, 115, 210, 223; French, 213, 223, 256n44. *See also* energy

oceans, 82
oil, 5, 7–8, 74; conversion of coal into, 174, 183; dependence of states on revenue from, 9–10; history of, 50–51, 114–15, 253n3. *See also* fossil fuels
Oliven, Oskar, 154–56; "Proposal for a European Superpower Grid," 155
Österreichichische Elektrizitätswirtschafts-AG, 198
Ostwald, Wilhelm, 73–74
Ötz Valley, 189
ovens, 100

paper manufacture, 3, 36–41, 84, 114, 117, 119–20
Papeteries Bergès, 120
Paraná River, 213
Paris, 1, 49, 51–54, 140, 198

Paris climate agreement (2015), 223
Paris-Lyon-Marseille railway, 120
Perrier, Leon, 140
Piedmont, 33, 63; Alpine waterpower of, 34, 204; lowering of the level of lakes in the, 106. *See also* Italy
pipes, 14, 39-40; iron, 2-3, 16, 40, 212
Poland, 30; coal exports to Germany of, 166; demand for the coalfields of Upper Silesia of, 129; industrial centers in, 200
pollution, 74; of coal consumption, 100; of the Rhine by the heavy concentration of industry on its banks, 220. *See also* environment
politics: and energy, 10; solutions to the problem of transport of energy in Italian, 63-64; of state intervention in reservoir construction in Bavaria, 110-12
popular movements: for large-scale exploitation of Alpine waterpower, 101-4, 112; for protecting nature from urban and industrial expansion, 105-7, 112. *See also* conservationists; environmental movements
Po River, 20, 63, 66, 146, 212; basin of the, 61
powerhouses, 14
power plants: economies of scale for, 65-66; electric illumination, 61; fossil-fueled, 99, 143, 152, 186, 190, 195, 210, 213; transmission of alternating current from, 66; wartime destruction of, 194. *See also* electricity; hydroplants
power transmission, 114. *See also* transmission
Prague, 49
precipitation, 20-21; abundant Alpine, 23, 134; increased Little Ice Age, 22; and runoff, 92. *See also* water
press: Bavarian, 102-3; French local, 43, 45; German, 109, 257n24
Pressel, Wilhelm, 72
Prussia, 29, 71, 86
Pyrenees, 19, 36

railroads, 28-29; Alpine, 91; Austrian nationalization of, 197; French, 37; German, 51, 85; hydroelectricity to power the state-run Bavarian, 90, 94, 100-102, 110-11; large-gauge mainline, 94-96, 139; narrow-gauge mountain, 94; power grids that electrified networks of, 237n5; Swiss nationalization of, 77, 94-95. *See also* electric trains; locomotives; nationalization
Rathenau, Emil, 54
Ratzel, Friedrich, 162
Rehbock, Theodor, 74
relief, 14, 23, 31, 34-38, 82, 91, 102, 117, 214, 229n41. *See also* mountains
renewables, 4-9, 14, 17-18, 41, 45, 56, 68, 82, 112, 213; advanced development reflected in, 72-74; inconstancy of, 224; transitions to, 10; waterpower as, 72-74, 188, 223-24, 226n7. *See also* energy; solar power; waterpower; wind power
Renner, Chancellor Karl, 166
reservoirs, 80-82, 84; artificial, 212; high-lying, 88-89, 169; high-pressure, 84; hydropeaking and, 219; impact on microclimates and regional climates of, 217; interconnected, 169; mountain lakes as, 91, 99, 105, 107-11, 172; politics of, 109-11; postwar building of Alpine, 245n14; postwar draining of Alpine, 195; power density of Alpine, 214; shifts in the seasonal distribution of water from, 219; storage of Alpine waterpower in, 137-43, 149, 209. *See also* glaciers; lakes; waterpower
Revue des Alpes, 46
Rheinisch-Westfälisches Elektrizitätswerk AG (RWE), 151-56, 169, 175, 182, 190, 222, 250n35
Rhine Falls, 44, 53
Rhine River, 20, 44, 49, 146, 167, 212, 227n12; anadromous fish species of the, 220; hydropower development of the High, 245n24

Rhône River, 20, 121, 125, 146, 212; Grand Dixence high-Alpine reservoir of the upper, 209, 213; hydropower development of the upper, 125, 209
Rich, Theodore, 149
rivers, 82, 227n12; Austrian non-glaciated, 181; dams on Alpine, 70, 84, 102–3, 215; exacerbated drought conditions on Alpine, 223; habitat heterogeneity of Alpine, 216; heavy sediment loads of Alpine, 103–4; impediments in Austrian, 217; impediments in Swiss, 216–17; morphology of mountain, 18, 103; potential energy of, 7; power development of navigable, 145–46; transport of timber by mountain, 108. *See also* channelization; dams; floodplain environments; incision; water
Romanche River, 125
Romania, 156
Rothschild, Baron Alfons, 57
Royal Bavarian Railway, 100
Rühlmann, Moritz, 16
Russia, 11, 156; oil resources of, 224. *See also* Soviet Union

salmon, 219–21
Salzach River, 219
Sautet Dam, 140–41
Savoy, 113, 120, 122. *See also* France
Scandinavia, 19, 199; hydro resources of, 94, 174, 187
science: of Alpine mountain formation, 19; and inventions, 14; postwar conservation, 221. *See also* technology
Schmick, Rudolf, 88–90, 102, 104
second industrial revolution, 4, 225n4. *See also* industrial revolution
Second World War, 4, 9, 11, 125, 160, 162; destruction of the, 194, 198; energy and the, 164–65, 184–92, 204; environmental history of the, 164. *See also* Austria; Germany

sedimentation, 19–21, 39, 104–5, 217; in reservoirs, 141. *See also* water
seismicity, 207–8; reservoir-induced, 208. *See also* earthquakes
Shelley, Mary: *Frankenstein*, 211
ships, 28; sailing, 26; steam-powered, 29
Siemens, Georg von, 94
Slovenia, 18; nuclear power in, 223
snowmelt, 21, 81, 92. *See also* water
social inequality, 103
Società Generale Italiana di Elettricità Sistema Edison (Società Edison), 60–61, 66, 70, 84, 235n29
Società Idroelettrica dell'Isarco, 158
Società promotrice dell'industria nazionale, 63
Societé électrométallurgique française, 44
Société métallurgique Suisse, 44
solar power, 6, 81–82, 223–24; coal as fossilized, 26; waterpower as product of, 23, 74. *See also* energy; renewables
South Tyrol, 136–37, 156–61, 204–6; annexation by Italy of, 204; campaign of "Italianization" in, 137, 157, 159–60; political history of, 247n47; submerging of the village of Reschen in, 206. *See also* Italy
Soviet Union: influence on postwar Italian politics of the, 204; invasion of eastern Austria by the, 196, 198; Nazi invasion of the, 203; oil resources of the, 172, 224; removal of economic assets from eastern Austria by the, 203; workers from the, 191. *See also* communism; Russia
Spain, 30, 137
Speer, Albert, 189–90, 194
Sri Lanka, 137
Stanley, William, 62
St. Blasien, 15
steam engines, 2, 5–6, 17, 26–28, 34, 55; coal as fuel for, 27–28, 222; electricity and, 54; generators and, 57; smoke and soot issued by, 74, 222. *See also* coal; fossil fuels; industrial revolution

steel, 195. *See also* coal

Stinnes, Hugo, 151

Study Commission for Electrical Railway Operation (*Studienkommission für den elektrischen Bahnbetrieb*), 95, 97–99. *See also* Switzerland

Swiss Eletkro-Watt Bank, 204

Switzerland, 18, 22, 24, 30, 44, 52–53, 59, 66, 91, 133, 150, 167, 213; Bernese Oberland of, 219; byzantine water law landscape of, 76; cheap postwar winter energy for, 205; export of hydroelectricity from, 146, 149, 187–88, 190; failure of the Sonzier reservoir in, 206; high-altitude waterpower exploitation in, 141, 209; hydroelectric railroads in, 94–97; lack of coal resources in, 173; lakes as power reservoirs for, 95–100, 138–39; nationalization of waterpower in, 76–77; northeastern, 33; southern, 84; western, 209. *See also* Frei-Land Society; Study Commission for Electrical Railway Operation (*Studienkommission für den elektrischen Bahnbetrieb*)

synthetic oil, 163, 165, 183, 191–92; bottlenecks of coal and electricity in the production of, 187

synthetic rubber, 163, 165

Tauernkraftwerke AEG, 201, 249n20. *See also* Tauernwerk

Tauernkraftwerke Glockner-Kaprun, 202. *See also* Tauernwerk

Tauernwerk, 170, *171*, 175–85, *177*, 180, *185*, 247n1, 249n20; halt to progress on the, 190–91; as Nazi project announced by Göring to Austrians, 250n38. *See also* Hohe Tauern; Kaprun/Kaprun Valley; Tauernkraftwerke AEG; Tauernkraftwerke Glockner-Kaprun

technology: electrical, 8, 42–47, 63; history of, 228n25; hydroelectric, 62; industrial, 60; miraculous, 14; obstacles for new, 43, 46–47; traditional waterpower, 17–18, 22–23, 25, 30; of the transport of power, 58–67. *See also* science

Tesla, Nikola, 33, 66, 232n40

textile industry, 28; factories of the, 33; northern Italian, 61; southern German, 86, 89. *See also* industrial revolution

Thermal Power Crash Program (*Wärmekraft-Sofortprogramm*), 190

tides, 55

timber, 24–25, 85, 108–9, 120

Todt, Fritz, 188–89, 251n62

tourism, 12, 20, 43, 83, 85, 106, 108

transformers, 62, 64, 66–67, 70, 78. *See also* electricity

transmission, 8, 14, 34, 45, 52–70, 76; of Alpine hydroelectricity to centers of consumption after the First World War, 135, 161; demonstration of, 60, 68–70; efficiency of, 58, 69; environmental impact of high-voltage, 221–22, 257n25; high-voltage, 152, 154, 174, 191, 193, 257n24; Lauffen-Frankfurt transmission with waterfall (1891), 69; long-distance, 61–70, 152, 154, 159, 168, 191, 193; overwater, 174. *See also* compressed air; electricity; hydroelectricity; infrastructure; power transmission

Treaty of St. Germain (1919), 166

Treaty of Versailles, 129, 152

tunnels, 169, 179; as water-powered wind, 189

turbines, 1–3, 9, 14–17, 25, 32–40, 47, 57, 70, 78, 88, 93, 212, 215. *See also* dynamos

Turin, 63–64, 158

Umberto I, King, 64

Union des Gaz, 61

United Nations, 199

United States, 11, 72, 103; administration of the Marshall Plan by the, 199; creation of national parks in the, 106; Lowell (Massachusetts), 53; oil resources of the, 172; Tennessee Valley Authority (Tennessee), 195. *See also* Niagara Falls

Verbundbetrieb, 152
Vienna, 18, 93, 165–66, 175–76, 182; power plants of, 197
von Clausewitz, Carl: *On War*, 113
von Hertling, Georg, 110–11
von Liebig, Baron, 71–72
von Ribbentrop, Joachim, 175

Walchensee. *See* Lake Walchensee
Walchenseewerk, 101–12, 116, 126–33, 218; painting of the, *132*; public movements related to the, 101–5, 237n4; as state-led, 110–11. *See also* Bavaria; Lake Walchensee
water: as a common good, 10; compensation for lack of fossil fuels through intense manipulation of, 11; inconstancy of Alpine, 93; storage of, 20, 41, 92–94, 98–99, 104, 217. *See also* glaciers; lakes; precipitation; rivers; sedimentation; snowmelt; waterfalls; waterpower
Water and Electricity Supply Office (Wasser- und Elektrizitätswirtschaftsamt), 167–69. *See also* Austria
waterfalls, 8, 15–16, 20, 30–33, 55, 112; destruction of Austrian, 179; glacial carving and, 38; in the Grésivaudan, 36, *42*; harnessing for a hydroelectric plant of, 67; height of, 40–41, 45; private sector control of, 77, 112. *See also* mountains; water; waterpower
watermills, 22–26. *See also* waterpower
waterpower, 2–3; belief in the superiority of Alpine, 3–4, 91; cost-effective, 31–32, 90, 126, 130–31; defining what constitutes acceptable uses of, 13; development of Alpine, 6, 10, 17, 19–20, 99, 121–33, 137–43, 187–88; generation of compressed air to run machines and tools by, 53; German wartime electricity supply and Alpine, 186, 189–91; high-altitude, 3, 39, 88–89, 151; high-pressure, 16–17, 30, 32–34, 36, 40; large-scale exploitation of Alpine, 4, 9, 53, 71, 90, 101–4, 106–7, 143, 152, 182; market considerations in the spread of Alpine, 9, 77; public-private partnership in the development of, 127; seasonal availability of Alpine, 91–93, 141, 177–78; state development of, 75–79, 90, 94, 106, 109–12, 116, 119, 128, 140, 197; storage of, 6–7, 81, 83, 88, 93, 96–100, 111–12, 127, 135, 178–79, 193, 201, 220, 224; traditional, 22, 24, 52, 230n5. *See also* batteries; electricity; energy; glaciers; hydroelectricity; hydroplants; nationalization; renewables; reservoirs; water; waterfalls; watermills; waterwheels
Waterpower of Bavaria, The (memorandum), 104
waterwheels, 22–25, 30, 32; diffusion of the vertical, 231n21; efficient industrial, 32; use of traditional, 234n6. *See also* waterpower
Watt, James, 28
Weber, Max, 74
Weisse Kohle, 72
Wells, H. G.: *Anticipations*, 49–50, 71
Westinghouse, George, 65–66
Westinghouse Company, 33, 62, 66, 232n40
"Why save electricity?" (Austrian socialist poster), 202
Wilhelm II, Kaiser, 72
Willibald, Georg, 108
wind power, 6, 55, 81, 223–24. *See also* energy; renewables
Wordsworth, William: "The Simplon Pass," 15
World Power Conference (WPC), 143–48, *144*, *147*, 154, 161, 245n20
World's Fair (Paris, 1889), 1, 3, 35, 44, 47
Württembergische Portland-Cement Werk, 68
Wyssling, Walter, 59

Yellowstone National Park, 106
Yugoslavia, 158, 180

Zinßmeister, Jakob, 72
Zurich, 18, 25, 146

The authorized representative in the EU for product safety and compliance is:
Mare Nostrum Group B.V.
Mauritskade 21D
1091 GC Amsterdam
The Netherlands
Email address: gpsr@mare-nostrum.co.uk

KVK chamber of commerce number: 96249943

The authorized representative in the EU for product safety and compliance is:
Mare Nostrum Group
B.V Doelen 72
4831 GR Breda
The Netherlands

www.ingramcontent.com/pod-product-compliance
Lightning Source LLC
Chambersburg PA
CBHW031758220426
43662CB00007B/450